Printed and bound in the U.S.A. Printed on acid-free paper.
10  9  8  7  6  5  4  3  2  1

ISBN 1-884015-76-X

Phone: (407) 274-9906
Fax: (407) 274-9927

$S^t_L$

Published by
St. Lucie Press
100 E. Linton Blvd., Suite 403B
Delray Beach, FL 33483

# ENVIRONMENTAL EFFECTS OF MINING

## Earle A. Ripley
## Robert E. Redmann
## Adèle A. Crowder

with
Tara C. Ariano
Catherine A. Corrigan
and
Robert J. Farmer
and
Technical Assistance from
L. Moira Jackson

Funding provided by the Centre for Resource Studies
Queen's University, Kingston, Ontario, Canada

$S^t_L$

St. Lucie Press
Delray Beach, Florida

# TABLE OF CONTENTS

# FOREWORD

Considering the importance of the resource industries to the Canadian economy, an understanding of their impact on the natural and human environment is vital to decision-making on the part of both corporations and governments. In both spheres, decisions must be made as to how to balance the requirements of a modern industrial society and the unavoidable reality that these will not be achieved without some effect on the environment. The challenge is to know when the cost of implementing a decision will outweigh the benefit to be gained—a monumental task when the state of knowledge and the scientific underpinnings concerning environmental issues are as imperfect as they are. In this context, it is understandable that the first edition of this book was the most popular publication since the Centre for Resource Studies was established in 1973.

The situation facing the mining and mineral-processing industry has been particularly demanding in recent decades, as the extent of its effects on the environment have become more clear and general public concern over this issue has increased. Over the period since the first edition of this book was published, important progress has been made both in awareness and comprehension of the problems involved in mineral production and in knowledge of how to prevent, mitigate, and correct them. These minerals are the materials which are the foundation of our contemporary way of life, and society expects to be able to continue to use them. In commissioning a new edition, the Centre aims to contribute to a growing understanding of the practical matters of protecting the environment in the course of activities that are intrinsically harmful.

As the authors note in their introduction, the environmental impact of mining is an extremely complex topic because it involves so many geographical, technical, scientific, and socio-economic issues. Each of these issues has its own specialties and sub-specialties, and the degree of detail within each is considerable. There is the added complication that each mining operation is distinct from every other, so that generalization is risky. Nowhere was this more evident than in the peer review process for this book. It was reviewed in whole or in part by dozens of individuals in the industry, in numerous government departments, and in university departments of mining engineering, biology, geology, and zoology.

We are very grateful to these many reviewers, whose comments and criticisms led to more sources of information and clarification.

George Hood
Centre for Resource Studies

## Acknowledgments

The authors would like to express their appreciation to the many individuals who helped us in the preparation of this book. The comments, questions, and contributions of reviewers were invaluable in our efforts to treat such a broad range of material as fully and accurately as possible. The research assistants, secretaries, and graphics experts were most obliging and professional. We are particularly grateful to Dr. Heather Jamieson of the Department of Geological Sciences at Queen's University for bringing her expertise and insight to a review of the final draft.

Finally, we would like to thank the Centre for Resource Studies for funding the project and for its support and patience over the extended period required to complete it.

Adèle Crowder, Queen's University
Robert Redmann, University of Saskatchewan
Earle Ripley, University of Saskatchewan

# INTRODUCTION

## Why Do We Need a Second Edition?

The first edition of this book was published in 1978 by the Centre for Resource Studies at Queen's University and reprinted twice, most recently in 1982. The objective of that edition was to provide an understanding of the operations and processes of Canadian mining operations and their environmental effects.

When that edition was published, public awareness of the potential for environmental damage from industrial activities was developing, and governments in Canada and abroad were turning their attention to controlling and correcting that damage and to requiring more from industry in all sectors in terms of both preventive and rehabilitative action. The high profile of the effects of acid deposition, then usually referred to simply as "acid rain," perhaps more than any other issue, crystallized awareness of the particular role played by the mining industry in terms of its potential impact on the environment. Both industry and government embarked on programs to deal with that problem; public pressure for change has continued, through environmental groups and the media.

Increased interest in the state of the environment in the ensuing period has created an even greater need for an overview of these issues for the nonspecialist reader, both inside and outside the mineral industry. Environmental research by and for the minerals industry has progressed rapidly over the past 15 years to the point that the first edition presented an incomplete and outdated picture of how the industry operates and what options are available in terms of environmental protection and rehabilitation. So much has changed, in fact, that this second edition is in reality an entirely new book. Its objective remains essentially the same: to present the general reader with an accessible source of information about past, current, and expected outcomes from mining activities in Canada.

By the early 1980s, renewed concern about the environment was mobilizing in the form of a better-coordinated effort to require governments to bring pressure to bear on polluters. The mining industry, for its part, was beginning to recognize its responsibilities and to work actively to fulfil them. Since the campaign to control acid deposition, many other problems have been recognized and many solutions proposed, tested, discarded, and supplanted, all in the quest for answers to complex questions that are still imperfectly formulated. Early mining activities were often based on certain assumptions later found to be inaccurate. In many cases, it has been impossible to take corrective or preventive action because the means to do so were not known. For example, understanding of acid drainage, arguably the most serious environmental problem facing the minerals industry, has developed only since the beginning of the 1980s. Cooperative research projects are now not only looking intensively at means of preventing or containing known problems and cleaning up old situations, but also examining processes and related activities to determine whether effects thus far unrecognized will surface. The mining industry itself has changed dramatically, in response partly to a more competitive marketplace and partly to increasing expectations and regulation. Contrary to the public's mental image of a miner as someone who works with a pick and shovel, mining has become very much a high-tech industry, and it is bringing this capacity to bear on environmental problems.

In the 1990s, the question is no longer simply one of counting fish kills and examining the composition of grasses. That is still a necessary approach from the point of view of measuring effects and devising solutions, but aims go beyond protecting fisheries or agriculture or recreation. Society has come to the collective recognition that it is not acceptable to cause irreversible damage and that damage must be repaired because that is the right thing to do. It is also the sensible thing to do, because many past activities were undertaken legitimately and in good faith, then later found to have created long-lasting harmful effects. Consequently, the current approach is to control and correct damaging changes as soon as it is scientifically and economically possible to do so.

Responses to the changes in public attitudes have come from governments at the federal, provincial and municipal levels, and from the mining industry. The federal response has had several components, including an environmental impact assessment process, laws which pertain to mining and other industrial sectors, the establishment of environmental standards, and policy reviews.

Revisions to federal environmental assessment (EA) procedures for new projects or expansions of existing projects have included the formalization of scoping (determining the limits of an investigation), the use of indicators or most-valued ecosystem components, and recognition of cumulative impacts (CEARC 1988a; Sonntag et al. 1987). Cumulative impacts may occur where several developments affect one ecosystem, or when an impact is repeated at different times (Beanlands and Duinker 1983; Palmer and Murphy 1984). Methods of resolving conflicts arising during an EA have been developed, including facilitation and mediation (Couch 1988). A future development was perhaps foreshadowed at a recent conference on natural resources law, when it was suggested that

EA should not just be a gateway for new projects but should function throughout their lifetimes (Kennett 1993).

The Canadian *Environmental Protection Act* (CEPA) became law in 1988; its provisions seek to prevent hazards from toxic substances. The CEPA applies not just to new projects but to all industrial activities. Assessing ecological hazards is an imprecise science: substances *may* be toxic, and organisms *may* be exposed. Protocols for hazard assessment have been developed by numerous regulatory bodies, but are difficult to apply in large ecosystems where a mixture of toxic substances may act simultaneously and where loading is uneven (Environment Canada and Health and Welfare Canada 1989). Federal policy has therefore concentrated on setting standards and making policy statements. In the case of the latter, the federal *Green Plan* (Environment Canada 1990a) was a general policy statement—its goal "to make Canada by the year 2000 the industrial world's most environmentally friendly country."

The federal government has been responsive to changes in international attitudes. The signing of the Montreal Protocol, which sought to establish global limits on the production of contaminants that destroy ozone, and such international agreements as the Great Lakes Water Quality Agreement with the United States, are illustrative. The United Nations Commission on Environment and Development was established in 1983 to follow up on the work started at a 1972 conference in Stockholm on the relationship between the world economy and the world environment. The Commission's visit to Canada inspired the Canadian Council of Resource and Environment Ministers to form the National Task Force on Economy and Environment, the members of which were leaders in industry, government, and academe. In 1987, the Commission's findings, published as *Our Common Future* (generally called the Brundtland Report), introduced to the world the expression "sustainable development." The main message of the Report was that economy and environment are indivisible and hence a sustainable balance must be achieved between environmental protection and resource development.

The Brundtland Report also inspired an international conference under the auspices of the United Nations in Rio de Janeiro in 1992—a symbolic twenty years after the Stockholm conference. The only outcome to date from Rio '92 has been the framing of an agreement on the preservation of global biodiversity. Canada has ratified this agreement, which seeks to maintain present diversity and to slow the pace of extinctions of animals and plants caused by human activity.

While the federal government is active in making policy and law related to the environment, resources fall under provincial jurisdiction in Canada. Ontario passed its *Environmental Assessment Act* in 1975, and all other provinces now have assessment procedures, which are described and compared by Couch (1988). They differ in procedures, ministerial powers of exemption, and funding to members of the public (e.g., in British Columbia separate procedures have been developed for mining and for energy projects). Some municipal governments have planning and assessment procedures that are as stringent as those of the provinces. Provinces also have acted similarly to the federal government in setting guidelines and making policies to guide their resource industries. An unanticipated result of these exercises in policy-making has been a considerable degree of overlap and duplication among jurisdictions. The Canadian Council of Ministers of the Environment has identified the harmonization of environmental regulation and management as its first priority

The past two decades have seen change in the environmental behavior of mining company management and personnel as well. Technical advances are being made with increasing rapidity, and it is on the technical realm that this new edition focuses. Equally important, but beyond the scope of this book, are the non-technical activities of mining companies. The job of a mineral company is to make money by extracting a mineral or minerals from the ground, but there is a great range in the way this can be done. Benefit/cost analyses must now include decommissioning and pollution controls, and many firms have environmental management systems and/or conduct environmental audits. A progression from "red-neck" to "green" management has been described by a mining executive (Brehaut 1991). Improving the surroundings of a mine is one reason for "green" management, but improving a financial situation may be another, since obtaining loans, insurance, and other financing may depend on meeting environmental criteria (Wolfe 1994). Full-cost accounting is still difficult. It is worth noting that the elimination of lead in gasoline—still a goal when CEPA was drafted—has been accomplished and is an example of societal full-cost accounting.

Since publication of the first edition, mining firms have entered into partnership with government in a number of environment-related initiatives, such as the National Task Force on Environment and Economy and Roundtables across the country. Referring to such initiatives, the Vice-Chairman of the National Task Force and Executive Vice-President of Inco Limited, the late Roy Aitken, observed in 1991, "Big business has responded as one would expect it to and has become very proactive in recognizing its role and the influence which it ought to exercise. That involves doing more than putting nice policy statements on the wall, or giving the engineering department a few codes of practice to follow." Such sentiments are gaining momentum in Canadian mining companies (e.g., besides their annual financial reports, Noranda and Falconbridge produce annual environmental reports describing the environmental effects of their operations, what they are doing about them, and what remains to be done). New management practices—environmental auditing, international standards (ISO 9000), environmental management systems (BSS 7750), and programs aimed at educating the workforce on environmental protection—have been adopted at the major mining firms. Falconbridge's comprehensive environmental management system is described in a case study of the company's management of the environment (Jackson 1995).

Some mining companies are involved in local environmental planning. Two new routes for involvement are the preparation of State of the Environment Reports and membership in Roundtables. Both of these activities occur at a range of scales from the federal to the municipal levels. Two federal State of the Environment reports have now been produced, in 1986 and in 1991. In the latter one, the part entitled *Environment and Human Activities* comprised 14 sections, one of which was on mining.

Joint efforts between the minerals industry and governments have burgeoned since the first edition was published. Together, they sponsor conferences for the purpose of sharing technical information about environmental matters, many of them international, because the problems faced by Canadian mining operations are common in many other countries. In addition to these opportunities for dissemination of the results of research, industry and government have individually and together sponsored numerous research programs. Perhaps the most familiar is the Mine Environment Neutral Drainage (MEND) program initiated by the industry and co-funded with the federal and some provincial governments.

The most recent example of joint initiatives, and certainly the broadest in terms of scope, is the Whitehorse Mining Initiative. This is a partnership among the federal and provincial governments and representatives of the mining industry, aboriginal groups, environmental groups, and labor, designed to ensure that mining can continue in Canada in such a way as to contribute to the economic welfare of all parties while minimizing the social and environmental costs.

## Scope of the Book

Mining and the environment is a complex and broad-ranging area of study in which research has been growing at an extremely rapid rate. It is clearly impossible to discuss it all in detail in a single book. The approach taken here is to provide a basic framework for understanding the issues and to direct the reader to the best-known sources for detail, supplying it where our own expertise exists, primarily in crop science and ecology. An extensive list of references is included at the end of the book, as is a glossary of mining and biological terms.

It is important to note that there is no such thing as an "ordinary mining operation" that can be referred to as an example of how things are always done. The techniques of mineral production and processing are very specific to the individual site due to differences in geology, mineralogy, terrain, and many other factors. We can generalize only up to a point, and what is true at one operation may not apply at another that appears to the non-miner to be identical. Also, and for similar reasons, we are limited in our ability to establish the economic context for mining and mineral-processing operations. Production varies greatly from one year to another, depending on commodity prices and demand, which are determined in the international marketplace and over which individual companies and operations have very little control. As a result, production figures for a single year provide little information about the industry as a whole. Reserves, too, are highly variable and the definition often depends on the source and the purpose for which they have been determined. Tables I.1 and I.2 are offered, therefore, as an overview of the contributions of various commodities produced in Canada from 1987 to 1993, in terms of both value and quantity. Table I.2 reduces all the amounts to thousands of tonnes, demonstrating the extent of range of volumes produced. The reader who is interested in annual production data is advised to consult the publications of the federal Department of Natural Resources, particularly the annual *Canadian Minerals Yearbook*, which places Canadian mineral activities in a more global perspective.

The material relates primarily to technical matters, sketching out policy issues and monitoring methods only to provide a context for indicating the basis for technical choices. Policy is a subject for another book. Federal laws that affect mining have been listed by Barton et al. (1988) and Barton (1993) and are not discussed in this book. Likewise, we do not examine issues of worker health and safety. Further, in the interests of controlling the length of this book, we discuss only those processing activities that take place on the same sites as mines. Off-

**Table I.1**  Percentage Contribution of Leading Minerals to Total Value of Production of Non-Petroleum Minerals, 1987–93

| Minerals in Canada | 1987 | 1988 | 1989 | 1990 | 1991 | 1992 | 1993p |
|---|---|---|---|---|---|---|---|
| Gold | 12.5 | 19.5 | 10.8 | 12.3 | 13.5 | 13.0 | 15.3 |
| Coal | 18.9 | 15.1 | 9.0 | 9.4 | 11.1 | 10.2 | 11.9 |
| Copper | 22.4 | 20.1 | 11.2 | 12.5 | 12.3 | 13.0 | 11.9 |
| Zinc | 17.2 | 18.8 | 12.8 | 11.6 | 8.0 | 11.1 | 8.2 |
| Nickel | 14.8 | 23.2 | 14.1 | 10.4 | 10.5 | 9.1 | 8.2 |
| Iron ore | 16.0 | 11.1 | 6.4 | 6.4 | 7.2 | 6.7 | 7.0 |
| Potash ($K_2O$) | 8.4 | 9.9 | 4.8 | 5.0 | 5.3 | 6.1 | 6.1 |
| Cement | 11.3 | 8.1 | 4.4 | 5.0 | 4.7 | 4.1 | 5.1 |
| Sand and gravel | 8.8 | 7.0 | 4.0 | 4.2 | 4.3 | 4.6 | 4.8 |
| Uranium (U) | 14.0 | 8.6 | 4.2 | 4.6 | 3.5 | 3.5 | 3.4 |
| Stone | 6.8 | 5.3 | 3.1 | 3.3 | 3.1 | 3.3 | 3.1 |
| Salt | 2.9 | 2.1 | 1.1 | 1.2 | 1.4 | 1.7 | 1.9 |
| Asbestos | 2.9 | 2.1 | 1.3 | 1.5 | 1.6 | 1.5 | 1.5 |
| Lime | 2.1 | 1.6 | 0.9 | 1.0 | 1.0 | 1.1 | 1.5 |
| Silver | 5.1 | 3.2 | 1.3 | 1.2 | 1.0 | 1.1 | 1.0 |
| Platinum group | 2.1 | 1.6 | 0.7 | 1.0 | 0.8 | 0.9 | 1.0 |
| Peat | 0.8 | 0.7 | 0.6 | 0.4 | 0.6 | 0.7 | 0.7 |
| Clay products | 2.5 | 1.6 | 0.9 | 0.6 | 0.6 | 0.7 | 0.7 |
| Lead | 4.7 | 3.2 | 1.3 | 1.5 | 1.2 | 1.5 | 0.7 |
| Sulphur in smelter gas | 1.2 | 0.7 | 0.4 | 0.4 | 0.6 | 0.4 | 0.7 |
| Cobalt | 0.4 | 0.4 | 0.2 | 0.2 | 0.4 | 0.9 | 0.5 |
| Gypsum | 0.4 | 0.7 | 0.4 | 0.4 | 0.4 | 0.4 | 0.5 |
| Other Minerals | 7.2 | 11.1 | 6.1 | 4.8 | 5.9 | 4.6 | 4.1 |
| Total | 100.0 | 100.0 | 100.0 | 100.0 | 100.0 | 100.0 | 100.0 |

p = preliminary
Source: Godin (1994).

**Table I.2** Contribution of Leading Non-Petroleum Minerals to Quantity of Production, 1987–93 (thousand tonnes)

| Minerals in Canada | 1987 | 1988 | 1989 | 1990 | 1991 | 1992 | 1993ᵖ |
|---|---|---|---|---|---|---|---|
| Gold | 0.12 | 0.13 | 0.16 | 0.17 | 0.18 | 0.16 | 0.15 |
| Coal | 61,211 | 70,644 | 70,527 | 68,332 | 71,133 | 65,612 | 68,600 |
| Copper | 794 | 758 | 704 | 771 | 780 | 762 | 699 |
| Zinc | 1,158 | 1,370 | 1,273 | 1,179 | 1,083 | 1,196 | 998 |
| Nickel | 189 | 499 | 496 | 195 | 188 | 178 | 181 |
| Iron ore | 37,702 | 39,934 | 39,445 | 35,670 | 35,421 | 31,582 | 31,720 |
| Potash ($K_2O$) | 7,668 | 8,154 | 7,014 | 7,345 | 7,087 | 7,040 | 6,970 |
| Cement | 13,371 | 12,350 | 12,591 | 11,745 | 9,372 | 8,484 | 9,842 |
| Sand and gravel | 278,546 | 287,653 | 274,848 | 244,316 | 216,264 | 240,616 | 229,940 |
| Uranium (U) | 12 | 12 | 11 | 9 | 8 | 9 | 9 |
| Stone | 119 | 120 | 118 | 111 | 88 | 89 | 79 |
| Salt | 10,125 | 10,690 | 11,158 | 11,191 | 11,871 | 11,088 | 11,371 |
| Asbestos | 665 | 710 | 714 | 686 | 686 | 587 | 509 |
| Lime | 2,330 | 2,518 | 2,552 | 2,341 | 2,375 | 2,384 | 2,447 |
| Silver | 1.4 | 1.4 | 1.3 | 1.4 | 1.3 | 1.2 | 0.9 |
| Platinum group | .. | .. | .. | .. | .. | .. | .. |
| Peat | 610 | 736 | 812 | 775 | 833 | 828 | 820 |
| Clay products | .. | .. | .. | .. | .. | .. | .. |
| Lead | 373 | 351 | 269 | 233 | 248 | 337 | 188 |
| Sulphur in smelter gas | 723 | 856 | 809 | 790 | 749 | 774 | 797 |
| Cobalt | 2.5 | 2.4 | 2.3 | 2.2 | 2.2 | 2.2 | 2.4 |
| Gypsum | 9,094 | 8,814 | 8,180 | 7,978 | 6,727 | 7,295 | 7,836 |

ᵖ = preliminary
" = too small to signify
Source: Godin (1994).

site smelters and refineries, therefore, are mentioned only to complete descriptions of the paths taken by the products of mines and their processing facilities. We also focus mainly on the major non-petroleum minerals. A number of minor metals are produced in Canada, but in such small quantities that their environmental impacts, even on the local scale, are negligible or are so unclear that they have not been included.

## Plan of the Book

Chapter 1 outlines current practices in Canadian mining. Exploration and development, extraction processes, beneficiation processes, metallurgical processing and refining, and decommissioning are described.

Chapter 2 summarizes ways of analyzing the environment, concentrating on aspects of ecosystems that can be measured before and after mining. Transport mechanisms between different parts of the environment, such as air to water, are described. Because of Canada's size, classification of regions is essential to provide a predictive framework carrying information about climates, vegetation, animals, soil processes, etc.

In Chapter 3 the effects of mining on the environmental factors and processes described in Chapter 2 are summarized. Dispersion of residuals is discussed.

Chapter 4 is devoted to protection, rehabilitation, and reclamation. This chapter includes sections on regulation, standards, and monitoring.

The second half of the book, Chapters 5 to 11, consists of chapters devoted to specific mined commodities. Each of the nine chapters describes a commodity, its Canadian production history, the situation of orebodies, ore extraction, ore processing, specific impacts on the environment, and short- and long-term remediation methods. The same general format is used in every chapter, but emphasis varies (e.g., human health is given more prominence in chapters dealing with uranium and asbestos). Some commodities are grouped within a chapter, such as gold and silver together, some eighteen metals as sulphide ores, and a variety of materials as structural and industrial minerals.

In the chapters on commodities, case histories have been used to provide detail. The case histories have been selected to cover a wide range of commodities, geographic regions, and types of impact. They include cases where there was a long history of mining before environmental controls existed, some relatively new mines, and an example of a mine planned to prevent detrimental environmental effects. Where the case histories are highly detailed (e.g., those on sulphide ores, noise, and dust), the same information is not repeated in the text.

# CHAPTER ONE

# MINERAL EXTRACTION

## Introduction

This chapter describes techniques and activities associated with exploration for mineral deposits, development of mines and processing facilities, extraction and beneficiation of ores containing desired minerals, and further processing of ores. The ultimate decommissioning of the mine facilities is also described here briefly; Chapter 4 deals in greater detail with reclamation and rehabilitation of mining properties. The descriptions of the various stages include brief summaries of their environmental effects, but it should be remembered that these effects are often quite specific to the mineral commodity being processed. Detailed explanations will be found in Chapter 3 and in the chapters dealing with commodity groups.

A major feature of these processes is that they result in the production of an extremely high volume of unwanted material, the disposal and treatment of which constitute one of the most serious environmental challenges to the industry. These wastes vary from approximately 30% of the mass of the ore—in the case of iron, gypsum and other non-metals—to about 50% for base metals (Godin 1991) and more than 80% for strip-mined coal (Dubnie 1972a). Since most base and precious metals are found in orebodies at concentrations, or grades, of only a few percent, even tenths of a percent, it is inevitable that a great deal of waste is produced when they are mined. An "average" Canadian metal mine rejects 42% of the total mined material immediately as waste rock, 52% from the mill as tailings, 4% from the smelter as slag, with the remaining 2% comprising the "values" for which the ore was mined (Boldt 1967; Godin 1991).

The effects of the mining and processing activities have a second characteristic that should be kept in mind for an understanding of their significance to the environment. Their unwanted by-products pose a problem not just because of

9

their volume, but because they are chemically reactive substances. Radioactivity may also be involved.

Mining activities take place in six sequential stages:

I. **Exploration**, which may involve geochemical or geophysical techniques, followed by the drilling of promising targets and the delineation of orebodies.

II. **Development**, i.e., preparing the minesite for production by shaft sinking or pit excavation, building of access roads, and constructing of surface facilities.

III **Extraction**, i.e., ore-removal activities that take place at the minesite itself, namely extraction and primary comminution (or crushing).

IV. **Beneficiation** (or concentration), which takes place at a mill usually not far from the minesite; at this point (except in the case of coal), a large fraction of the waste material, or gangue, is removed from the ore.

V. Further processing, which includes **metallurgical processing** and one or more phases of **refining**; it may be carried out at different locations.

VI. Since every orebody is finite, a final **decommissioning** stage is required through which the disturbed area is returned to its original state or to a useful alternative.

These stages are not necessarily discrete, but are likely to be taking place simultaneously on different parts of the same site.

Some minerals, such as many structural materials, are ready for market after the completion of only the first three of these stages. Others, such as asbestos, potash, and graphite, require physical separation from gangue, or waste, materials (stage IV). The remainder, which includes most of the metals except iron, require pyrometallurgical separation (stage V) to isolate the desired material to a marketable degree of purity. In many cases, further processing beyond stage V is required to produce a more highly refined product for specific applications. This additional processing is usually carried out away from the mine, mill, and smelter sites, and thus is not covered in this book.

Each of the six stages involves several different processes, the choice of which is governed by such factors as the type of ore and orebody, the geographical location of the orebody, the availability and cost of ancillary materials, energy, manpower and markets, and environmental considerations. Each stage produces something required by the next stage, whether materials or information. At the same time, each produces some effects on the environment, to one or more of the land, water, or air.

## Exploration

Exploration activity is closely related to the general economic climate and to the demand for minerals. This results in a considerable variation in activity from year to year, which is reflected in the resulting environmental effects.

Mineral exploration usually begins with a reconnaissance to locate likely mineral-deposit areas. Explorationists can often rely on geological information already compiled. For ground-based work, government maps and, in some cases, regional mapping are available. The early phases of exploration can be carried out using remote-sensing methods from satellites and aircraft: as promising areas are located, low-flying aircraft and ground-based exploration methods are used. This is followed by more detailed exploration surveys using geological, geophysical, and geochemical methods (Boyle et al. 1992; Green 1991). If an area is sufficiently promising, these will ultimately lead to stripping, trenching, or diamond drilling. To provide some sense of the relative effort put into each phase, expenditure in Canada in 1992 for both on- and off-property non-petroleum mineral exploration was $385 million. Ignoring overhead costs, 46% of this total was used for drilling (underground and surface), 18% for geological work, 10% for geophysical, and 5% for geochemical (Bouchard and Cranstone 1993). Three-quarters of the geophysical work was ground surveys and the remainder was airborne.

### Geophysical Exploration

Geophysical exploration may be carried out from the surface as well as from aircraft, using such current tools as seismic, gravity, resistivity, magnetic, electromagnetic, radar, and induced polarization. Radio-wave and electrogeochemical methods are recent developments.

### Other Forms of Exploration

Pits and trenches are dug in order to examine the underlying geological formations and to acquire samples for chemical analysis. These tasks, as well as the removal of overburden (surficial material that impedes access to the orebody), are usually accomplished by either manual or machine digging, hydraulic methods, or blasting. Hydraulic and blasting methods are probably the most significant in terms of the area affected.

Exploration technology has advanced in recent years. Prospectors have increased their use of remote-sensing and computer technology and their employment of deep drilling and "down-hole" geophysical methods for locating deeper orebodies (Green 1991). As a result, large quantities of geological, geophysical, and geochemical data are being accumulated, each of which on its own provides a perspective on the mineral-bearing potential of the underlying rocks. Integrating these types of data—most recently by using geographic information systems (GIS)—creates a powerful new tool for probing the earth's mineral resources more widely and deeply than ever before (Watson et al. 1989; Green 1991).

Another interesting method is drift prospecting, the study of glacial sediments to identify minerals and trace them to their sources. Research on this topic began in the 1950s and has been escalating in recent years (DiLabio 1990). It has been used effectively in discoveries of a rare-earth-niobium-beryllium deposit in Labrador and on gold deposits in Québec, Newfoundland, and Saskatchewan. The Canadian government's Lithoprobe program uses seismic and many other methods for exploring the depths of the lithosphere. A radar-imaging system is expected to be launched within the next ten years (Whiteway 1990).

There are several biological methods of mineral exploration that have not yet gained wide acceptance in Canada. Generally non-invasive techniques, they are expected to have relatively minor environmental effects. They use indicators such as biotic species or chemical compounds (Zonneveld 1983), including *geobotany*, which is based on the examination of an area's vegetation cover or the soil flora; *biogeochemistry*, which is based on the chemical analysis of the vegetation and soil; and *geozoology*, which uses animals to locate areas of mineralization.

With the development of satellite remote-sensing of vegetation (Tucker et al. 1985), this methodology has been incorporated into GIS-based integrated prospecting systems.

Computer modelling technology is also finding increasing use as a tool for transforming exploration data into orebody models, which can be of considerable assistance in assessing ore reserves and in planning extraction strategies (Green 1991). It might be noted that this technology may someday be used to control robot miners, if the work of such organizations as the Canadian Centre for Automation and Robotics in Mining comes to fruition, this research organization was established in 1987 as a joint venture between McGill University and L'École Polytechnique de Montréal.

## Transportation and Services

Transportation is not a major issue for the substantial amount of exploration activity that is directly adjacent to existing mines. There is also an occasional discovery like Hemlo, conveniently located just off the Trans-Canada Highway in Ontario. For the many mineral exploration programs carried out in remote areas, however, access becomes a problem and may require the construction of roads or runways. These may be only temporary needs, but if a mine is developed, such infrastructure may be required for its lifetime.

The services used in support of exploration vary with the number of personnel involved and with the nature of the work. They usually comprise campsites and other facilities for personnel and facilities for the storage of equipment and fuel in remote areas. Local services in nearby communities are incorporated in other cases.

## Environmental Effects of Exploration

The exploration phase may affect far more terrain than do subsequent mining operations, but for a shorter time period and usually with less severity. Barker and Curtis (1991) estimated that only one prospective mining site in 5,000 actually produces profitably. There are environmental effects associated with each type of exploration activity, except for satellite-based methods. Low-flying aircraft affect people and wildlife; exploration grids leave visible scars that may also disturb wildlife and lead to considerable erosion; trenching and stripping may have long-term effects on adjacent watercourses; and drilling can lead to contamination of both surface water and groundwater. In areas of permafrost (which includes the northern half of Canada), any exploration activity that disturbs the vegetative cover may cause thermokarst features to develop.

Overall, transportation is likely to be the source of a major portion of the environmental effects of any development—including mining operations. The impacts may include construction of railways or roads, and once they are built, their effect on wildlife depends upon traffic density and types of users. Roads built into isolated areas can have major impacts upon game animals such as caribou, moose, or sheep, which are exposed to vastly increased hunting pressure. Fish also become more accessible to sport fishermen, and hence populations may be diminished.

Reclamation of roads is generally possible, although it may be difficult in certain situations. The road-construction material itself may present problems, as in cases where roads were built of waste rock that contained sulphides. The rapid oxidation of this material causes serious problems of acid drainage, which is discussed in detail in later chapters.

The primary goals during the planning stage should be to minimize the area that will be cleared or stripped, to minimize the disturbance to native vegetation and soils when disturbance cannot be avoided, and to reduce or prevent erosion by means of physical manipulation of the soil and/or the seeding of adapted plants. A detailed guideline for the reclamation of roads and landing strips built for exploration purposes has been given by McDonald and Dick (1973). Such infrastructural projects have special problems in Arctic regions, since soils and vegetation there are particularly vulnerable to disturbance (Aho 1966; Heginbottom 1973; DIAND 1983).

## Site Development

The site development phase consists of four main activities: a feasibility study to evaluate the deposit and the best method to use in its mining; the design of the mine and its environmental control structures; an environmental impact assessment and associated public enquiries; and construction.

The completion of each of these activities may require two to three years, although some may run concurrently. The impacts of exploration discussed in the first section, particularly transportation and services, are generally typical of the development stage as well. Since surface disturbance and waste generation—due to construction activities and operation of machinery—will likely be at maximum levels during site construction, pollution control is particularly important at this time.

It is in the course of site development that most of the environmental protection plans must be formulated and incorporated into the operation. Careful attention to the choice of methods and devices for this purpose from the outset is now a fundamental consideration in the design of mining facilities and is a major component of the environmental policy statement adopted in 1989 by The Mining Association of Canada (Miller 1990). If properly implemented, the design and siting of buildings, waste disposal systems, stockpiles and services both for the mine and for the local population (workers and supporting services) will minimize the extent of environmental effects. Proper design of mills and waste disposal methods and good selection of sites for them will prevent irreversible damage to adjacent terrestrial and aquatic ecosystems. Indeed, responsible efforts in this regard are now statutory obligations, as discussed in Chapter 4.

## Extraction

Most ore deposits may be broadly classified into one of three types on the basis of general physical characteristics. *Massive stockwork or vein types*, such as those associated with most non-ferrous metalliferous ores, are of highly variable size, shape, and inclination. *Bedded types*, associated with coal, iron ore, some sulphide ores, and a number of salts, tend to be of much greater horizontal than vertical extent and of quite uniform thickness. *Loose, unconsolidated types* of deposits, such as alluvial gravel and mineral sands, range in size from very small placer deposits to many square kilometers in the case of some mineral sands deposits; this type includes placer gold deposits as well as industrial sand and gravel.

Mining methods are often divided into surface and underground operations. The line between the two is gradually blurring, however, as many orebodies are mined by a combination of the two methods. We also consider a third category: non-entry mining. As Table 1.1 indicates, the environmental effects of each of the three categories depend on different methods of extraction, which are generally dictated by the type of orebody.

Although only one sixth of Canadian ores, in terms of tonnes hoisted, exclusive of industrial minerals, were mined by surface methods in 1950, by 1970 the fraction had risen to two-thirds (Dubnie 1972a) and even somewhat higher in recent years (Godin, various years). The change has been even more dramatic with regard to coal. Underground production diminished from about 25% in 1970 to less than 5% by 1986, almost all of which came from submarine mines in eastern Nova Scotia (Romaniuk and Naidu 1987).

**Table 1.1**  Environmental Effects of Various Methods of Mineral Extraction

| Mining Method | Environmental Advantages | Environmental Disadvantages |
| --- | --- | --- |
| **Underground** | | |
| open stoping | less waste rock than with surface mining | high subsidence potential; oxidation of exposed materials |
| filled stoping | lower risk of subsidence; disposes of some waste material | possibility of oxidation and even combustion of back-filled material; slurry drainage and water disposal; may affect aquifers |
| **Surface** | | |
| open pit | few, other than accessibility and lower worker risk than underground | large amount of waste rock, dust, noise, mine drainage, ore oxidation |
| alluvial | relatively easy to control environmental damage although not often done | high potential for particulate emission to atmosphere and hydrosphere; surface disturbance |
| **Non-entry** | | |
| auger | minimal surface disturbance and worker risk | low extraction efficiency |
| *in-situ* leaching; solution | reduction of solid wastes, mill tailings, surface disturbance, worker risk | requires disposal of large quantities of soluble salts, possible groundwater contamination, and surface subsidence |
| *in-situ* utilization | minimal surface disturbance, worker risk and solid residuals | difficulty of controlling and containing underground process; high potential for underground contamination; explosions |

The extractive process generally involves the removal of the ore from its deposit and includes some primary separation of waste material from the ore, together with enough comminution to facilitate the removal and transportation of the ore from the mine to the mill. Ores hosted by crystalline rock (that is, gold and most base metals), whether mined by surface or underground methods, are usually fractured at the mine face using an explosive. This is most commonly a mixture of ammonium nitrate and fuel oil, which is safer and more versatile than dynamite. In his *Pit Slope Manual*, Murray (1977) discussed the type and extent of surface disturbance as indicated by satellite imaging in the mid-1970s.

## Underground Mining

Figure 1.1 illustrates a typical underground mine. From the perspective of this book, it is appropriate to classify underground mining methods into two main types: open stoping and filled stoping. The former, which is used where the rock is sufficiently strong, involves leaving underground cavities after the ore is removed by drilling and blasting. When the latter type is used, portions of the mined-out areas are filled with waste materials, either to facilitate further mining

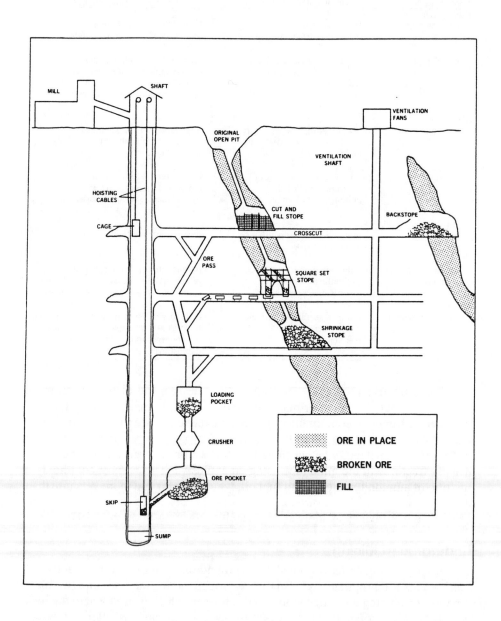

**Figure 1.1** Illustration of an underground mine. Source: Marshall (1982).

activities in the area or to reduce compressional collapse and subsidence after the area is mined out. About one-third of the base metal ore removed by underground methods in Canada is mined using backfill methods. The choice of whether or not to use backfilling depends on the strength and shape of the orebody, as well as the costs involved. Although using backfill has been constrained by the extra cost (Roberts and Masullo 1986), the practice could become more widespread as environmental pressures mount and backfilling technology improves. In any case, other methods of disposing of wastes will always be necessary because crushing and milling processes cause the volume of material to increase dramatically, partly because the ground material is less dense than the original ore and partly because of the addition of reagents.

Brunswick Mining and Smelting Corporation (BMSC) recently installed a semi-automatic system for backfilling cut-and-fill and modified open stopes at a mine in northeastern New Brunswick. Crushed rock from a nearby basalt quarry is mixed with a cement slurry before being deposited underground. The operation is part of BMSC's "effort to unfold new mining innovations aimed at optimizing productivity." In some backfill operations, no cementing is used, while in others fly-ash or smelter slags are added instead of portland cement to bind the fill (Kheok 1986; Churcher 1989).

Backfilling with mill tailings (the waste product of the milling process, described in the section on beneficiation), transported underground as a slurry, is quite widespread, despite difficulties with water, such as the potential for contaminating groundwater. There have been recent improvements in the technology to eliminate these problems (Churcher 1989) as well as the development of alternative methodology, such as pneumatic transport (Roberts and Masullo 1986). A very recent alternative is to form the waste material into a paste rather than a slurry (Landriault 1992); this has a higher solids content (75%–90% compared with 60%–70% for a slurry), is less expensive to distribute and has a stiffness approaching that of rock fill. Although still rather costly, the use of high-velocity air to transport materials underground is becoming more economically competitive.

## Surface Mining

Surface mining is now the dominant extractive method used in Canada for the mining of coal, base metals, iron, and asbestos, as well as most other industrial minerals (Godin 1992). In general, open-cast surface methods are most suitable for large orebodies located at shallow depths, whereas underground methods tend to be used for deeper, richer ores (Brown 1990). In addition, certain orebodies, especially those that are large, low-grade, and disseminated, are more appropriately mined from the surface.

For many mines, a combination of surface and underground methods is used in order to reach the entire orebody more efficiently. An orebody is usually mined from the top down, initially by surface methods, until it becomes more economical to go underground than to remove further waste material. As the pit deepens, increasing amounts of waste material must be removed in order to avoid dangerously steep pit-wall slopes.

Although open-cast mining methods disturb greater surface areas and produce a larger amount of solid waste, their use continues to increase. Because surface mining is more readily adapted to large-scale mechanization, thus reducing the number of workers required and enhancing industrial safety, the development of larger-scale extraction methods and machinery is making surface mining more economically attractive.

## Massive and Bedded Deposits

Surface mining may be classified into two types: open pit and strip mining (Dubnie 1972a). The essential difference between the two is that the former is designed to exploit massive orebodies, and the latter bedded deposits. Both types of mining require the removal of large quantities of overburden and waste rock. The open pit method requires the removal of large amounts of waste material while the mine is being developed and operated; in strip mining, by contrast, the overburden is stripped off, and the deposit is mined—and the wastes replaced—in a cyclic manner, so that the waste fraction is almost constant during the developmental and operational stages of the mine.

Strip mining is used in Canada almost exclusively in the extraction of relatively shallow coal deposits. The coal is removed either by dragline—as in the plains of Alberta and southern Saskatchewan—or by shovel-truck operation—as in the cordilleran region of Alberta and British Columbia. The generally flat, horizontal shape of such bedded deposits requires the disturbance of a large area when these are mined from the surface. Present-day operations generally strip off the overburden and stockpile it, producing a series of parallel ridge-like waste piles, between ten and fifteen meters above the original terrain. When mining activities are completed, the stockpiled overburden is used to reclaim the site to previous or other suitable use. At present, some of the best examples of these techniques for simultaneous mining and reclamation are being used in the mining of mineral sands in Australia and coal in the United States.

Somewhat different mining methods must be used in the more rugged terrain of the cordilleran areas of Alberta and British Columbia. These vary considerably with surface and seam topography, and with depth and composition of the overlying material. In sloping areas, the surface must be "benched" to level it for the operation of equipment for removing the overburden. If it is not feasible to replace the overburden, because of severe erosion potential, it is hauled away to a waste dump.

A cross-section of a typical open pit mine is depicted in Figure 1.2. After the removal of overburden, the orebody is mined by making successively deeper cone-shaped cuts. The cone shape of the excavation facilitates access to the

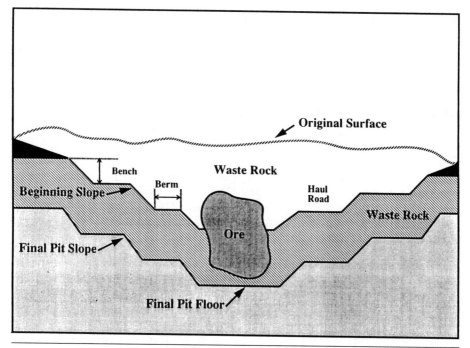

**Figure 1.2** Cross-section of open-pit mine. Source: Marshall (1982).

orebody and, more importantly, avoids the potential instability of steep pit walls. On average in Canada, for each tonne of useable ore extracted from an open pit, 630 kg of waste rock and 50 kg of overburden must be removed (Godin 1991). The ratio of waste to ore, called the stripping ratio, varies from 0.13 to 6.0 for individual mines (Dubnie 1972a). The amount of waste is highly dependent on the wall slope required for safety, and this in turn depends mainly upon the mechanical strength of the rock (Coates and Gyenge 1972). Slopes ranging from 20 to 67 degrees are found in Canadian mines; the most common is approximately 45 degrees (Dubnie 1972a).

### Alluvial Deposits

The almost ubiquitous distribution of sand and gravel deposits, combined with their low unit value and largely urban markets, has resulted in a wide distribution of pits and quarries across the country, with the greatest concentration near larger urban areas. Sand and gravel typically comprise approximately two-thirds of the total mass of end-product materials extracted from the ground by all mining operations in Canada (Godin, various years). In fact, in terms of finished product, sand and gravel comprise 57% of the mass of all mined materials in Canada, followed by structural stone at 24%, with all the remainder making up only 19%. It might be noted that a mine may have sand and/or gravel pits for its own on-site use.

Placer mining in Canada dates back to the mining of copper by native peoples (Gilbert 1989). The discovery of gold in the Yukon in the late nineteenth century led to the Klondike gold rush of 1898, during which placer mining methods were used to recover considerable quantities of the precious metal, reaching a peak of more than 30,000 kg in 1900. Since then, as primarily manual extraction methods gave way to giant dredges, decreasing demand led to a decline in production to less than 200 kg by 1972 (LeBarge and Morison 1990). The number of dredging operations decreased from 13 in 1917 to a few by the early 1970s. More recently, a strengthening market has prompted a rise in production to over 5,000 kg in 1989 (Van Kalsbeek et al. 1991), while the number of operations rose to 90 in 1979 and to well over 200 in 1989 (Gilbert 1989; LeBarge and Morison 1990). In recent years, production has been from 82 streams, with 16 of them accounting for 84% of the total.

## Non-Entry Mining

Non-entry mining of underground deposits is conducted from the surface and does not require much underground access by either personnel or large machinery. The main types of non-entry mining are auger mining, solution mining, *in-situ* leaching, and *in-situ* utilization by means of underground gasification or combustion of coal deposits.

Some orebodies containing metals—most commonly those containing copper and uranium—may be mined *in situ* by treating them with a chemical or bacterial solution (McCready 1986a) that dissolves the desired metals. The solution is then processed above ground to remove the solute. The method is best suited to deposits that are too deep, too small, and of too low a grade to be mined economically by other means and, of course, those for which the method is chemically suitable. The method is, in effect, a combination of the stages of extraction, concentration and, to a certain extent, metallurgical processing.

While leaching (discussed in the section on metallurgical processing and refining) has been used quite successfully above ground, with both mined ores and low-grade waste material, *in-situ* leaching has had only limited success underground (Thomas 1973). Some of the reasons for this are the difficulties experienced in rendering the ore sufficiently permeable to disperse the solvent uniformly, for example, and in containing and recovering the pregnant—or mineral-bearing—solution (Ward 1983; Lastra and Chase 1984). Technical innovation has spurred a resurgence of interest in this method (Hiskey 1986). While it avoids many of the environmental problems associated with extracting an ore and processing it above ground, it introduces the possibility of contaminating an aquifer if the leaching solution is not adequately contained (Osiensky and Williams 1990). Although few operations in Canada employ *in-situ* leaching in conjunction with other conventional mining methods, one has been using it to extract uranium in the Elliot Lake region of Ontario since the 1950s (Hardcastle and Sheikh 1988). More detail may be found in Chapter 7.

Solution mining is similar to *in-situ* leaching, except that the solvent used is water or steam. The best-known example of solution mining is the Frasch process for sulphur, in which steam or superheated water is pumped down and melts the sulphur, which rises to the surface as a foam. In Canada, true solution methods are used only for the mining of salt and potash, described in Chapter 10.

It is possible to utilize fuel minerals *in situ*, thus avoiding some of the environmental problems associated with the extraction and subsequent processing of the materials at the surface. Although some use of the method has been made for the gasification of low-quality coal in Great Britain, the United States, and the former Soviet Union, it has its own difficulties, including lack of control in the combustion zone (Gregory 1983).

### Environmental Effects of Extraction

The environmental effects of the extraction stage tend to be mainly local, associated with surface disturbance, the production of large amounts of solid waste material, and the spread of chemically reactive particulate matter to the atmosphere and hydrosphere. These effects broaden out to the regional scale only in the case of very large open pit operations or closely spaced facilities such as gravel pits. The thousands of sand and gravel pits across the country pose environmental challenges. Many of these operations produce quantities of particulates that are carried away by drainage water and the atmosphere. While water-quality effects will probably be similar for both above- and below-ground operations, surface disturbance, water movement and air quality effects are likely to be greatest in the case of surface mines. Table 1.2 summarizes these effects and traditional and more advanced methods of mitigating them. Acceptable reclamation practices are being developed (Browning 1987; Kryviak 1987), as demonstrated by the Redland case study in Chapter 4.

The safe storage of the waste rock from primary and secondary comminution and of mill tailings, often referred to as "spoils," represents one of the most challenging tasks facing the mining industry. Some of these wastes are suitable for use in highway construction (Emery and Kim 1974) and in the building of tailings dams (Dunbavan 1990). Much of the waste contains metal and sulphur compounds. These are sources of acid drainage that may need treatment or control for hundreds of years and, together with sulphur dioxide emissions from smelting, pose the most serious environmental problems for the industry. In some cases, higher metal prices make it worthwhile—and current technology makes it possible—to re-mine old waste heaps that were considered worthless a few years ago (Cristovici and Leigh 1986). In most cases, however, mining wastes must be covered to reduce oxidation, toxicity, erosion, and unsightliness, as well as to

**Table 1.2** Abatement Procedures for Some Environmental Effects of Mineral
Extraction

| Effect | Abatement Procedures | |
| --- | --- | --- |
| | Traditional | Advanced |
| surface disturbance and waste dumps | reclamation, backfilling, and slope engineering | greater use of waste material for mine backfilling, roads, construction |
| | physical stabilization: covering with inert material such as slag, soil, concrete | greater use of non-entry methods of mining and alternative methods of disposal |
| | chemical stabilization: spraying with oil-resin emulsion, for example | better waste-dump siting |
| | vegetative stabilization | |
| hydrospheric effluents | settling ponds, recycling, lime neutralization | use of wet drilling or enclosure and dust collection, more recycling |
| | chemical treatment: neutralization, coagulation, precipitation, oxidation, reduction, ion exchange | biological polishing |
| atmospheric dust emissions | water spraying; road surfacing; covered storage and transport facilities | more effective use of dust collection equipment, particularly for smaller (<10 mm) particles |
| | low-efficiency dust collectors, such as cyclones | |
| | moderate-efficiency dust collectors, such as wet scrubbers | |
| | high-efficiency dust collectors, such as fabric filters and electrostatic precipitators | |

allow the area to be used for other purposes. These issues are discussed in more detail in Chapters 3 and 4 and in the chapters dealing with specific commodities.

Hydrospheric emissions from the extraction stage are all considered under the term "mine water." They include water from rain, snowmelt, surface waters, and aquifers, as well as that used in mining operations (e.g., to dampen dust underground). When this water comes in contact with broken rock and ore, it picks up particulate matter, as well as products of oxidation or reduction and dissolution, including acids or alkalis and metals. When it reaches receiving water bodies, such as streams or lakes, mine water can cause undesirable turbidity and/or sedimentation, its chemical composition can have toxic effects on vegetation and animals, and it may alter temperative regimes. These problems are addressed through the use of settling ponds and through chemical treatment of the water before release. One useful aspect of mine drainage can be its heat content. The Macassa gold mine, in northeastern Ontario, installed a system that uses the waste heat from mine water to warm the underground ventilation air in winter (De Ruiter et al. 1989), offering potential savings of more than $100,000 in annual fuel costs.

Atmospheric emissions attributable to the extraction stage of mining come mainly from the action of wind on disturbed land and stockpiles of ore and waste material. Other sources include underground mine ventilation systems, machinery movement and exhaust, and reactions such as oxidation or radioactive decay of ores. The substances emitted to the atmosphere are mostly those present in the ore and waste materials, along with any reaction products mentioned above. The extraction stage primarily produces larger particles with limited dispersion, which have major effects on mine workers and, occasionally, on local residents. The most serious consequence in atmospheric terms is acid deposition, the collective term used to refer to acid rain, acid snow, acid mists, and acid fog. This is discussed in Chapter 3.

Although many devices are available for the collection of atmospheric particulate emissions, the removal of undesirable substances from air and water emissions does not end the problem. The materials removed—usually solids— must still be discharged or contained and may be chemically reactive. This is often difficult because of the very small particle sizes involved.

The potential environmental effects of non-entry mining methods are relatively minor, the most undesirable being land subsidence due to removal of underground material without backfill or leaving of pillars, air pollution from underground coal gasification and combustion operations, and water pollution from *in-situ* leaching and solution mining due to leakage into aquifers and discharge of wastewater. Interest in more benign methods may well give impetus to development of more effective non-entry technologies.

As more accessible ores are depleted, mining may shift to orebodies in more remote areas—particularly the Arctic (Gerard 1976; Udd 1989), the ocean floor (Goblot 1981), and the sea itself (Pearson 1975; Platzoder 1979; Doggett and Mackenzie 1995). Manganese nodules represent one of the more promising examples of these orebodies; they are found on the ocean floor at moderate to great depths, and contain sizeable reserves of cobalt, nickel, copper, and manganese. Another example is sediment basins, which may hold a wealth of metals (Loncarevic 1974). The exploitation of marine mineral deposits will result in many new environmental problems of international concern (Archer 1970; Goodier and Soehle 1971). Some indication of these challenges has been given by studies of the effects of coastal mining in British Columbia (Kay 1989), and the marine disposal of terrestrial mill wastes. Littoral dredging operations, besides damaging the site itself, may also affect a much wider area through the creation and movement of suspended particulates and may even accelerate coastal erosion. Machinery used in the operation will likely produce enough noise to disturb waterfowl and marine life, and any leakages of petroleum fuels or lubricants will produce oil-slick contamination. Offshore mining in estuaries or shallow coast areas will have the potential to cause severe disruption of sensitive ecosystems, while the mining of manganese nodules through deep-water oxygenation may lead to the leaching of toxic substances from the nodule ore. While there has been little exploitation of ocean mineral reserves, due to the considerable financial and logistical problems involved, potential impacts of deep seabed mining on the global economy have been assessed (United Nations 1986).

The concept of mineral extraction as an interim land use is beginning to gain acceptance. This is particularly true in the case of gravel pits and quarries, which are most numerous on the outskirts of large cities. Resistance remains very high elsewhere, however, as demonstrated by recent decisions like that of British Columbia to create a wilderness preserve in the Alsek-Tatshenshini region rather than allow development of the rich Windy Craggy deposit.

## Beneficiation

The earliest metallic-mineral deposits exploited were of sufficiently high grade to be smelted directly or to need no beneficiation (e.g., copper in the Keewanawan area and the Coppermine River in Ontario). Most of these rich deposits have long since been exhausted, so that typical ores in the present day average only a few percent of base metals by weight, and considerably less of precious metals. Because these ores are too dilute to manage directly, except by hydrometallurgical methods (see the section on processing and refining below), they are usually prepared for further processing using an operation called *beneficiation*. Other ores, such as salts and many industrial minerals, are of higher grade, but require the removal of various impurities before they are suitable for marketing.

Beneficiation is the dressing of an ore in preparation for subsequent stages of processing such as smelting, leaching, and refining. It serves to remove unwanted ore constituents, thus increasing the concentration of the desired mineral (that is, improving the grade of the ore) and/or to alter the physical properties of the ore, such as particle size and moisture content. The beneficiation stage is carried out at a mill, which is usually located near the minesite, mainly because of the prohibitive cost of transporting large quantities of low unit-value ore.

The input material to the beneficiation stage is the mined ore, and the output may be either a finished product ready for market—as in the case of most of the non-metallic minerals—or a concentrate requiring further processing—as is the case for most metallic minerals. Present-day beneficiation (depicted in Figure 1.3) usually consists of three steps: preparation, in which the ore is comminuted by crushing and/or grinding; concentration, to separate the desired ore mineral

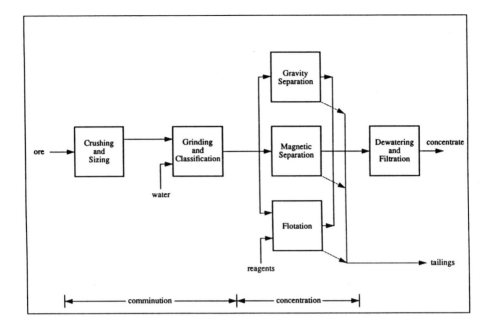

**Figure 1.3** Simplified flow-chart of the beneficiation stage of mining, in which an ore is comminuted and separated into its desired (concentrate) and gangue (tailings) fractions.

from other ore material, or gangue, which is discarded as waste; and a final dewatering of the concentrate.

Concentration often involves several steps of its own, as when the ore contains more than one economically attractive commodity and it is necessary to separate them and process each individually. In some cases, other treatment may follow the concentration step and be included in the beneficiation stage. One example is the agglomerating of iron ore to render it more physically suitable as blast-furnace charge for the smelting stage.

Three main methods are currently used for concentrating an ore (see Figure 1.3 and Table 1.3): gravity separation, often used for coal, iron ore, and asbestos; flotation, the most common method for the base metal sulphide ores and for potash; and magnetic and electrostatic separation, used for iron ore, titanium, tungsten, and others (Smith et al. 1972; Sirois and MacDonald 1983; Parekh et al. 1991).

As noted in the previous section, most ore undergoes primary comminution as part of the extraction process. Secondary comminution, involving the use of crushers and grinders, is usually the first operation on the ore after it reaches the

**Table 1.3** Concentration Methods Used in Canada

| | | Gravity separation | | | | | | Magnetic | Flotation |
|---|---|---|---|---|---|---|---|---|---|
| | general | cyclones | heavy-media | jigging | pneumatic | spirals | tabling | Separation | |
| **Metals** | | | | | | | | | |
| antimony | ◆ | | | | | | | | ◆ |
| gold | ◆ | | | ◆ | | | | | ◆ |
| iron | ◆ | ◆ | | | | ◆ | | ◆ | ◆ |
| lithium | | | | | | | | | ◆ |
| molybdenum | | | | | | | | | ◆ |
| nickel | | | | | | | | ◆[1] | ◆ |
| niobium | | | | | | | | | ◆ |
| silver | ◆ | | | ◆ | | | ◆ | ◆ | ◆ |
| sulphides[2] | ◆ | | | | | | | | |
| tantalum | ◆ | | | | | | | | |
| tin | ◆ | | | | | | | | ◆ |
| titanium | ◆ | | | | | | | | |
| tungsten | ◆ | | | | | | | | ◆ |
| **Non-metals** | | | | | | | | | |
| asbestos | | | | | ◆ | | | | |
| barite | | | | | | | | | ◆ |
| calcite | | | | | | | ◆ | ◆ | |
| coal | | ◆ | ◆ | ◆ | ◆ | | | | |
| dolomite | | | ◆ | | | | | | ◆ |
| fluorspar | | | ◆ | | | | | | ◆ |
| graphite | | | | | | | | | ◆ |
| gypsum | | | ◆ | | | | | | |
| mica | | | | | ◆ | | | ◆[1] | |
| nepheline syenite | | | | | | | | | ◆ |
| potash | | | ◆ | | | | | | |
| talc-soapstone | | | | | | | | | ◆ ◆ |

1. to remove iron
2. includes copper, lead and zinc

Sources: Energy, Mines and Resources Canada (1990); Godin (1991).

concentrator. The crushed rock is separated into various size classes by screening. It is then fed to a milling circuit—consisting of one or more dry or wet mills of the ball, pebble, roll, autogenous, or semi-autogenous (SAG) types—where the ore is ground to a sufficiently fine size to liberate most of the desired minerals.

The final particle-size distribution is controlled by the use of classifiers, such as hydrocyclones, operating in a feedback circuit. After concentration by one of the above-mentioned methods, the ground ore is sent either to the next stage of processing or to market. The finely ground waste materials from the concentration stage are called mill tailings, and their disposal presents one of the most significant environmental problems facing the mining industry. They will be discussed in the section on environmental effects of beneficiation.

## Crushing and Grinding

The crushing and grinding stage of comminution is done to break the ore up into particles as small as the grain size of the desired mineral component. This amounts to "liberation" of the mineral, which is then separated from the gangue component by a concentration process.

Crushing is carried out in one or more stages, using jaw or gyratory crushers for the larger rock, followed by cone, roll, or reduction-gyratory crushers that reduce the ore to the 5–25 mm size range. Grain size is further reduced by means of grinding, which usually involves tumbling the crushed ore in a rotating cylinder along with some grinding medium. The cylinder is lined with steel, rubber, or a combination of the two. The grinding may be carried out using the dry ore itself as the medium, as is the case with autogenous milling; it can also be conducted using wet pebbles, steel balls, or rods as grinding media. As in crushing, it is common to have several stages, with the ground ore being sized and routed by devices such as centrifugal cone classifiers. The use of steel liners and grinding media introduces into the ore a certain amount of iron contamination that must eventually be removed.

While comminution equipment has not changed very dramatically during the past fifty years, there have been considerable improvements in its control and in the optimization of its use (Parekh et al. 1991). Crushing and, particularly, grinding operations must be tailored to the composition and grain-size distribution of each particular ore, which means that the type and timing of each operation must be adjusted whenever the ore shows a change in these characteristics. Computer models of grinding circuits (Laguitton et al. 1981) are proving to be useful tools for monitoring ore characteristics and making these adjustments automatically, thus increasing efficiency and reducing waste and use of chemicals. There have been other recent improvements in the technology of jaw and cone crushers (Parekh et al. 1991), where increased efficiency is reducing media wear and power requirements.

Some improvements have recently been made in equipment design, control and optimization of use. Parekh et al. (1991) concluded that a reduction of 40% in energy consumption is possible by exploiting new comminution technology. As an example, compressive comminution—such as cone crushing—is more energy efficient than impact comminution—such as rod and ball mills. Separation methodology is also less effective, generally, for very small particle sizes; this problem is being addressed through the development of such techniques as high-gradient magnetic separation (see section immediately following), combined magnetic/gravity separation, magnetized fluidized beds, selective flocculation, and column flotation (Parekh et al. 1991). High-efficiency classifiers alone have been estimated to provide 24%–30% power savings for industrial-mineral comminution (Cienski and Doyle 1992). Collectively, these improvements in beneficiation technology should serve to raise recovery rates and to reduce the quantities of potentially toxic substances that must be discharged along with wastewater and tailings.

## Gravity Separation

The separation of materials of different densities, through the use of the force of gravity, is almost as old as mining itself. Most gravity separation, such as jigging, tabling, and sink and float methods, involves a water medium and the use of spirals. One exception is the use of air—called aspiration—for the separation of asbestos. Other gravity concentration devices include sluices, the Reichert cone, and the Bartley-Mozley concentrator (Smith et al. 1972; White 1985). Although gravity separation has been largely replaced by flotation, it is still used in the concentration of coal, of some industrial minerals, and of metallic oxide ores, such as magnesium.

## Magnetic and Electrostatic Separation

Magnetic fields may be used to separate substances that exhibit ferromagnetism, such as those containing iron, cobalt, or nickel, from those which do not (Kelly and Spottiswood 1982). The method is most widely used for concentrating iron ores and for removing iron-containing minerals from ores of the industrial mineral nepheline syenite, of nickel, and of other metallic sulphides. A high-gradient magnetic separator, capable of reducing the ash and sulphur contents of a variety of Canadian coals, was developed by CANMET (Mathieu 1981). One approach, which uses a combination of magnetic and gravity separation methods (Martinez and Spiller 1991) has proven effective in concentrating such magnetic and weakly magnetic substances as iron ore, chromite, wolframite, and the rare-earth minerals.

The electrostatic method uses a high-voltage electric field to separate materials of different electrical conductivities. This method is limited to dry materials and has been most widely used with titanium-bearing alluvial deposits and iron ores. The need for the total feed to be dry is a disadvantage; however, it is not necessary to dry the finely divided finished products, as is required after flotation. A combination approach, using electrostatic sorting with a fluidized bed, was developed by Burgess et al. (1970) for the beneficiation of coal.

## Flotation

The flotation method for concentrating mineral ores is based on principles of surface chemistry and has become the most widely used method of ore beneficiation in Canada. It has universal use for metal sulphide ores and other metallic ores, coal, and industrial minerals (see Table 1.3), often in combination with other concentration methods.

The reagents used in the flotation process are very numerous; they may be grouped into three general types: frothers, collectors, and modifiers.

According to Scott and Bragg (1975), the recycling of mill water presents few problems for most gravity and magnetic concentration circuits. There is also little difficulty with bulk flotation circuits. There are some circumstances, however, where the physico-chemical characteristics of the water are not compatible with the metallurgical processes. For instance, the water may contain substances that clog or otherwise damage equipment or machinery. Of the gross water use in Canadian mineral extraction and milling, an average of 70% was recycled in 1986 (Table 1.4). Recycling efficiencies depend on the mineral; in Quebec, for example, mines producing ferrous metals and industrial minerals recycled 78% of the water used, gold mines 51% and copper and zinc operations 40% (Mining Association of Quebec 1993). Less recent (but more specific) milling data of Scott and Bragg (1975) indicated that Canadian copper mills recycled approximately 80% of their water, and nickel-copper mills in the Sudbury region almost 100%. However, difficulties persist with the most demanding differential flotation circuits. Mills in northeastern New Brunswick, processing zinc-lead-copper ores, were unable to recycle more than 34%–70% of their water (Montreal Engineering Co., Ltd. 1973); in addition, they faced considerable challenges in reducing the concentration of their effluent (Rolia and Tan 1985). Scott (1987) describes further research on water treatment, combined in some cases with the segregation and separate treatment of water.

To allow solid particles to settle out, most water, whether recycled or discharged to the environment, is held in a pond or a series of ponds for a period of time. This containment also facilitates treatment to neutralize acids and to remove metals and radioactive substances, as well as the partial destruction of contaminating flotation agents (Bragg 1975). The more rapidly and efficiently these ends can be achieved, the better. Poling (1973) found that the addition of salt (sea) water accelerated the sedimentation rate of mill effluent considerably. He went as far as to propose the use of seawater in place of fresh water in the flotation circuits. The challenge of purifying wastewater sufficiently so that it

**Table 1.4** Water Use in Canadian Mining, 1986 (Mt)

| Category | # sites | Total intake | Recycled water | Gross water use | Total discharge | Minewater | Net discharge | Consumption | Gross water use by purpose | | | |
|---|---|---|---|---|---|---|---|---|---|---|---|---|
| | | a | b | a + b | c | d | c - d | a - c + d | Processing | Cooling | Sanitary | Other uses |
| **Extraction/Milling** | | | | | | | | | | | | |
| gold | 44 | 32.7 | 67.1 | 99.8 | 45.5 | 15.7 | 29.8 | 3.0 | 38.2 | 58.8 | 1.6 | 1.3 |
| Cu/Cu-Zn | 26 | 96.8 | 128.9 | 225.7 | 181.2 | 109.1 | 72.1 | 24.8 | 196.3 | 20.0 | 3.0 | 6.3 |
| Ni-Cu | 13 | 19.8 | 44.5 | 64.3 | 25.9 | 7.2 | 18.8 | 1.0 | 31.8 | 19.5 | 1.3 | 11.7 |
| Ag-Pb-Zn | 13 | 24.2 | 5.6 | 29.8 | 36.3 | 20.3 | 16.0 | 8.3 | 26.3 | 1.3 | 1.3 | 1.0 |
| uranium | 7 | 24.8 | 19.0 | 43.8 | 42.7 | 22.1 | 20.6 | 4.2 | 35.9 | 7.0 | 0.9 | 0.0 |
| iron | 8 | 230.0 | 685.8 | 915.8 | 248.3 | 12.8 | 235.5 | -5.5 | 883.9 | 28.3 | 3.7 | 0.0 |
| other metals | 6 | 3.0 | 6.8 | 9.8 | 4.0 | 0.5 | 3.5 | -0.5 | 9.7 | 0.0 | 0.0 | 0.0 |
| Sub-total | 117 | 431.4 | 957.7 | 1389.0 | 583.8 | 187.7 | 396.1 | 35.2 | 1222.1 | 134.9 | 11.8 | 20.3 |
| | | | | | | | | | | | | |
| asbestos | 7 | 2.5 | 1.4 | 3.9 | 15.8 | 12.1 | 3.7 | -1.2 | 0.4 | 2.8 | 0.6 | 0.2 |
| peat | 16 | 0.0 | ? | 0.0 | 0.0 | 0.0 | 0.0 | 0.0 | 0.0 | 0.0 | 0.0 | 0.0 |
| gypsum | 10 | 0.5 | 0.4 | 0.9 | 8.2 | 8.2 | 0.0 | 0.5 | 0.8 | 0.0 | 0.1 | 0.0 |
| potash | 12 | 14.0 | 83.9 | 97.9 | 20.8 | 6.1 | 14.8 | -0.8 | 22.4 | 60.6 | 1.2 | 13.8 |
| salt | 10 | 34.3 | 3.7 | 37.9 | 32.5 | 0.0 | 32.5 | 1.8 | 25.7 | 11.4 | 0.8 | 0.0 |
| other non-metals | 28 | 12.9 | 85.5 | 98.4 | 10.9 | 5.5 | 5.4 | 7.4 | 96.7 | 1.0 | 0.2 | 0.5 |
| Sub-total | 83 | 64.1 | 174.9 | 239.0 | 88.3 | 31.8 | 56.5 | 7.6 | 146.1 | 75.7 | 2.8 | 14.4 |
| | | | | | | | | | | | | |
| bituminous coal | 18 | 11.2 | 30.9 | 42.1 | 12.6 | 5.2 | 7.4 | 3.8 | 41.0 | 0.6 | 0.5 | 0.0 |
| sub-bituminous coal | 5 | 0.6 | 0.3 | 0.9 | 0.3 | 0.0 | 0.3 | 0.3 | 0.7 | 0.0 | 0.2 | 0.0 |
| lignite | 5 | 0.1 | ? | 0.1 | 6.0 | 5.9 | 0.1 | 0.0 | 0.0 | 0.0 | 0.0 | 0.0 |
| Sub-total | 28 | 11.9 | 31.2 | 43.1 | 18.9 | 11.0 | 7.8 | 4.1 | 41.7 | 0.6 | 0.7 | 0.0 |
| | | | | | | | | | | | | |
| Total | 228 | 507.3 | 1163.7 | 1671.1 | 691.0 | 230.5 | 460.4 | 46.9 | 1409.9 | 211.2 | 15.3 | 34.8 |
| | | | | | | | | | | | | |
| **Smelting/Refining** | | | | | | | | | | | | |
| all non-ferrous | 20 | 388.6 | 336.0 | 724.5 | 364.5 | 0.0 | 364.5 | 24.1 | 258.2 | 445.1 | 11.2 | 10.0 |

Sources: Tate and Scharf (1989a, b).

may be safely discharged to the environment—or reused in mining processes—can usually be met by using a combination of treatment methods such as neutralization, oxidation, reduction, precipitation, and ion exchange (Kemmer and Beardsley 1971).

A recent development has been the redesign of the conventional shallow flotation cell into a columnar tank. These column flotation tanks were used initially for the beneficiation of coal, but are now finding application with other ores, such as talc and base metal sulphides (Parekh et al. 1991). Combined with new bubble-generating systems, the column cells have advantages, including an increase in concentrate grade, higher output, and greater moisture content.

The complexity of the flotation process, coupled with the variable nature of the ores being processed, means that optimum operation requires a mixture of art and science. Combining operating experience and data into a computer-controlled "expert system" makes it possible, for example, to maximize chemical recycling while minimizing spills. A prototype expert system has been operated successfully on a copper-flotation circuit in British Columbia (Edwards and Mular 1992).

Some wet processing of asbestos-bearing rock has been carried out in the United States and is now being introduced in Canada. Most current Canadian operations use an air-separation method based on the density difference between the lighter fiber and the host serpentine rock (Sirois and MacDonald 1983). This dry-milling process is extremely dusty, and thus requires the use of an adequate air-filtering system to capture the dust at source so as to keep the fiber content of the air below two fibers per cubic centimeter, the acceptable level for human health (Somcynsky 1986). The filtered air is either discharged to the environment or, during cold weather, recirculated to the mill to conserve energy. Properly applied, present technology can reduce atmospheric emissions by 92%. The main emissions to the atmosphere are through the venting of the mill ventilation air and through wind erosion of tailings piles. Greater use of wet-separation methods would reduce emissions to air, but would replace them with increased hydrospheric emissions. More detail may be found in Chapter 11.

## Environmental Effects of Beneficiation

The main environmental effects of the various beneficiation processes discussed above, along with traditional and advanced abatement procedures, are summarized in Table 1.5. Practices vary from mine to mine, depending on orebody characteristics and company policy. Many already employ most of the advanced procedures. Table 1.6 summarizes the environmental advantages and disadvantages of the different concentration methods discussed in sections above.

Most of the atmospheric emissions resulting from the beneficiation stage arise from the transportation, crushing and grinding of the ore, and from dry separation methods. Such emissions may be considerably reduced by minimizing

**Table 1.5**  Abatement Procedures for Major Environmental Effects of the Beneficiation Phase

| Effect | Abatement Procedures | |
|---|---|---|
| | Traditional | Advanced |
| Solid waste (mill tailings) | Use of mine backfill<br><br>Industrial use | Removal of iron and sulphur before discharge to land or water<br>Greater use as backfill<br>Greater industrial use |
| Water use and liquid waste (process water) | Some recycle and reclaim<br><br>Settling pond and lime treatment before discharge to waterways | Greater recycle and reclaim percentages<br><br>More effective treatment before release<br>Reduction of amount of water used |
| Atmospheric emissions (crushing, grinding and transportation) | Minimal | More use of dust collection devices, covering of transporting vehicles, and sheltering of storage piles |
| Atmospheric emissions (air separation) | Dust collection devices | Virtually complete control of emissions of such harmful substances as asbestos, combined with an efficient servicing and monitoring program |

the handling—and exposure to air currents—of fine material, and through the use of totally-enclosed process lines (Asklof 1974). Once created, atmospheric particulates may be removed from the air through the use of such devices as wet scrubbers, electrostatic precipitators, mechanical collectors, and fabric filters (Down and Stocks 1977). Although this is relatively easy to do for point-sources, it becomes much more difficult in the case of dispersed-source emissions arising principally from the handling of finely broken dry ore, and the transportation of concentrate and mill tailings from the mill all contribute to dust production and consequent fugitive emissions. Although assessing and managing fugitive emissions can be difficult and costly, considerable reductions can be achieved through proper design and simple "housekeeping" measures (Helming 1976).

In contrast to the air separation of asbestos, the main contaminants of flotation are hydrospheric in nature, rather than atmospheric. While some materials are passed on to the metallurgical stage as components of the concentrate (such as iron, sulphur, silicate), the bulk is waste rock sent to the tailings pile. A large fraction of the reagents and water used in the process may be recycled.

**Table 1.6** Environmental Effects of the Mineral Concentration Methods Used in Canada

| Concentration Method | Environmental Advantages | Environmental Disadvantages |
|---|---|---|
| flotation | process water can be recycled | large water requirement; hydrospheric residuals contain reagents, particulates, and often heavy metals |
| gravity separation | | |
| dry | no hydrospheric residuals; no reagents involved | large airflow needed; atmospheric residuals require extensive control measures and subsequent disposal |
| wet | no atmospheric residuals; media are generally recycled | possible hydrospheric residuals if heavy media used |
| magnetic and electrostatic separation | no reagents involved; water requirements negligible with dry process or small with wet process and water can easily be recycled | some release of particulates to atmosphere with dry process, or hydrosphere with wet process |

## Metallurgical Processing and Refining

Metallurgical processing and refining include all treatment an ore receives after its extraction and beneficiation. Most of this treatment involves changes in the chemical nature of the mined minerals.

With few exceptions, this further processing of mined and beneficiated ores involves the isolation of a metal from its sulphide, oxide, silicate, carbonate, or other compound. These types of processes are called extractive metallurgy , and may be broadly divided into three groups: *pyrometallurgy*, in which elevated temperatures are used to assist the extractive reaction; *hydrometallurgy*, in which a liquid solvent is used to leach out the metal from its ore; and *electrometallurgy*, in which electrical energy is used to effect the dissociation of the metal in aqueous solution.

While some metals are extracted predominantly by one or other of the above methods, others involve combinations of two or even three of the methods. Table 1.7 summarizes the methods used at present in Canada. Many of the rarer metals are produced as by-products of the processing of other more common metals, and are extracted from smelter slags, flue dusts, or electrolytic slimes, as described in Chapters 5 –11. On average, more than 90% of the metal values contained in the concentrates are extracted during metallurgical processing (Lemieux et al. 1992). Recovery rates range from approximately 88% for lead, zinc, and molybdenum to 95% for copper and 98% for gold.

**Table 1.7** Extractive Metallurgical Methods Used for Different Types of Ores in Canada

| Ore Type | Metal | Pyro-metallurgy | Hydro-metallurgy | Electro-metallurgy | Comments |
|---|---|---|---|---|---|
| quartz | Au | | x | | |
| | Ag | x | x | x | mainly by-product of base-metal refining |
| oxide | Ca | x | | | electric retort-Pidgeon process |
| | Cu | | x | | |
| | Fe | x | | | sintering |
| | Mg | x | | | electric retort-Pidgeon process |
| | Si | x | | | electric furnace |
| | Th | | x | | by-product of uranium refining |
| | Ti | x | | | electric arc smelting |
| | U | | x | | |
| sulphide | Sb | x | | | by-product of Pb smelting |
| | Bi | x | | x | by-product of Pb refining |
| | Cd | | x | x | by-product of Zn smelting & refining |
| | Co | x | x | x | by-product of Ni-Cu refining |
| | Cu | x | | x | |
| | In | x | x | x | by-product of Pb-Zn smelting |
| | Pb | x | | x | |
| | Hg | x | | | by-product of base-metal refining |
| | Mo | x | | | $MoO_3$ from $MoS_2$ by Pidgeon process |
| | Ni | x | | x | |
| | Pt metals | | x | | by-product of Ni-Cu refining |
| | Rh | | x | x | by-product of Cu-Mo roasting |
| | Se | | x | x | by-product of Cu refining |
| | Te | | x | x | by-product of Cu refining |
| | Zn | x | x | x | |

The number of multiple entries in Table 1.7 provides evidence of the limitations of the metallurgical classification presented above. Zinc concentrate is an example of a metal recovered by a combination of all three processes: it is first roasted to convert the insoluble sulphide to an oxide, then leached with dilute sulphuric acid to produce a zinc sulphate solution that is purified, and, finally, electrolyzed to produce metallic zinc.

## Pyrometallurgy

Pyrometallurgy is the oldest form of extractive metallurgy and still accounts for more than half of non-ferrous metal production (Kellogg 1981) and, in Canada, usually involves sulphide ores. The purpose of the process is to convert a metal from its sulphide or oxide form to a form close to the pure element through the elevation of temperature. The heat required to raise the temperature may be derived from the combustion of a fossil fuel, the oxidation of a component of the ore, electricity, or other source. This definition, although not universally used, may be justified on the basis that the primary concern in a pyrometallurgical process is the high temperature itself, rather than the means of achieving it. The main present-day use of pyrometallurgical extractive methods is in the smelting of copper and copper-nickel sulphide ores.

Pyrometallurgy involves four stages: preparatory treatments, such as drying, calcining, and roasting, and such agglomerating processes as sintering; smelting in blast furnaces, reverberatory furnaces, retorts, etc.; converting; and fire refining.

The most widely used pyrometallurgical process is smelting, in which the ores or concentrates are combined with flux materials and heated to sufficiently high temperatures to cause melting.

The preliminary roasting stage burns off 20%–50% of the concentrate's sulphur content, releasing it as sulphur dioxide. The roasting stage is frequently omitted, particularly if the sulphur content of the ore is low, in which case more sulphur will be driven off during smelting. The concentration of the sulphur dioxide released to the atmosphere by multiple-hearth roasters is quite low (2%–4% by volume) because of the large volume of air involved in the combustion process, making it expensive to recover the gas. Fluidized-bed roasters, on the other hand, use less air and produce sulphur dioxide concentrations in the range of 10%–14%; however, they require finely ground, dry concentrate feed. While the roasting process is autogenous (i.e., the heat is provided by the combustion), the smelting stage usually requires supplementary heat.

In the smelting operation, either raw or hot roasted concentrate is melted, along with a siliceous flux, to produce matte, which is approximately 50% pure. Since most of the sulphur has been removed during roasting, the concentration of released gases from smelting and converting may be as low as 1%–2%. Several types of furnaces and converters, including flash and electric furnaces and top-blown rotary converters, avoid combustion gas dilution and release sulphur dioxide at much higher concentrations. The conversion process raises the metal purity as high as 98% by injecting air or oxygen to oxidize most of the remaining iron and sulphur. The dust and fumes from this operation are collected and recycled into the process flow. Sulphur dioxide concentrations from this stage—usually 3% to 8%—may be recovered, although the variable concentration and high exhaust temperature introduce some difficulties. The metal value of slags produced by the conversion process is usually so rich that slags are recycled into smelter feed.

The output from the converter may undergo a fire-refining process before it leaves the smelter. This involves adding a flux to the molten metal and blowing air through it to oxidize the remaining impurities; through this process, impurities are reduced by about half in preparation for electrolytic refining.

Since, as noted earlier, most of the ores processed by pyrometallurgy in Canada are sulphides, which contain large fractions of pyrite or pyrrhotite, the main contaminants are sulphur and iron. Both of these are oxidized during the smelting process, with the bulk of the sulphur becoming gaseous sulphur dioxide, and most of the iron ending up in the slag as iron oxide. Minor contaminants derive from other constituents of the mined ores and from introduced reagents such as fluxes and reducing agents.

A materials-flow chart indicating typical gaseous and other contaminants is presented in Figure 1.4. Atmospheric emissions are in the form of gases— primarily sulphur dioxide—and particulates, which include a wide range of metallic and non-metallic compounds. Once in the atmosphere, these substances may undergo, in very complex ways, transportation, chemical reaction, dilution, transformation, and deposition. Some of these mechanisms are described in Chapter 2, and more detailed information on emissions from non-ferrous metal mining and processing are given in Chapters 3 and 7.

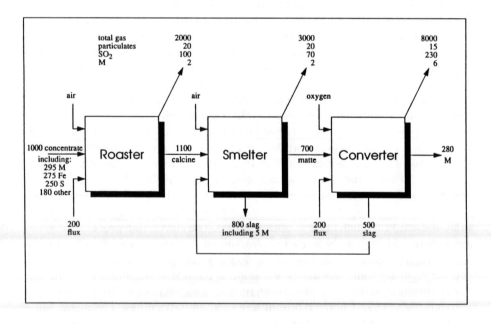

**Figure 1.4** Materials-flow diagram for a typical base-metal (M) pyrometallurgical operation. All quantities are in kg and relate to 1 t of roaster concentrate feed. Sources of data: Environment Canada (1981, 1982a, b).

## Hydrometallurgy

The term hydrometallurgy describes the use of solvents to extract metals from a wide variety of ores, including gold, copper, nickel, zinc, and uranium. In comparison with pyrometallurgy, this method is relatively new, dating back only a few centuries (Van Arsdale 1953). Some of its early uses were metal recovery from waste ore heaps in Hungary during the sixteenth century and, a century ago, from mine water in the United States (Hiskey 1986). The method has an environmental advantage over pyrometallurgy in that no sulphur dioxide is produced, and an economic advantage in that the sulphur content of the ore is more easily extracted in its more valuable elemental form (Sirois and MacDonald 1983). In many large operations in the United States, and elsewhere, this method is used to process the oxidized fraction of an ore such as copper, the main part of which is processed by pyrometallurgy (Kilburn 1993). Its widest use in Canada, on the basis of the weight of ore processed, is in the processing of zinc sulphide ores, which was discussed at the begir        ~f the section on metallurgical processing and refining.

The most commonly used hydrometallurg
ous solutions of acids or bases. The overall h
1.5) comprises four steps: a preparation ph

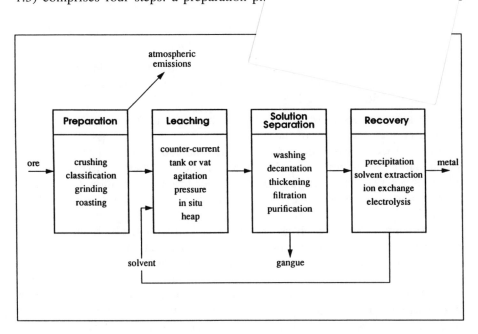

**Figure 1.5** Process-flow diagram for a typical hydrometallurgical operation, illustrating main throughputs and residuals.

leaching; solid-liquid separation and treatment; and concentration and deposition of the metal from the pregnant—that is, metal-bearing—solution.

The preparation phase serves two distinct purposes: to render the ore soluble in the desired solvent, usually through a procedure (such as oxidation, reduction, sulphitization, etc.) that effects a chemical change in the ore, and to modify the particle size distribution of the ore in order to optimize the subsequent leaching operation. For example, zinc sulphide, which is insoluble in all common solvents, may be rendered soluble by means of an oxidizing roast. Normally, the smaller the particle size, the greater the rate of extraction; however, particles that are too fine will impede the flow of the solvent and will require agitation for proper leaching. In many cases, very little preparation is used, with slightly more for leaching dumps and heaps than for *in-situ* leaching (Hiskey 1986). The use of more elaborate preparation and leaching methods is usually justified only for larger high-grade orebodies (Lastra and Chase 1984).

After the ore is prepared, it must be brought into contact with the solvent in one of several different ways, including:

- *in-situ* leaching and solution mining, which are carried out underground (see previous section on non-entry mining);

- dump leaching, which involves the leaching of rock in waste dumps;

- heap leaching, in which run-of-mine ore is stacked on an impermeable pad on the surface and sprayed periodically with a solvent;

- several other variants, such as vat (described in the Nova Gold case study in Chapter 6), percolation, thin-layer, agitation, differential, and pressure leaching (Hiskey 1986; Gilchrist 1989).

The last of these is a relatively new development that uses hydraulic pressure to speed up the dissolution of the ore.

In the processing of gold ores, a common method used to separate the gold is cyanidation. A cyanide compound is added to the crushed ore to dissolve the gold particles. Once the gold has been precipitated from the solution, the solution must be treated to remove the cyanide or to destroy it to prevent it from leaching into the environment.

Following the solution phase, the pregnant solution must be separated from the gangue by means of decantation, thickening or filtration. Except in heap or *in situ* leaching, the gangue is often washed with water to remove traces of the solvent. This may be carried out by counter-current decantation, in which the slurry flows through a series of three to six thickeners in the opposite direction to the wash water. The discharge flow is approximately 60% solids, so that two-thirds of a tonne of water is discharged with each tonne of solids. The use of density gauges and variable-speed pumps can improve wash efficiency and minimize the amount of water discharged (Scott and Bragg 1975).

The final step of the hydrometallurgical process is the concentration and/or purification of the pregnant solution, and the deposition of its contained metals from solution. The desired metal values from a pregnant leach solution are usually recovered by one of the following methods:

1. chemical precipitation, which involves the introduction of another metal into the leaching solution where it dissolves, replacing the original metal in solution and so causing it to precipitate out;

2. ion exchange, which is the diffusive exchange of ions between a pregnant solution and an added crystalline or resinous solid, without any change in the structure of the solid;

3. solvent extraction, by which one metal is separated from another through the use of an organic solvent that is immiscible with water.

A fourth method, electrolysis, will be discussed in the following section, since it is properly an electrometallurgical process.

The quantities of water required for leaching operations vary greatly with the fraction of solids in the leaching stage, which ranges from 30%–70% (Scott and Bragg 1975). Some uranium operations using steam for heating require, for this purpose, approximately 100 kilograms of water per tonne of ore processed. There are also large water losses in the form of vapor in agitation leaching. When counter-current decantation is used, wash water requirements may reach 2.5 tonnes per tonne of ore. Close control of pulp water content in counter-current decantation can improve wash efficiency and reduce water use, thus minimizing the volume of contaminated water.

## Electrometallurgy

Electricity may be used to aid in the [*quantities of water required for leaching operation*] ores either by electrolysis, including electrowi[nning and electroplat]ing, or by electrothermic methods, such a[s ... furnaces] and induction furnaces. Electrolysis meth[ods ...]. The others are pyrometallurgical methods and[ ...] only in terms of those special factors that diff[er from com]mon pyrometallurgical processes.

Electrolysis for metallurgical processing was developed toward the end of the nineteenth century (Kellogg 1981; Gilchrist 1989). It involves electrochemical reactions in either aqueous or fused-salt solutions.

For *electrorefining*, the impure metal acts as the anode; a solution of high electrical conductivity is used as the electrolyte, and the pure metal is plated out onto the cathode. Electrolysis in aqueous solution is most widely used for the electrorefining of copper, lead, and nickel (Environment Canada 1987). Impurities either remain in solution, adhere to the anodes or fall to the bottom of the cell.

Many metals—such as bismuth, cadmium, cobalt, gold, silver, selenium and tellurium—may be recovered from these impurities.

In contrast to the soluble anodes used in electrorefining, the *electrowinning* process uses insoluble anodes. The metallic source is, instead, the electrolyte, which may be an aqueous solution or a molten chemical compound of the metal, such as a chloride or an oxide. The main example of the former process is the electrowinning of zinc from purified zinc sulphate solution obtained from the leaching of roasted zinc concentrate. The latter method is used for metals that cannot be electrolyzed from aqueous solution, such as aluminum, magnesium, and the alkali earths (Sirois and MacDonald 1983). Energy requirements for electrowinning are even higher than for electrorefining, ranging as high as 8.6 $GJ \cdot t^{-1}$ for copper in aqueous solution, and 86 $GJ \cdot t^{-1}$ for aluminum in fused-salt bath. Sirois and MacDonald (1983) noted that there was scope for improving the cell design and efficiency of the fused-salt process for the recovery of lead and associated metals from lead chloride electrolyte. A step in this direction was recently taken by Murphy and Chambers (1991), who were able to reduce the energy requirement for the process from 4.7 $GJ \cdot t^{-1}$ to less than 2.5 $GJ \cdot t^{-1}$ by means of a number of technical modifications to the cell design. There is still room for research directed toward a better control of impurities in both electrorefining and electrowinning.

Electrochemical methods may be included under the pyrometallurgical heading, since it is the heating effect—and not the electric current—that is of importance. However, it is pertinent to list here the types of electrochemical processes and to indicate the extent of their use in Canada. Electric heating has a number of advantages over fuel firing for high temperature capability, including more accurate temperature control, no requirement for air to support combustion, and greater thermal efficiency. These result in many benefits from the environmental perspective: less heat loss, less gas emission, and the elimination of contaminants from fuel combustion. The main types of electric furnaces are the arc furnace—used extensively in the steel industry and in the manufacture of ferroalloys—and induction and resistance furnaces.

## Environmental Effects of Further Processing

The disparate issues of technology, economics, and environmental responsibility encompassed in decisions involving such trade-offs are drawn together under the rubric "total accounting." What is implied is that evaluations of the costs of using a particular technological approach should not be restricted to the company's direct capital and maintenance costs and operating expenses. Rather, they should also include the expected costs to society of the company's having used that technological approach, in the form of repairing environmental damage, restoring the site to a useful state, even the imputed costs of "free" commodities such as air and water. In total accounting, for example, the method of

**Table 1.8**   Abatement Procedures for Major Environmental Effects of the Further Processing Phase

| Effect | Abatement Procedures | |
| --- | --- | --- |
| | Traditional | Advanced |
| solid wastes (tailings from hydrometallurgy, smelter slag) | disposal on land, in freshwater bodies, and in the sea; some use as mine backfill and for other industrial purposes | more efficient disposal to minimize environmental impact; greater use as mine backfill and for other purposes; removal of potentially harmful substances, such as sulphur and iron before disposal |
| disturbance of aquifers (by *in-situ* leaching) | after-the-fact restoration | adequate mine design and planning before the fact |
| smelter gases (particulates, $SO_2$, etc.) | sulphur extraction (elemental and acid forms); dust collection devices; production cutbacks or shutdowns; higher stacks | further sulphur extraction; greater use of dust collection devices; development of new processes with lower emissions or more useful residuals |
| energy wastage | some of the energy released in pyrometallurgy is used in the process | many opportunities exist for reducing energy consumption in hydrometallurgy and electrometallurgy, and for capturing more of the waste heat from pyrometallurgy |

producing electrical heating would be considered in making the choice of a furnace or similar technology.

The major environmental effects of the further processing stage of mining in Canada are summarized in Table 1.8. Whatever method of extractive metallurgy is used, the fact that most Canadian metallic ores are sulphides is inescapable. The processing of sulphide ores leads inevitably to the production of either elemental sulphur or sulphur compounds, such as sulphur dioxide, or sulphates dissolved in waters, all of which are potential contaminants. Undoubtedly, the atmospheric emission of sulphur dioxide from pyrometallurgical processing plants is the best known and by far the most significant effect because, if not controlled, it leads to acid deposition. Although smelter emissions have been reduced considerabl_____ _____ __ ___ _____ __ emissions per unit of ore processed), they stil *atmospheric emission* _pogenic sulphur dioxide emissions in Canada *of sulphur dioxide* .990). That most Canadian base metal ores are *leads to acid* of sulphur inevitable, but most of it can be di *deposition* nment as gaseous sulphur dioxide. An increas____ _____ _____ and marketed in the forms of elemental sulphu___ _____ _____ other sulphur compounds such as calcium s_____ _____ _____ 1994). In the past, most sulphur products w_____ sulphur, but an increasing proportion is now being produced as by-products of the oil and gas and mining industries. This has resulted in periodic surpluses of sulphur which, in the words of Vroom (1971), are a "challenge and an opportunity" for humanity to manage. Some of the newer uses for sulphur are found in pre-formed and foamed products; in sulphur concrete and asphalt; as an ingredient in paints and coatings and as an impregnant; in plant, soil, and waste treatments; as a reaction-medium; and in petrochemical processes and fuel cells.

Sulphur dioxide has been a major pollutant, acting at regional and international scales. It is produced naturally by such events as volcanic eruptions and by certain industrial activities, primarily smelting and the combustion of fossil fuels, particularly coal, with a high sulphur content. There is some evidence that anthropogenic emissions of sulphur dioxide have levelled off (Brimblecombe and Lein 1989) and that atmospheric concentrations may even be declining (Environment Canada 1990). The goal of Ontario's *Countdown Acid Rain Program*, which required the province's four main sources of sulphur dioxide (Inco, Ontario Hydro, Algoma Steel, and Falconbridge) to reduce their total emissions from 2,194 kt in 1980 to 885 kt by 1994, has been met.

The other main atmospheric contaminant attributable to this stage of mining is particulate matter, which amounted to approximately nine per cent of the national total of anthropogenic emissions in 1985 (Kosteltz and Deslauriers 1990). Some of the particulates are chemically inert, but others are chemically reactive. A number of other substances are released to the atmosphere, but in such small quantities that only workers in the processing plant are likely to be affected; these include electrolytic mists (Stern 1968), metal and metal-oxide fumes (Stern et al. 1984), and phosphine and arsine from hydrometallurgical and electrometallurgical processes (Habashi and Ismail 1975).

The main lithospheric contaminant of this further processing stage is smelter slag, followed closely by tailings from hydrometallurgical processing. While some of this material is being recycled or re-mined for secondary values and while other practical use is being made of more of it, there is still a considerable amount left that is discharged by dumping. The industry now recognizes that returning waste by-products to the smelting process is environmentally sound and has the financial advantage of recovering primary metal values (Kosub 1991). Potential new uses, such as substitutes for certain industrial minerals, have been the subject of study (Emery 1975). The iron content of slag might be exploited if an economical way could be devised to extract it. When disposal is the only feasible course, the disposal site must be chosen carefully, and a program of stabilization, followed by reclamation of the area, is the logical procedure (Bradshaw and Chadwick 1980). While slag is not completely benign—iron can leach from slag heaps under some conditions—it is much more stable and does not present the environmental problems associated with acid-generating tailings and waste rock dumps.

There are few direct hydrospheric emissions from the further processing stage. While water is involved in most hydrometallurgical and electrometallurgical

**Table 1.9** Extractive Metallurgical Methods Used in Canada

| Metallurgical Method | Environmental Advantages | Environmental Disadvantages |
|---|---|---|
| pyrometallurgy | minimal liquid wastes; lower energy requirements | more demanding beneficiation requirements; sulphur dioxide emissions; generally less pleasant working environment |
| hydrometallurgy | little discharge to the environment if done on a closed-circuit basis; tailings generally less finely ground than for pyrometallurgy; greater potential for integration with ancillary operations | waste solutions may be released to environment; large energy requirements |
| electrometallurgy | produces high-purity product; high degree of control possible | large amount of electrical energy required; production of electrolytic mists and some toxic gases; possibility of toxic chemical wastes to hydrosphere; generally used together with one of the above methods |

processes, it is recycled to a large enough extent (Tate and Scharf 1989a) that there is little discharge to the environment, except through spills and leakage. In 1986, almost 50% of the gross water used in non-ferrous smelters in Canada was recycled (Table 1.4). In some cases, solid wastes are dumped into water bodies for disposal, and may produce siltation and other effects. A study of the effects of smelter wastes in the Columbia River showed that, while slag dumped into the river had little effect on water quality, smelter wastewater increased the water hardness as well as the loading of calcium, sulphate, fluoride, phosphate, and zinc (Reeder 1971). Many of the concentration increases were slight, due to the large magnitude of river flow in comparison with the quantity of effluent, but natural mixing processes did not produce full dilution until 5 to 15 km downstream from the discharge point. This issue is discussed in greater detail in Chapter 3.

The three methods of extractive metallurgy have been compared with one another in Table 1.9 from the perspective of their effects on the natural environment. In practice, electrometallurgy is always used in conjunction with one of the other two, so that the choice is usually between pyrometallurgy and hydrometallurgy. In situations where physical factors do leave this choice, there are a number of tradeoffs between the two methods, from both the economic and the environmental perspectives. The choice is a complex one, involving many factors. Hawley (1974) discussed the two processes in detail and concluded that "from an environmental standpoint, hydrometallurgy is the route to follow." By contrast, the U.S. National Academy of Sciences (Committee on Mineral Resources and the Environment 1975) has advised, "Rather than a substitution of hydrometallurgy for pyrometallurgy, the design of more energy-efficient processes must seek out the special advantages of both approaches." Although hydrometallurgy is finding wider use, an estimate by Kellogg (1981) attributed more than 95% of world production of all metals—including more than 50% of non-ferrous metals—to pyrometallurgy.

Emissions of particulates and sulphur oxides can be—and in many cases have been—reduced to a small percentage of the uncontrolled values. The energy requirements of pyrometallurgy are largely supplied by the oxidation of the sulphur content of the ore and are much less than those necessary for hydrometallurgy. There is evidence that innovative design can reduce process-gas volumes and fugitive emissions and can lead not only to less environmental stress, but also to more efficient operation. Four practices that would achieve these ends were recommended by Kellogg (1981) and Environment Canada (1982b):

1. maximum use of oxygen in place of air, for more complete oxidation;

2. use of intensive reactor design, such as flash, injection and cyclone smelting;

3. full use of the fuel value of sulphide concentrates as an energy supply for the process; and

4. design of continuous processes to replace batch processes.

Oxygen enrichment reduces the requirement for the use of additional fuel, eliminating the contaminants produced by its combustion. At the same time, it produces less off-gas with higher sulphur dioxide concentration (simplifying conversion to sulphuric acid). Research is also progressing in such areas as the recovery of trace elements from smelter fumes and dusts and the application of mathematical modelling and expert systems to improving the efficiency of metallurgical operations (Kosub 1991; Parekh et al. 1991).

The large amount of heat generated by pyrometallurgical processes is often vented to the atmosphere; however, some operations are able to recover much of the heat from smelting (Snelgrove and Taylor 1989), as well as from sulphuric acid plants (Bond 1989). The heat is usually used to make steam and to generate electricity.

Problems associated with hydrometallurgy are also being addressed. A major area of research is the development of leaching methodology to reduce costs and to convert undesirable contaminants to forms that are marketable. Sulphur dioxide pressure leaching, which produces elemental sulphur as a contaminant, continues to receive considerable attention. The use of other oxidants, such as hydrogen peroxide and oxygen, is also being investigated for use with lead, gold, and silver. Bacterial oxidation, by means of the bacterium *Thiobacillus ferrooxidans* (McCready 1986b) has already been used to break down iron and sulphide compounds in refractory gold ores in South Africa; the method is now being developed for use at similar mines in the United States and Canada (Scott 1992a; Kennedy 1992).

This bioleaching approach will circumvent the need for roasting (and its concomitant sulphur dioxide production) when processing these difficult ores (Scott 1992b). It also has the advantage of requiring much less cyanide, when used with an autoclave, than was previously necessary. The ferric chloride leach process, developed by CANMET, has the potential to treat relatively low-grade bulk base metal concentrates, obtaining very high recoveries of the contained metals (Craigen 1991). Substantial reductions in energy use are possible through the use of light composite impellers in leaching and solvent extraction applications, and by adding oxidants such as hydrogen peroxide or sodium chlorate to the leaching process to reduce retention times (Lakshmanan et al. 1989). Electrowinning and electrorefining processes could be rendered more energy-efficient through the use of periodic current reversal and higher liquor temperatures.

## Decommissioning

The mining of a finite orebody leads inevitably to its exhaustion. In some cases, this takes fewer than five years; in others, more than fifty. The life of a mine may be extended if new technology and/or higher market prices permit the exploitation of lower grades of ore or if on-site exploration finds additional mineralization, but inevitably the operation will come to an end, and the mine will close. Mills and smelters also have finite lives, although generally longer than those of the mines themselves, since processing facilities usually service several mines at various stages in their life cycles. The rehabilitation of areas disturbed by mining activities has been mentioned briefly in other sections of this chapter, particularly with respect to mineral exploration activities. This section discusses, in general terms, the rehabilitation of disturbed lands and aquatic areas at the termination of mining activities. Chapter 4 and the various case histories consider these issues in some detail.

Historically, when mining operations reached the end of their useful lives, they were abandoned; disturbed sites, waste dumps, and mine structures were seldom reclaimed. As a result, altered surface topography, surface and sub-surface drainage, disturbed vegetation and soil, and abandoned roads and buildings were left behind after the operation ceased. This approach is no longer acceptable; moreover, it has become clear in recent years that many exhausted underground mines may be suitable for other uses, such as the storage of liquids or dry goods, horticultural nurseries, industrial facilities, or even tourist attractions (Boivin 1989).

Legislation often requires that the mine workings, building sites, and other disturbed areas be returned to their original state, or at least to "a state which will support plant and animal life or be otherwise productive or useful to man at least to the degree it was before it was disturbed" (Bratton 1987). The Mining Association of Canada, in its *Guide for Environmental Practice* (Miller 1990), exhorts its member companies to "return disturbed land to a safe, stable and productive condition." In some cases it is impossible to return land to its former state; in mountainous terrain, for example, where steep slopes often lead to instability and severe erosion, a return to the original contouring is not attempted (Bell et al. 1989). In these circumstances, as well as for many gravel pits and surface coal mines, moulding new topography, using a landscape architecture approach, is more practicable (Hall 1987). Of an estimated 2,700 gravel pits in Alberta (Kryviak 1987), approximately 100 have been reclaimed by means of recontouring, regrading for slope stability and positive drainage, and revegetating. Some reclaimed mining areas have been remodelled for recreational use, even to the extent of including lakes stocked with fish (Milliken and Mew 1970; Maneval 1975; Browning 1987).

In the case of agricultural land, the goal is to return the site either to its pre-mining productivity or to a higher level: where mining disturbance caused the mixing of subsurface limestone with surface glacial outwash material, the disturbed area can be more productive agriculturally after mining than before (Ripley et al. 1982).

The restoration of disturbed sites is often complicated by acid drainage and high concentrations of toxic substances. In many cases, proper application of existing methodology is successful. A process for reclaiming land disturbed by surface mining has been developed by Sweigard and Ramani (1986). It is based on the assessment of several natural factors—including topography, climate and soil characteristics—in relation to cultural factors such as site size, shape and accessibility. The process leads to the development of alternative scenarios and their evaluation and selection.

Land reclamation is a process that promotes soil conservation and/or soil improvement, as well as the productive use of disturbed land. According to Peterson and Etter (1970), this involves three phases: prior land-use planning; physical operations to achieve a suitable topography if necessary; and natural or assisted revegetation and subsequent management of reclaimed land. In some areas, such as tundra, it is important to restore the pre-mining condition and not try to improve its production.

While most discussions of reclamation have centered on the damage done to land surfaces, it is important to remember that mining also has reclaimable environmental effects in other areas. These include damage to water bodies, hazards to human health and property and to fish and other wildlife, and aesthetic damage. Hogg (1971) defined reclamation as the amelioration of any of these deleterious effects. This approach provides a necessarily comprehensive view of the subject, emphasizing the interrelations between the various spheres of influence. Polluted land and water, for example, affect one another through various dispersal processes (see Chapters 2 and 3); reclaiming one disturbed site may well have benefits elsewhere (Jackson 1995).

Repairing the damage caused by mining activities—whether the area is returned to its original condition or to an acceptable alternative—is usually a fairly expensive proposition. In addition to one-time clean-up activities, it may require monitoring and ongoing remedial action—such as neutralization of acid drainage—for many years to come. Monitoring for radioactivity will also be a long-term commitment. Since decommissioning is now considered part of the overall mining operation, companies should budget for the associated expenses from the outset.

Because environmentally responsible decommissioning activities are a relatively recent practice, compared to mineral extraction, this stage of mining is still in its infancy. As it matures, it will undoubtedly become more sophisticated and yield better results. Detailed examples of current reclamation and abandonment practices will be presented in Chapters 4 through 11.

## Summary

The mining process generally involves four activities: exploration, development, extraction and processing, and decommissioning. Each of these phases has the potential to affect the environment to varying degrees.

The search for mineral deposits becomes more difficult as those closer to the surface and in known areas of mineralization are discovered and exploited. Development of more powerful prospecting methods is essential; otherwise, it will be necessary to use current methodology more comprehensively and intensively just to satisfy current needs. The inevitable result of this would be increasing pressure on the environment as surface exploration probes wider and deeper and pushes the geographical frontiers.

It is clear that proper design of mining facilities and infrastructure helps to minimize the impacts of operation. Most of the environmental protection plans are created and implemented during the development stage. In fact, ultimate decommissioning plans are often part of the initial planning.

Underground and surface mining are the main methods of ore extraction utilized in Canada. In 1992, over 70% of mineral production (exclusive of industrial minerals) came from surface mining operations. For both types of operations, waste rock is the primary source of potential environmental concerns. Associated with the disposal of waste rock are the risk of acid drainage, which can lead to heavy metal contamination, and problems with atmospheric dust emissions.

The processing practices—beneficiation, smelting, leaching, and refining—may create other sources of environmental concerns, such as solid wastes (tailings), contaminated water and liquid wastes, and dust emissions. These concerns will be dealt with in Chapters 3 and 4 in general terms and in the chapters related to the commodities for which there are specific environmental issues.

# CHAPTER TWO

# ECOSYSTEMS

## Introduction

Burton (1991) describes mining and other resource development as "concerned with the deliberate modification of existing ecosystems and their associated environment." He notes:

> Some ecosystems, often characterized by physical and biological diversity and complexity, are essentially stable in character and can be substantially altered without significantly impairing their integrity. Others, often characterized by structural simplicity or monoculture biology, are potentially unstable....

The mining industry typically faces the latter variety, and Burton points out that managing them effectively "requires a systems-oriented approach and requires some understanding of, or at least concern for, the dynamics and stability of the system." While Burton was speaking in the context of reclamation and rehabilitation of minesites, it is valid to extend his observations to all six stages described in the previous chapter.

All organisms, including people, live in ecosystems, which are communities of plants, animals, and micro-organisms interacting with their environment and each other. Thus, an organism's environment includes not only the physical and chemical factors that affect it, but other organisms with which it interacts.

Before it is possible to evaluate the impact of an activity, it is important to understand the system that is being affected. What factors would be measured before and after mining activity for purposes of environmental assessment? Are these factors different on land and in water? The purpose of this chapter is to describe attributes of ecosystems that may be altered by mining and delineate what aspects of them can readily be measured. Chapter 3 then relates mining activities to these elements and explains their impacts on ecosystems.

In Canada, mining is carried out in three broad types of ecosystem: urban areas, managed areas of agriculture or forestry, and unmanaged areas, such as boreal (i.e., northern) forests or tundra. The factors to be measured to examine the effects of mining may vary from one type of ecosystem to another, particularly between terrestrial and aquatic ecosystems.

The scale of definition of ecosystems is arbitrary and depends on the immediate purpose. A prairie slough or a small Arctic island can be considered as a separate ecosystem, while the circulations of the atmosphere and the oceans demand consideration of the biosphere or the entire globe as a unit. In management of natural areas, it is now customary to work at the larger, "landscape" scale, such as a watershed. A forest, for example, is considered holistically with all its rivers, streams, lakes, and wetlands. While recognizing the linkages between them, it has generally been found easier for our purposes to divide regions into terrestrial, wetland, freshwater aquatic, and marine aquatic areas. In discussing the impacts of mining, focus necessarily shifts, for example, among contaminant uptake by an individual salmon, the population to which the salmon belongs, the river and ocean communities in which the population moves, the animals that eat salmon, and the physical and chemical characteristics of their habitats. Close and distant views are both necessary.

Most of the effects of mining are stresses of various kinds. The responses of an ecosystem to stress include five general possibilities:

1. after some disruptions, it resumes operation as before;

2. it operates at a different level (e.g., lower crop production);

3. it operates with some new structures (perhaps with some new species or food pathways);

4. a new ecosystem with different structure emerges; or

5. it collapses.

According to Keith (1994), "The complexity of ecosystems makes it impossible to predict precisely how the self-organization and regenerative processes will occur. The challenge of ecosystem management is to protect or restore the capacity to self-organize and therefore maintain or regain integrity. The objective of maintaining 'ecosystem integrity' should be the overall purpose guiding human relations with the environment."

In this chapter, attributes of ecosystems are grouped under the following headings and ways of measuring each attribute are outlined: (1) Measurement of Energy Flow, (2) Food Webs, (3) Structure and Niches, (4) Communities and Diversity, (5) Populations, (6) Biogeochemical Cycling, (7) Succession, and (8) Stability and Resilience.

## Measurement of Energy Flow

To understand the impact of mining on ecosystems, we need to know whether it is affecting the flow of energy through the ecosystem; we therefore need to know where and how to measure it. All organisms use energy. We are most familiar with human energy exchanges, in which we burn food calories and absorb and lose heat (through radiation, convection and conduction) to air, water, or solids with which we are in contact. The energy derived from food is used to maintain body heat, for growth and tissue renewal, and for activity such as movement. Lack of food or extremes of heat or cold kill us. Other animals similarly derive their energy from the plants they eat or from eating each other.

Green plants can "fix" a small part of the electromagnetic spectrum in the visible light wavelengths to make carbohydrates from carbon dioxide and water. The energy is fixed as chemical bonds, primarily as sugars, and the necessary catalysts are pigments such as chlorophyll. Factors that can limit photosynthesis include shortfalls in light, water, carbon dioxide, essential nutrients for cell structure and for other metabolic processes, and the quantity of chlorophyll that plants can deploy over a certain area. If a river becomes turbid because of sedimentation from, say, alluvial mining, lack of light or water for plants and algae may make photosynthesis impossible. Some bacteria obtain energy for their use by catalyzing exothermic reactions in mud or water, some of which involve sulphur compounds. In addition to having energy-fixing green plants and bacteria, some ecosystems are net importers of energy fixed elsewhere. For example, large quantities of dead plants and animals are washed into both marine and freshwater estuaries, and in some lakes fallen leaves boost the energy inputs in autumn.

Energy fixation varies greatly in Canada, as plants are adapted to different climatic regimes. In this book, therefore, we consider mines and their impacts in the context of ecoregions; this topic is discussed later in the chapter.

How is energy fixation measured? On land, either the fixed energy per unit area of ground per unit of time is measured or the mass of the dry matter produced can be estimated. The rate is called the *primary productivity*; the sugar or dry matter is called *primary production*. In practice, these parameters are not used in many environmental assessments, but regional averages provide expected values that can serve as guides to management of recovery from stress. In practical terms, the most likely measurements include crops (average live weight per unit area, or livestock) or annual increments in forests. In urban areas, primary production is not of great interest, while in farmland or managed forests, it is vitally important.

In aquatic ecosystems, either productivity or primary production can be measured. Both microscopic algae and larger plants—such as waterweeds and seaweeds—may be important. Rates of net production in Canadian ecosystems range from less than 100 g•m$^{-2}$•yr$^{-1}$ in tundra or arid areas to over 1,200 g•m$^{-2}$•yr$^{-1}$ in wet forests (Whittaker 1975).

## Food Webs

Some of the energy fixed by a plant may be acquired by an animal that eats it; this transaction forms part of a food chain or food web. A familiar example of a food chain is grass→cow→person. Solar radiation is fixed by the plant, some part of which is eaten, providing energy for the cow, and the cow in turn provides some energy for humans. Each level—the grass, the cow, and the person—uses energy for respiration, in order to provide for its own maintenance, growth, movement, etc. These levels are known as trophic levels.

In natural grazing ecosystems, stability is maintained when only a certain portion of the plants are eaten and only a certain proportion of grazers are eaten by the carnivores. Short-term balance is not a characteristic of all grazing food webs. Herbivorous populations—such as those of spruce budworm, army worms, tent caterpillars, and grasshoppers—can destroy populations of host plants in large areas. In such systems, environmental assessment must be based on long-term observations. In Canada, mammalian terrestrial grazing food webs include herbivores such as voles, hares, deer, and moose in the south, and arctic hares, lemmings, caribou, and muskoxen in the north. Mammalian carnivores range in size from shrews to martens, fishers, timber wolves, and grizzly bears. Groups of organisms such as invertebrates, reptiles, or fish include both herbivores and predators.

Aquatic food webs, based on tiny phytoplankton, often have as many as five levels of energy transfer, such as phytoplankton, zooplankton, small fish, large fish, and birds (such as ospreys) or mammals (such as seals). Terrestrial and aquatic food webs are often linked, for example, by omnivores such as bears, which eat at more than one trophic level.

Since the transfer of energy between trophic levels is never complete—that is, not all the plants or animals get eaten or digested—upper trophic level animals tend to be less numerous than lower ones and are often solitary animals with large territories.

Dead animals and plants and wastes such as faeces are the basis of many food webs, called *detrital* food webs. Organisms that derive energy from materials in dead cells include bacteria, fungi, and a great variety of invertebrates in water, muds, and soils. Grazing and detrital food webs are generally not distinct from each other. Many animals eat both live and dead matter, and some fisheries are based on detrital webs; complex, interwoven grazing and detrital energetic pathways are features of estuaries and riverine systems.

In environmental assessment, it is usually only economically important animals, such as salmon or caribou, which are considered. To maintain such animals at an upper trophic level, their food web must be maintained. Measurements at lower levels of a food web may therefore become necessary. To do so, it is necessary to understand what is eaten and during which seasons, particularly if the animals migrate, as salmon and caribou do.

## Structure and Niches

Ecosystems are commonly recognized by their structure; for example, prairie is dominated by grasses and forbs, forest by trees, marshes by emergent aquatic plants, bogs by dwarf shrubs, and so on. In Canada, vegetation generally determines structure, except in the case of occasional features made by animals, such as beaver dams. Types of Canadian ecosystems include forest, prairie, desert, tundra, and alpine areas.

Where conditions for growth are difficult, only one thin layer of vegetation—composed, for example, of tundra lichens and mosses, or small forbs and sparse grasses on arid badlands—is usually possible. Water is generally the factor that limits the number of plant layers; when more is available, there may be several layers of chlorophyll on top of one another. In temperate forests, these can comprise large canopy trees and their epiphytes, small understorey trees, shrubs, and one or more layers at ground level. In the coastal rainforests of British Columbia, the canopy can be more than 40 meters high.

In aquatic ecosystems, structure is important in shore zones, where layering is provided in freshwater systems by submerged macrophytes (large aquatic plants) and in marine systems by macrophytes and large seaweeds. Reefs, shoals, and islands increase the number of hiding places and refuges for invertebrates and fish. As the complexity of structure increases, more microclimatic differences and more types of hiding places and refuges are created. The opportunities for animals with varied ways of living increase, so that, generally, the more complex the structure, the greater is the diversity of animals.

Each species of organism has its own niche, which is defined partly by where it lives and partly by how it lives. Organisms with specialized niches—e.g., birds that eat only seeds—are called *specialists*. Organisms with broad niches—such as omnivorous birds—are called *generalists*.

Global systems for classifying regional vegetation generally define structure first: Is the region savanna, forest, semi-desert? Is a wetland a swamp, a marsh, or a peatland with shrubs? Is a shoreline a smooth sandy beach without vegetation or rocky with seaweed in bays? Subjective judgment or measurements can be used in describing structure. The section later in this chapter on dispersion of residuals describes the ecoregions of Canada and explains how their structure is defined.

## Communities and Diversity

Biodiversity has been defined in Canada as referring to all animals, plants, and micro-organisms in terrestrial, freshwater, and marine environments. It includes four levels: landscape, community, species, and genetic diversity within species. The United Nations Convention on Biological Diversity, an international agreement ratified by Canada in 1992, has three objectives: conservation of biodiversity, sustainable use of biological resources, and equitable sharing of benefits from the use of genetic resources.

Two of these objectives are of interest from the perspective of the mining industry; namely, the protection of biodiversity from potentially adverse effects of activities at industrial and extractive sites, and possible future uses of micro-organisms in mineral extraction after genetic alteration.

Extinction of species has always occurred, but the natural rate has been greatly accelerated by human activity. In Canada, the Committee on the Status of Endangered Wildlife in Canada (COSEWIC) has estimated that the country has some 140,000 species, half of which have not been identified. As an example of a well-known group, there are about 1,000 species of fish; 53 of these are listed as endangered, threatened, or vulnerable (COSEWIC 1994). The degree of our ignorance is shown by the fact that the Committee evaluates only vertebrates and vascular plants. The status of invertebrates, fungi, and bacteria—which largely control nutrient cycling and soil processes—is unknown (Federal/Provincial/Territorial Biodiversity Working Group 1994).

Every species has a unique set of requirements, including microclimate, nutrients, water balance, favored microsites, and interrelationships with other organisms. The way an organism lives, and where it lives, together define its unique niche. There is, however, a large overlap in the requirements of some groups of species, and so they tend to occur together. For example, sugar maples thrive on fairly moist soils and are limited to the north of their range by cold and to the west by water availability. Tree species that co-occur with them include eastern hemlock, basswood, red and white oaks, white pine, ironwood, and about twenty others. They tend also to co-occur with mycorrhizal fungi,[1] with insects that eat their leaves, their flowers or their wood, with squirrels and birds that eat their seeds, and with woodpeckers that nest in their dying trunks. Underneath a stand of sugar maples is an assemblage of mosses, ferns, and forbs that thrive in their shade. All of these organisms may be said to make up a community; one maple stand tends to resemble another. Also part of the community are all the animals living in the maple stand, and the deer, cottontails, and coyotes that use it, moving through it sporadically.

Similarly in aquatic ecosystems, the same species of phytoplankton, the same species of zooplankton, the same invertebrates, and the same fish tend to co-occur where conditions are similar, making up a freshwater community.

One of the effects of stress can be to lower the capacity of an ecosystem to support specialists—for example, predators that eat only one type of prey, or hummingbirds that require supplies of nectar (Mathews et al. 1971; Odum 1985; Smith 1990). Generalists, which have broader niches, may be more successful. Adaptation to stress is in itself recognized as a specialization in some plant species. A group of plants that generally profit from disturbance have yet another special strategy for survival: they are the weedy plants that appear first on disturbed ground such as road verges, railroad tracks, or areas that have been bulldozed. They are generally short-lived, with a large reproductive capacity (Grime 1977; Begon et al. 1986).

Stress in ecosystems may result in the loss of top predators. Loss of carnivores or piscivores may be the result of disruption of food webs or a change in physical habitats, but these animals may decline because of biomagnification of toxic substances in their tissues. Loss of piscivorous birds because of mercury or lead accumulation has been well documented, but is not generally related directly to mining. Rather, lead in waterfowl usually comes from sinkers and shot and mercury from chloralkali plants and the paper industry. Biomagnification and bioaccumulation are discussed below.

Diversity has two components. First, how many species occur in a given area or community? This is called *species richness*. Second, how uniform is their distribution? This is called *evenness*. To use a simple example, assume only two species, A and B, co-occur. A could occupy from 1%–99% of the area, and B could occupy 1%–99%. If A and B each occupied 50%, evenness would be maximal.

The simplest measurements used are lists of species in an area, such as a list of birds or fish species. For a better understanding of an area, however, richness and evenness are combined as a diversity index. Relative numbers of individuals belonging to different species are measured by the evenness component of diversity indices. In environmental assessments, it is usually not practical to work with all species; those selected may include rare species, "characteristic" species that contribute to structure, or species of particular economic interest. Generally, the continuing presence of rare species indicates that a whole ecosystem is functioning well because they tend to have specialized niches; under stressful conditions, the rare species may decline before more abundant organisms do. Characteristic species must be retained if structure is to be maintained (e.g., the pine trees in a pine forest or the kelp in sea otters' hunting areas). Species of particular economic interest can include invertebrates, fish, wildfowl, game birds, fur-bearing animals, and animals used as food by aboriginal peoples, as well as crop plants, fibre plants, and lumber-producing trees.

Diversity indices are extremely useful for comparing a particular area at different times, especially for purposes of environmental assessment.

## Populations

Within communities, breeding groups of organisms form populations. Populations are a useful size at which to study impacts on ecosystems, since they have measurable properties, including birth rates, mortality rates, sex ratios, age structures, and, sometimes, size. Within a population, there are usually genetic variations, and genetic variance is a valued component of biodiversity. Variance between populations of a species is usually larger than that within a population.

The size of populations varies enormously. The small extreme is represented by whooping cranes, which have been hovering on the brink of extinction. At the other extreme are multitudinous populations such as those of some black flies or mosquitoes. Some populations in Canada are known to be decreasing—bullfrogs, for example—and others increasing, such as purple loosestrife.

Some populations maintain a fairly steady total number or density, while others undergo peaks and crashes. In some populations, fairly regular cycles in abundance and density are normal. These cycles may be controlled climatically, by food abundance or quality, by predation or disease, or by genetic changes. Examples of cycling populations include lynx and snowshoe hares in northern Canada, and meadow voles and raptors such as owls and hawks in the south.

The net reproductive rate or the intrinsic rate of growth of a population can be measured in the field. Populations in which additions of individuals exceed the death rate grow, in some cases exponentially, until the lack of an essential element or of some other resource, or competition or predation causes them to crash. Local extinction (called *extirpation*) may result. Other populations grow until they reach a fairly steady state, and the environment is then said to have a particular "carrying capacity" for that population. It is common for a population to overshoot and then undershoot its carrying capacity, especially in an organism that does not breed frequently so that it is not finely attuned to its environment. Some populations grow and crash in fairly regular cycles (e.g., meadow voles or lynx).

It is possible to construct life tables, similar to the mortality tables used by actuaries for life insurance and pension purposes, for some organisms and to use them to forecast future trends in a population. In some populations—e.g., in sponges or creeping plants—definition of an individual organism is difficult. Faced with this type of population—a grass, for instance—density or cover measurements may be more practical choices than work on population dynamics (i.e., counting individuals before and after an impact).

Populations may be territorial, remaining in one area all year. They may be migratory, moving between summer and winter areas, as caribou do, or they may migrate out of Canada in winter, as many bird species do. Seasonal habitats and staging sites for migrating animals are of international concern, and some are protected by international agreements, such as the North American Waterfowl Management Plan for migratory birds.

Populations tend to evolve over time. The genotype of an organism is the whole of its genetic properties, determined by the genes in its chromosomes, half of which are derived from each parent. The phenotype of an individual is the expression of some or all of these genes; for example, a tree may have the genetic ability to reach a height of fifteen meters, but because of poor conditions, its phenotypic expression of its genes may only give it a height of 10 meters. Its offspring will again contain the genotype for a potential height of fifteen meters. Recessive genes are not phenotypically expressed unless the genes from both parents are for the recessive trait.

A population contains a range of genotypes, and most breeding systems allow a combination of genes in every generation. Natural selection is the process determining relative success of different genotypes in a population. Examples are the appearance and spread of bacteria with resistance to antibiotics, insects with resistance to pesticides, and plants with metal tolerance. Unsuitable genotypes die off, while successful ones have more offspring. Selection is not so much "survival of the fittest" (in the sense of battling stags) as selective deaths from insufficient food, unsuitable habitat, disease, or from being eaten. Success for an individual organism consists of passing its genes on to subsequent generations.

In addition to genetic recombination, which "shuffles the pack" of genes, mutation also occurs. The rate of spontaneous mutation for a single gene is about once in $10^6$ to $10^8$ replications. Mutations may be harmful or helpful to individuals. If they are useful mutations, they will spread through the population with each succeeding generation. If they are sufficiently harmful, the individuals carrying them may not produce offspring or may produce unfit offspring.

Genetic recombination and mutation both offer new genotypes for selection, and the intensity of that selection differs from species to species in different regions and over different periods of time. Fecundity ranges from a few offspring in a lifetime to perhaps two hundred nestlings for a bird or a lemming or thousands of fry for a fish. The number of individuals in a species does not generally alter greatly between generations, and the majority of these young organisms die.

The capacity to tolerate a stress from any source, either in the environment or within the organism itself, may be a genetic one. Within some populations, genotypes capable of, say, metal tolerance may occur at a low frequency—in perhaps one individual in every five hundred. Under the selection pressure of a contaminated environment, these individuals grow and reproduce better than all the rest of the population. Their offspring include some individuals with the tolerant genes, and the number of tolerant individuals gradually increases in each generation. Appearance of a tolerant population is more rapid if the population breeds frequently—for example, if there is more than one generation every year. Because of differences in intervals between reproductive episodes, the types of organisms that are known to show metal tolerance include grasses and clams rather than trees and whales. Bacteria, with a different organization of genetic information than that in higher plants and animals, can produce tolerant populations rapidly.

## Biogeochemical Cycling

The physical activities described in Chapter 1 have the effect of altering distributions, concentrations, and forms of elements, including metals. These disruptions affect organisms that are made up of about 15 essential elements, together with traces of other elements such as copper. The essential elements, in order of their abundance on land, are oxygen, carbon, hydrogen, nitrogen, calcium, potassium, silicon, magnesium, sulphur, aluminum, phosphorus, chlorine, iron, manganese, and sodium. The specific requirements of organisms differ— some, such as diatoms for example, needing silicon, and others needing selenium. The demand for a particular element can vary with time, at different stages of a life cycle. Some of the elements that are essential become toxic if they are overabundant in the organism's environment, such as copper and selenium. The same element can be deficient, balanced in supply, or toxic in excess, both in the environment and inside an organism. Sodium in the human diet is an example of an element that can be deficient, balanced, or toxic.

Plants obtain carbon and oxygen from the atmosphere, and their other essential elements through roots, from the interstitial water in soil. In addition, aerial parts of plants receive wet and dry fall-out and can absorb some substances through their leaves (Treshow 1984). Aquatic plants can take in elements through both aerial and submerged leaves, through stems or roots from water, and through roots from interstitial solutions in sediments. Contaminants follow nutrient pathways in some cases; for example, arsenic is likely to enter a plant through the roots following a metabolic pathway for phosphorus (Campbell et al. 1988).

Animals obtain essential elements through eating plants or each other. They also absorb elements such as oxygen through respiratory surfaces exposed to the ambient medium, such as lungs or gills, and by drinking water. In addition, they can absorb nutrients and toxins through their skins or external surfaces. Some animals are filter feeders that process particles in water or mud, both organic and inorganic, and are therefore strongly affected by changes in sedimentation; large animals often eat non-organic material in a search for essential elements, at salt licks, for example.

When a substance is taken into an organism and retained within its tissues at a higher concentration than that of the surrounding medium—whether water, soil, or interstitial water—bioaccumulation occurs. Bioaccumulation can be expressed as a concentration factor. If organisms that contain a contaminant are eaten or ingested by another species, the predators may contain a higher concentration than the prey. Two processes can be involved. First, if the predatory organism is larger and longer-lived it may acquire the body burden of large numbers of prey during its lifetime. Second, if the predator has a higher lipid content, it may store more of a lipid-soluble compound. An increase in concentration of a contaminant up the levels of a food chain is called biomagnification. The classic case of metal biomagnification is that of mercury, which can cause Minamata Disease in humans, having worked its way up the food chain from methyl mercury in mud. Most metals, in fact, are not biomagnified. If upper trophic levels in a food chain accumulate a contaminant at lower concentrations than their prey, biominification has occurred (Campbell et al. 1988).

Elements are cycled out of organisms back to the atmosphere, water or soils as wastes or as dead tissues. A wolf, for example, cycles elements in its breath, in urine, in faeces, in shed hair or claws, and after its death. Recycling generally takes place through the activity of detritivores or decomposers, which use the wastes or dead material to obtain their energy and essential elements. Gradually, organic molecules containing the essential elements, such as proteins and carbohydrates, are broken down to simpler inorganic molecules, with releases of energy. Nutrient cycling occurs largely between inorganic forms of elements outside organisms—such as carbon dioxide or ammonia—and organic molecules inside organisms or in detritus. A nutrient cycle is shown in Figure 2.1. Nutrients are carried along with much of the hydrological cycle, which is shown in Figure 2.2.

The time that an element takes to cycle from inorganic to organic and back to inorganic molecules can vary enormously. Phosphorus can move in and out of phytoplankton cells in minutes; it can also be locked into the cells of a tree-trunk, remaining there for centuries. After death, phosphorus release from material such as teeth and bones can be slow, depending on their rate of decay.

In urban ecosystems, biogeochemical cycling is usually considered only in the context of management of wastes. In farmland or managed forests, such cycling is very important because shortages of essential major elements restrict crop growth. The elements that are most frequently limiting are nitrogen, phosphorus, and potassium. Concentrations are measured by testing soil samples, either with field kits or in labs. In some areas, the availability of a trace element— such as selenium—may limit growth of farm animals, but not plants. The same nutrients are generally limiting in natural ecosystems, but natural vegetation accommodates better to the levels of nutrients present, with sparse growth where one or more elements are in short supply.

In aquatic ecosystems, element concentrations are relatively easily measured in solution. Responses of aquatic ecosystems to supplies of nutrients tend to be rapid because of the short reproductive cycles of plankton. When an element— such as phosphorus—ceases to limit phytoplankton growth, rapid expansion of some populations results, forming the algal blooms of eutrophication. Algal blooms generally have secondary effects reverberating throughout an ecosystem as their necromass affects oxygen concentrations in sediment and water, thereby changing conditions for fish.

## Succession

All ecosystems are in a state of flux. In central Canada, southern ecosystems were disrupted until about 12,000 years ago by the last continental glaciers. In the high Arctic, by contrast, periglacial conditions still continue. Paleoecological

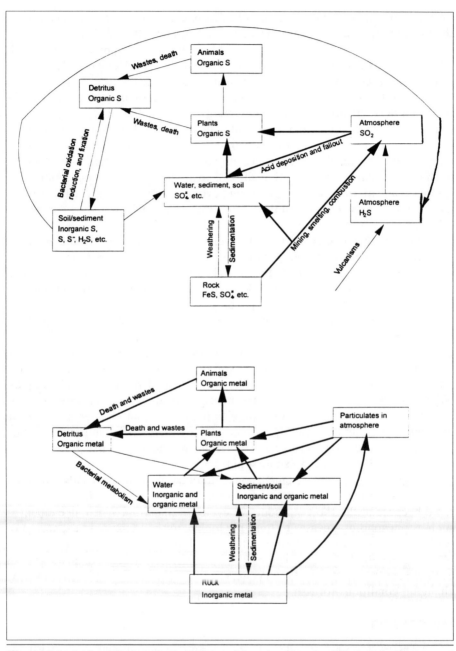

**Figure 2.1** A simplified cycle for sulphur (above) and a simplified cycle for a metal without a gaseous phase (below). The bold arrows indicate flows most strongly affected by mining.

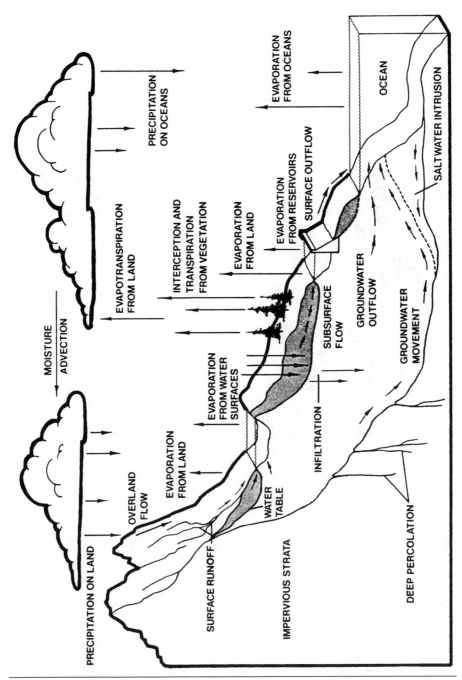

**Figure 2.2** The cycling of water through the biosphere. Source: Ripley (1992).

records show major changes in vegetation since glaciation, as vegetation zones and their associated animals migrated north (Ritchie 1987). There has not been a steady progress; the northern limits of many species have, at times, been further north than at present, and there have also been colder periods than the present, such as the Little Ice Age during the seventeenth century.

Within each vegetation zone, succession—sequential replacement of one community type by another through a series of stages—occurs naturally. An area of vegetation dies, and it is replaced either by similar plants or by different ones. Causes of disturbance include natural aging, drought, flooding, herbivory, parasites, and disease. The scale of disturbances varies; the smallest takes place when, for example, an individual forest tree dies and makes a gap, and the largest occurs when forest fires destroy thousands of hectares. The scale and frequency of disturbances tend to be characteristic of different ecoregions. In the boreal forest, fire frequency ranges from as little as twenty years to centuries. In deciduous forests, the fire frequency is less, as it is in rainforests. In the unfarmed parts of the prairies, the disturbances of greatest importance are drought, fire, and herbivory.

Succession occurring on a pristine surface, such as land exposed by a retreating glacier, is called *primary* succession. Succession after disturbance, where soil has already formed, is called *secondary* succession. When a plant community maintains a mix of species in equilibrium with climate, it may be called a *"climax"* stage. The set of stages before this equilibrium is reached is called a *"sere."*

In aquatic ecosystems, succession is generally on long timescales—e.g., as lake basins become sedimented. In wetlands, succession depends largely on water levels and sedimentation patterns, and ranges from abrupt changes in beaver ponds to gradual changes occurring at the scale of centuries in bog basins. Some ecosystems, such as Great Lakes wetlands, are "pulse stabilized" in relation to irregular changes of water level; that is, there may be sudden changes of state, followed by long periods of slow change. For example, a swamp forest may be flooded and become a lagoon, returning to a forested state after some centuries.

Overall diversity generally increases with an ecosystem's age, but maximal diversity of part of an ecosystem can occur in younger stages; for example, in abandoned fields, maximum diversity of plants can occur when both field and forest plants co-occur. Consequently, when measuring diversity indices for environmental assessment, the stage of an ecosystem's succession must be taken into account. A field that is changing back to forest, for example, may have a high species richness due to imported weeds, but this will decline as native trees succeed. In short, high biodiversity is not necessarily a good thing.

Long-term succession can be traced in paleoecological evidence such as lake or bog sediments. Short-term successional data are derived from forest surveys, tree ring analysis, fire scars, etc. For forests, several successional computer models, based on the life histories of tree species and on known frequency of disturbances, are available (Shugart et al. 1992).

## Stability and Resilience

Stability is the obverse of disturbance; ecosystems with few disturbances or only minor disturbances are relatively stable. To some extent, stability and resilience are not related, because a simpler ecosystem may be more resilient than a complex one that has been developing longer. The scale at which an ecosystem is considered also gives different views on stability and resilience. If the boreal forest is considered as a whole, every year some areas will be burned and will return to an early successional stage. This does not mean that the boreal forest is not resilient, because the vegetation that is reestablished is also boreal forest vegetation. All the successional phases—from grasses and fireweeds, aspen or chokecherry scrub to old white spruce, etc.—are part of boreal forest, as much as old stands of black spruce or mixed stands with balsam fir. On the other hand, if permafrost is destroyed under tundra vegetation, the scars can remain without returning to the previous vegetation within decades.

Complex food webs are generally considered more stable than simpler systems, because they contain alternative pathways for energy and nutrients (Odum 1983; Smith 1986). For example, if a forest has ten dominant tree species, the loss of one may not be a fundamental change, whereas in a forest with two dominants or one dominant species, loss of one such species is a major upheaval.

## Regions in Which Mining Occurs

In considering environmental relationships of a mine, it is necessary to have a very thorough understanding of local conditions. Identification of all its elements by direct observation would be extremely time-consuming, so being able to categorize the ecosystem in which it will be located can provide a headstart. This is of particular interest when an environmental impact assessment is required. Various approaches to categorization are detailed in this section, with an emphasis on systems currently used by Canadian governments.

### Biomes

Very large ecosystems, generally present in more than one continent, are called *biomes*. Biomes in Canada include tundra, boreal forest, mountain and conifer forest, deciduous forest, and prairie. The largest biome in Canada is the boreal forest, stretching from the Yukon to Labrador and Newfoundland; most Canadian mines are in this region.

Net rates of productivity and mean biomass for biomes found in Canada are summarized in Table 2.1. Mean rates and mean biomass are not very meaningful when ranges are so large, but the table emphasizes the differences in sparseness

**Table 2.1** Means and Ranges of Net Primary Productivity and Biomass in Selected Biomes

| | Net Primary Productivity g·m⁻²·yr⁻¹ | | Biomass kg·m⁻² | |
|---|---|---|---|---|
| | Range | Mean | Range | Mean |
| Temperate forests | | | | |
| Evergreen | 600–2,500 | 1,300 | 6–200 | 35 |
| Deciduous | 600–2,500 | 1,200 | 6–60 | 30 |
| Boreal forest | 400–2,000 | 800 | 6–40 | 20 |
| Prairie | 200–1,500 | 600 | 0.2–5 | 1.6 |
| Tundra, alpine | 10–400 | 140 | 0.1–3 | 0.6 |
| Rock, sand, ice | 0–10 | 3 | 0.0–2 | 0.02 |
| Cultivated land | 100–3,500 | 650 | 0.6–12 | 1.0 |
| Swamp, marsh | 800–3,500 | 2,000 | 3–50 | 15 |
| Lake, stream | 100–1,500 | 250 | 0–0.1 | 0.02 |
| Continental Shelf | 200–600 | 360 | 0.001–0.04 | 0.0 |
| Estuaries | 200–3,500 | 1,500 | 0.01–6 | 1.0 |

Source: Whittaker 1975.

or lushness found between northern or alpine tundra, prairies, and forests. Agricultural ecosystems are most like prairies, with high productivity rates, but with relatively little overwintering biomass, because it is removed as a crop.

Generalizations usually made about consumer biomass are that it is higher in forests than prairies and higher in prairies than tundra. The proportion of net production consumed is higher in lakes and estuaries than in terrestrial ecosystems, and probably higher in grasslands than forests. Few studies have compared all trophic levels on the same basis, and the values generally cited for different ecosystems are largely based on global data collated in the 1970s by Whittaker (1975). Other generalizations include comparisons of nutrient turnover, which can be much more rapid in aquatic than terrestrial biomes. Litter mass, the debris that accumulates before being incorporated into soil organic matter, is generally largest in the boreal forest in Canada, but is relatively large in some Arctic ecosystems because its decomposition is extremely slow in dry, cold areas. Decomposition rates in grasslands and forests depend so much on water regimes and microclimate that generalizations may not be helpful.

Nutrient cycling is slow in Arctic and alpine areas. In boreal forest, decay of plant matter in the soil is largely controlled by fungi, and is also relatively slow; the soil generally has a thick cover of plant litter and tends to be acidic. In the mixed coniferous/deciduous forest and in the deciduous forest, nutrient cycling is more rapid, and soils are largely brown forest types rather than the podzolic types of the boreal biome. Prairie soils tend to be alkaline or saline toward arid regions because of strong evaporation drawing solutions upwards in summer.

Predictions about disturbance, stability, and resilience will differ between different biomes. As well, the scale of processes such as fire differs between them, so for prediction of effects of mining it is necessary to understand the structure and function of biomes.

Some geographic processes such as a spring melt become hazards when they impinge on structures or settlements. Hazards in the Canadian mountain biomes include earthquakes, volcanic eruptions, lava flows, ashfalls, rock falls, slumps and avalanches, torrents and debris flows, snow and ice avalanches, floods and flashfloods, and forest fires (Gardner 1993). Mines have been affected by these hazards. Gardner (1986) described the effects of avalanches on mines in the Rocky Mountains, mostly before the 1930s. As recently as 1965, however, an avalanche onto an access road near Stewart, British Columbia, killed 26 people and injured 23 others (Gardner 1993). Flood damage in the Rockies was exemplified by the destruction of most of Anthracite, a small coal-mining settlement near Banff, Alberta, in 1904 (Gardner 1993).

Most landslides in the Rocky Mountains have occurred in areas of limestone or dolomite (Ford 1993). In these soluble rocks and in areas of gypsum, anhydrite, and salt, aqueous solutions of the minerals can cause a landscape called *karst*, first described from eastern Europe. The term applies to both geomorphic and hydrological systems. Karst areas, often steep and dramatic but also sometimes flat, develop sink holes, limestone pavements, depressions, fissures, and caves, and may have drainage systems that are partly underground. These areas can present hazards for construction of dams, roads, and railways (Ford 1993).

Estimates of hazards in Canadian biomes must take account of the coldness of the country; geographers consider three-quarters of the country cold, including mountainous, flat, inland, and maritime regions (French and Slaymaker 1993). One of the most important components of cold landscapes is permafrost, which is a layer of ground that does not thaw completely during the summer. The upper boundary of permafrost is called the *permafrost table*, which may or may not reach the surface. Even in the Far North, a surface layer, the "active layer," freezes in winter and thaws in summer, and an unfrozen layer may exist between the permafrost table and the bottom of this active layer. Thawing of the surface is affected by the type of rock or soil, by cover of vegetation or snow, and by the aspect and the albedo. Also, permafrost may be a reflection of past rather than present climate. Where the mean air temperature is now less than –6°C, permafrost is generally continuous in Canada, reaching as far south as 55° in Hudson Bay (French 1976). It may exceed 500 meters in depth. Further south, discontinuous or patchy permafrost stretches across the area bounded by a mean temperature of –1°C.

Even where permafrost is undisturbed, it is not a static condition as its name implies. (The term "geocryology," while not a precise synonym, is preferred by researchers in this field.) The nature of this type of terrain has both physical and chemical ramifications. As a dynamic system, it can move objects around with great force, a serious consideration for the construction of mining and other operations, like pipelines. Further, it is actually the sediments, or soils, which are frozen, and water can still move through the frozen ground, the freezing point having been lowered by electrochemical activity surrounding soil particles. Little is known about the rate of movement of water through frozen soils, but it is reasonable to expect that tailings or other wastes may leach even into ground that has been viewed as a natural barrier to pollutants.

French (1976) describes geomorphic features associated with permafrost. Where it is developing, ice wedge polygons, pingos, palsas, and peat plateaus may be found. Where it is decreasing, a different set of features are formed, including ice depressions, slumps, thaw lakes, sinks, and beaded drainage systems and mounds. Because these latter features resemble those produced by karst processes, this type of landscape is called *thermokarst*. It is the kind of terrain that has resulted where roads or runways in the Arctic have destroyed underlying permafrost.

Glaciation has affected some landscapes in Canada that were karstic. Ford (1993) described the area at the Nanisivik lead-zinc mine on Baffin Island as having a jumble of boulders of ore and dolomite embedded in ice. He considered that water had been injected at least 60 meters under a glacier, dissolving dolomite. Following ice recession, the water froze; in the mine, the present mean annual ground temperature is −11°C.

Models of climatic change all agree that global warming tends to be greater in high latitudes. Consequently, given current predictions of a new warming trend, changes in permafrost can be expected. Such melting could affect the stability of tailings dams, tracks or roads, and buildings (French and Slaymaker 1993).

## Ecoregions

Because of the vast areas in biomes, it is useful to be able to "place" a mining operation within some smaller geographic unit about which predictions can be made. Within biomes, ecoclimatic regions are defined as broad areas of the earth's surface characterized by distinct ecological responses to climate, as expressed by vegetation and reflected in soils, wildlife, and water (Ecoregions Working Group 1989). In the 1980s, a comprehensive classification of the ecosystems of Canada was drawn up to transcend previous systems designed for specific resource sectors and to describe the ecological setting in which all resource uses take place. The resulting system (Wiken 1986; Burnett 1991) is hierarchical, consisting of 15 ecozones (shown in Figure 2.3), 47 ecoprovinces, 177 ecoregions, and 5395 ecodistricts.

The ecoregion level of this hierarchical system has been used, for example, to examine the effects of mining in northern Canada (Eedy et al. 1979), acid rain sensitivity in Québec (Gilbert et al. 1985), and impacts of other types of developments (Rubec 1979). The federal reports on the State of the Environment also use this system. The necessity to incorporate wildlife information into landscape

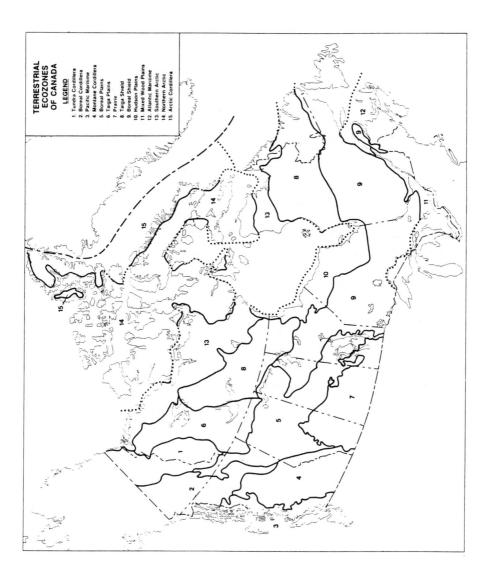

**Figure 2.3** Terrestrial ecozones of Canada. Sources: Wiken (1986); Burnett (1991).

classification systems was emphasized by Stelfox and Ironside (1982; 1988), and the frequent inadequacy of information on aquatic systems by Welch (1978).

Using Figure 2.4, the sites of mining operations in Canada can be located within ecoclimatic provinces (Ecoregions Working Group 1989; EMR 1991). Most are within the Boreal ecoclimatic region, including nearly all of the gold and uranium mines. Most of the remaining mines are concentrated in the Cool Temperate, Grassland, Cordilleran, and Interior Cordilleran regions, while mining of potash and other salts occurs mainly in the Grassland region. Table 2.2 describes the characteristics of the ecoclimatic provinces.

One of the values of knowing the ecological region in which a mine is placed is that the classification provides information on topics such as rare species. For examining trends in wildlife populations and endangered species, Canada has been divided into seven life-zones (Hyslop and Brunton 1991). The Great Lakes/ St. Lawrence zone, after three hundred years of settlement, has the highest number of endangered and threatened species. If, however, the numbers are related to total numbers of plant and animal species present in the various ecoregions, the prairies have a high proportion of threatened birds and terrestrial mammals (Environment Canada 1991).

## Wetlands and Aquatic Ecosystems

The hierarchical systems described above are based on terrestrial ecosystems, although larger landscape categories include water bodies. To determine a mine's effects on a wetland, a river or lake, or the ocean, background information can be sought from separate data sets and categorizing systems for wetlands, freshwater habitats, and coastal sites.

### Wetlands

Wetlands are defined as "land which has the water table at, near or above the land surface or which is saturated for a sufficiently long period to promote wetland or aquatic processes as indicated by hydric soils, hydrophytic vegetation and various kinds of biological activity that are adapted to the wet environment" (Tarnocai 1980).

Peat formation occurs where the production of plant matter exceeds the rate of decomposition (generally under cold conditions). A wetland is defined as a peatland when the peat is 40 cm or more in depth (Moore and Bellamy 1973). Wetlands are classified at three levels: as classes, forms, or types. Classes are general categories based on vegetation physiognomy, hydrology, and water quality: they comprise bog, fen, swamp, marsh, and shallow open water. Forms are based on surface form, basin characteristics and proximity to water bodies. Types are based on vegetation morphology.

Twenty wetland regions have been identified in Canada (see Figure 2.5). Their geographic distribution generally resembles the pattern of terrestrial

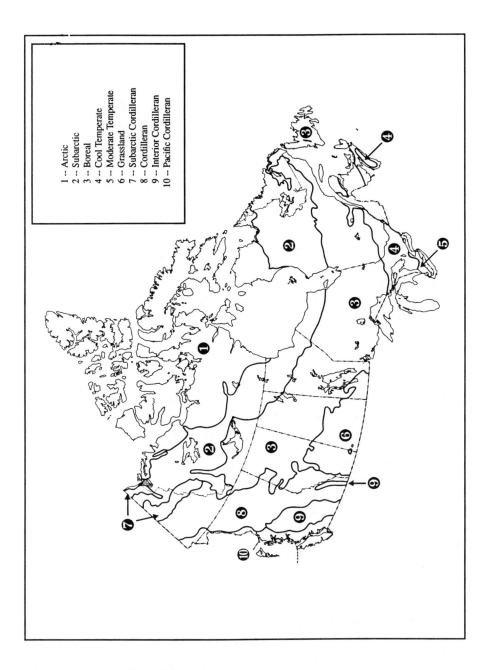

**Figure 2.4** The ecoclimatic provinces of Canada. Source: Ecoregions Working Group (1989).

**Table 2.2**   Ecoclimatic Provinces of Canada and Their Vegetation

| Ecoclimatic Provinces | Vegetation Development |
|---|---|
| Arctic | Treeless, with tundra, polar semi-desert, or polar desert |
| Subarctic | Open-canopied conifer woodlands, with tundra patches |
| Boreal | Closed-canopied forests of conifer or mixed conifer-hardwood |
| Cool Temperate | Mixed forests of shade-tolerant hardwood-conifer |
| Moderate Temperate | Deciduous forests |
| Grassland | Grassland with or without small groves of hardwood trees |
| Subarctic Cordilleran | Open-canopied conifer woodland and alpine tundra in elevational zones |
| Cordilleran | Closed-canopied conifer or mixed-wood forest, open-canopied conifer woodland, and alpine tundra in elevational zones |
| Interior Cordilleran | Grassland (with or without scattered trees), closed-canopied conifer or mixed-wood forest, open-canopied conifer woodland, and alpine tundra in elevational and rain shadow zones |
| Pacific Cordilleran | Closed-canopied conifer forest, open-canopied conifer woodland, and alpine tundra in elevational zones |

Source:  Ecoregions Working Group (1989).

ecoregions, but the vegetation of bogs and fens may resemble upland vegetation of more northerly regions (Zoltai 1988) because the location of wetlands in depressions coincides with cold microclimates.

The region with the most mining activity is the Boreal wetland region, which has cold winters and warm summers. Precipitation ranges from moderate in the center of the country to high in maritime coastal areas. In the subhumid western area, wetlands occur only in depressions and fens are abundant, while in the humid east, raised bogs are more common.

At a smaller geographic scale than th [*Region with the most mining activity*] ure several regional classification systems in use lar (1976) described a classification for the prai oth and duration of flooding; in Ontario an eva to municipal planning rates wetlands according ze, occurrence of rare species, and human use.

## Freshwater Ecosystems

Freshwater systems include rivers, lakes, and parts of estuaries. Twenty-five river basin regions are generally recognized in Canada, flowing into five ocean basins: the Pacific, Arctic, Gulf of Mexico, Hudson Bay, and Atlantic, as shown in Figure 2.6.

Flow rates, the reliability of streamflow, and some water chemistry data were summarized by Pearse et al. (1985). The most obvious chemical difference

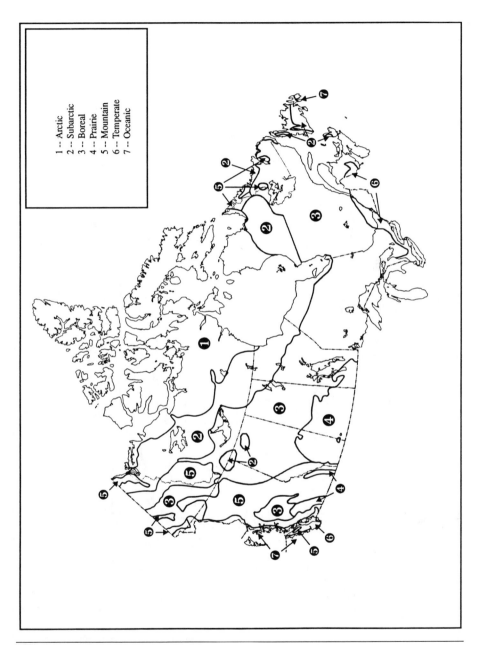

**Figure 2.5** The wetland regions of Canada. Source: National Wetlands Working Group (1986).

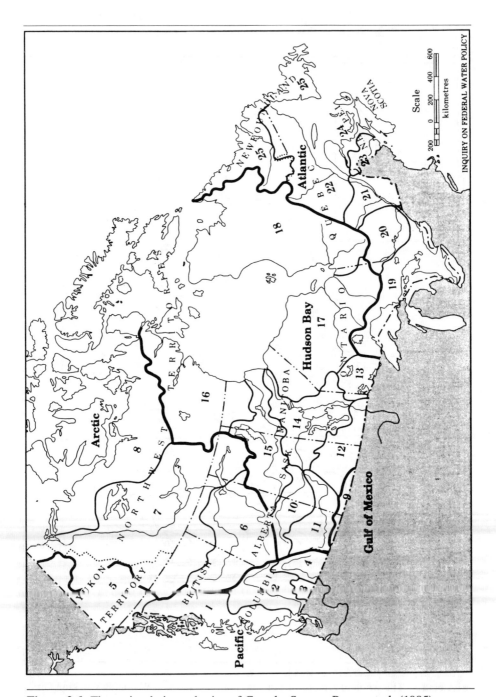

**Figure 2.6** The major drainage basins of Canada. Source: Pearse et al. (1985).

between basins is in water hardness, which is greatest in the Mackenzie and Assiniboine-Red systems.

Studies on Canadian river systems range from small-scale, conducted for example by local conservation authorities, to very large-scale investigations such as that of the watershed of the Mackenzie Valley. A current large-scale study is proceeding in the St. Lawrence Valley as part of the international response to the Canada-U.S. Great Lakes Water Quality Agreement (1972 and 1987). Descriptions include runoff, bedforms, channel morphology, habitats, valley forms, and drainage networks.

Lakes are analyzed using morphometry, water regimes, and water quality. General analysis of phenomena such as thermal stratification and turnover have been described in limnology texts such as those of Hutchinson (1994) and Wetzel (1983). The literature on Canadian lakes is so large that it has been summarized in a series of bibliographies compiled by Environment Canada. Some bibliographies are issued for regions, from Newfoundland west to the Yukon (e.g., Lindsey and Patalas 1981). Provincial data are also copious and readily available. Studies on the effects of acid deposition necessitated a large program of analysis and monitoring, starting in the 1970s (Charles 1993).

The Great Lakes have an extensive literature and a series of databases of their own. A popular introduction to their ecology can be found in *Great Lakes: Great Legacy?* by Colborn et al. (1990). For nearly a century, matters of international interest, such as lake levels and water chemistry, have been considered by the International Joint Commission dealing with boundary waters between Canada and the United States. Pursuant to the Great Lakes Water Quality Agreement, research has been focused on areas of the Great Lakes that had deteriorated because of contaminants or other anthropogenic impacts. In some of these areas—the harbor of Port Hope on Lake Ontario, for example—a source of the problems was mining or refining. Cleanup activities have been started in some of these areas, such as the Bay of Quinte on Lake Ontario and Toronto harbor.

Data on wetlands are available federally and provincially; hydrological data are now being kept by the Atmospheric Environment Service of Environment Canada.

### Marine Coastal Areas

Types of coast include steep rocky areas, such as fjords, and gentler shores covered with a range of substrates from solid rock, through boulders and cobbles to sand and mud. All types have valuable ecological features, such as bird colonies on the cliffs of Labrador and British Columbia, but the most highly productive (for organisms of all kinds) and diverse areas are generally estuaries, tidal flats, and saltmarshes. Saltmarshes form less than 5% of the Arctic coast, 10%–20% of the Pacific coast, and 80%–90% of the Atlantic coast.

Detailed biophysical descriptions are available for some areas, such as a Pacific estuary (Hunter et al. 1982) and parts of the coast of Labrador (Lopoukine and Hirvonen 1979).

One of the most useful environmental data sets on Canadian coastlines is maps of sensitivity to oil spills. A national program began in the 1970s, and large-scale detailed coverage was partly computerized by the 1980s. The mapping, coordinated by Environment Canada, uses geophysical, hydrographic, meteorological, biological, and economic data. Valuable fisheries are mapped in great detail. This sensitivity mapping system has been extended to the freshwater coast of Lake Ontario.

Because of the potential transport of residuals, a knowledge of currents and tides throughout the year is necessary for work at coastal mining sites.

## Dispersion of Residuals

Residuals, or waste materials, from mining and mineral processing can be in the form of solids, liquids, or gases. In each of these forms, they may have the potential to affect the atmosphere, the hydrosphere (fresh or salt water or groundwater), or the lithosphere (soils, rocks, or sediments). As indicated in Chapter 1, each stage of mining and processing and each mineral commodity has its own distinctive set of environmental effects through dispersion of wastes. Clearly, all of these effects are not equal in their significance, but at every level, there is the possibility of uptake and exchange of residuals by organisms. This section is intended to describe the processes by which the dispersion takes place. The physical and chemical properties of the substances and their impacts are discussed throughout the following chapters. Table 2.3 lists the various dispersal processes by which waste substances enter the atmosphere, hydrosphere, and lithosphere. Methods for preventing the dispersion of these undesirable substances are discussed in later chapters.

Solid wastes include waste rock, tailings, and particulate matter, all of which may be chemically reactive and consequently generate soluble acids or alkalis, soluble metal salts, and radionuclides. Liquid wastes can be produced by such reactions or directly by the processes themselves. Gases are usually by-products of the metallurgical and refining stages. Radon is released by waste rock.

Although most lithospheric residuals are not available to organisms unless dissolved in water (Krenkel 1973), they may be easily picked up by wind or water erosion (becoming air or water pollutants) and transported to other land or water areas (Beamish 1974; Hartley and Schuman 1984).

### Waste Rock and Tailings

Surface waste piles are subject to wind and water erosion as well as to chemical reactions resulting from exposure to air and water. The products of these reactions, as well as the waste materials themselves, may be spread far from

**Table 2.3**  Biospheric Dispersion Processes

*Dispersion in the Atmosphere*
- transport of vapor-phase material by wind
- transport of particles by wind
- vertical diffusion
- fall out (dry deposition) of particles
- rain out (wet deposition) of particles
- dissolution of vapor in the ocean
- chemical changes, such as oxidation and hydrolysis

*Dispersion in the Lithosphere*
- diffusion into soil as vapor or in water solution
- volatilization from soil by vaporization or by distillation
- volatilization from incinerators and reprocessing plants
- leaching into groundwater
- absorption and desorption from soil particles
- wind and water erosion of soil particles
- metabolism by soil organisms
- uptake by plants and animals

*Dispersion into Fresh Water*
- chemical transformations, including precipitation
- absorption/desorption from suspended particles and sediments
- transport on sediments
- resuspension and transport of shallow-water sediments
- diffusion into sediments
- evaporation from surfaces
- distribution of water and its contaminants by irrigation
- uptake and metabolism by plants and animals

*Dispersion in Oceans*
- chemical transformations, including precipitation
- absorption/desorption from suspended particles and sediments
- sedimentation of suspended particles
- resuspension and transport of shallow-water sediments
- diffusion into sediments
- uptake and metabolism by marine plants and animals
- transport by ocean currents
- vertical mixing and diffusion in the surface layer
- concentration in surface films
- evaporation of wave spray

Adapted from Study Panel on Assessing Potential Ocean Pollutants (1975).

the source location by the erosive forces. Thus, containment of the waste piles implies isolation of the material from air and water, in order to minimize chemical transformation and erosion.

Tailings and other mining wastes have high susceptibility to wind erosion if the surface remains bare and dry, if the particles are of fine texture, and if their aggregation or clod size is small. Chepil and Woodruff (1963) devised a measure of soil erodibility based on the percentage of non-erodible dry particles of diameter greater than 0.84 mm. The units of soil erodibility are tonnes of soil loss per hectare per year. By this measure, Québec iron ore tailings have, for example, an erodibility of approximately 30 and British Columbia base metal tailings 150, while others exceed 400. All but the first are highly susceptible to wind erosion if the surface is dry and unprotected.

The actual amount of wind erosion from a soil surface has been found to depend upon the soil erodibility, the soil surface roughness, the wind fetch,[2] the amount of vegetative cover, and a climatic factor (Chepil and Woodruff 1963). While the first of these factors is related to the nature of the soil material, as described above, the second, third, and fourth are local factors that are subject to modification—e.g., by surface treatment, planting of shelter belts, or revegetation. The final factor depends upon wind speed and soil moisture, and summarizes the climate of a geographical area in terms of its potential for wind erosion. A map of the average values of this wind erosion climatic factor for North America (Chepil et al. 1962) shows that the only areas of Canada subject to appreciable wind erosion are the southern portions of the Prairie provinces. A model for agricultural land, however, does not necessarily apply to constructions like tailings impoundments, which may be quite atypical of the surrounding terrain. The broadest Canadian study of wind erosion, conducted by a Senate committee, documented more widespread occurrences, for instance, in Québec (Sparrow 1984). Aggregate quarries also produce dust and in many cases create an urban nuisance across the country.

Rain and snowfall also erode material from waste rock, slag piles, and the outer walls of tailings impoundment areas, washing them onto adjacent land areas and into water bodies. The amount of erosion may be very large during periods of high rainfall intensity, and can be estimated by using the universal soil loss equation (Wischmeier and Meyer 1973; Bhowmik 1985). Similar to the wind erosion expression described above, this equation expresses water erosion as a function of rainfall, soil erodibility, surface-slope length and steepness, surface cover (vegetation), and erosion control measures.

Mine waste heaps are also susceptible to collapse, often triggered by earthquakes, but also under static conditions, usually because of "liquefaction" resulting from water buildup in the embankments. Some of the factors that may contribute to this sudden flowing of waste materials are poor drainage, high confining pressure, and large and frequent cyclic stresses on the embankment. An additional factor is the nature of the material; the lower the bulk density of the material, the more likely it is that liquefaction will occur.

Fairly typical was the failure of the southwest wall of a tailings dam at the Matachewan mine, near Elk Lake, Ontario, in October 1990. This resulted in the discharge of 130,000 m³ of tailings and water into the Montreal River system (Heffernan 1992). Although the mine was inactive at the time and the failure was related to the overflow of an adjacent lake, the mining company was held responsible and convicted under the Ontario *Water Resources Act*.

A great deal of study has been devoted to improving slope stability of embankments through proper design and the effective use of drains (Coates 1972; Williams 1973; Das et al. 1990). A study of tailings-embankment stability in the Northwest Territories (Roy et al. 1973) found that the Arctic climate caused many problems not found in more temperate regions. These were related primarily to permafrost and to the freeze/thaw cycle of soils.

The solid residuals include mill tailings and smelter slag as well as eroded material from disturbed land and waste and tailings heaps. While the disposal of solid wastes directly to water bodies was once common practice, the procedure has gradually lost favor as a result of increasing environmental pressure (Williams 1975; Ellis 1987). However, the difficulties associated with other forms of disposal for these wastes have led to a resurgence in its use as the "lesser of several evils" (Scott 1987).

Safe practice of this method of disposal requires a knowledge of water systems and their variability in terms of such factors as sensitivity and absorptive capacity. The ultimate destination of effluents must be taken into account in determining standards of their composition before discharging them. For the country as a whole, river drainage accounts for approximately 60% of the precipitation (Ripley 1987), so that most of the remaining 40% must be returned to the atmosphere through evaporation, sublimation, transpiration, or respiration. There is a wide variation of water residence times within and between drainage basins, ranging from minutes in coastal areas to years in the Great Lakes and some parts of the Prairies. As water drains through streams and rivers toward the sea, it carries materials in solution, suspended solids, and entrained sediments. The Canadian National Committee for the International Hydrological Decade (Slivitsky 1978) has estimated the average annual sediment discharge of Canadian rivers to be 300 Mt, most of which derives from natural sources.

The most obvious effects of the disposal of mining wastes to the hydrosphere are sedimentation and turbidity. A study of three British Columbia operations, which discharged between 3,000 and 26,000 tonnes of tailings in the Pacific Ocean each day, found that these direct physical effects were responsible for considerable destruction of habitat of many benthic and other aquatic organisms (Littlepage et al. 1972; Goyette 1975). The turbidity effects were produced by large quantities of soluble materials and fine particulates (also referred to as slimes), which scattered incoming light; this reduced the light intensity and, ultimately, the overall biological productivity of the area.

Upon discharge to the hydrosphere, substances from both anthropogenic and natural sources may be acted on by diffusive and mass-flow dispersal processes. While a smoothly flowing stream represents nearly pure mass flow, a stagnant pond has only diffusion to stir its waters. Water bodies contain currents due to the inflows and outflows of water, to the flow around rough objects, and to wave-action produced by atmospheric winds. Temperature and density changes generally ensure a turnover of lake water in spring and fall in southern Canada, whereas Arctic lakes and ponds may not thaw completely (Wetzel 1993; Hutchinson 1994).

Dispersion is a relatively rapid process within the hydrosphere, particularly in small bodies of water such as rivers and lakes. However, it is limited to discrete areas which represent a small fraction of the land masses. A further constraint is that much of the mid- and high-latitude hydrosphere freezes in winter, virtually closing down its dispersive capabilities. Mass transport is the major transfer mechanism—particularly in the direction of flow—in moving water bodies. Density gradients, winds, tides, and currents cause mixing and movement of solutes, which can also be altered during their progress by chemical processes. Even in the quietest waters, and at the water-sediment interface, some mixing is provided through the activity of organisms. Nevertheless, the hydrosphere is far from homogeneous. The water of every stream, pond, river, and lake differs in its content of dissolved and suspended materials (Mrowaka 1974).

The dispersion of residuals from the hydrosphere to other spheres is far more complex than that of those released to the lithosphere. The hydrological cycle (Figure 2.2) describes the flows of water, whether solid, liquid, or gas, among the three spheres, through the processes of precipitation, runoff, and evapotranspiration. The sun supplies the energy for ice and water to escape the confines of the surface, be cleansed of their contaminants and to soar briefly (approximately ten days, on average) as water vapor in the atmosphere. The water molecules, however, may be accompanied by volatile loss of contaminants. This moisture condenses again when cooled by atmospheric processes, and immediately starts to collect impurities, mainly from natural sources. Gravity causes it to fall and continues to draw it and its contained substances down to the lowest possible level.

The water may return to the atmosphere almost immediately or continue on a long journey through soil, stream, river, and eventually to an ocean in which circulations may carry it thousands of kilometers and to great depths before releasing it to the atmosphere again. Some water passes through the ice phase and may remain immobile for months—or even millennia, if incorporated into a glacier. More than 71% of the Earth's fresh water is tied up in the form of ice, and 28% as groundwater (Ripley 1992). Of the remainder, only a small part is contained in the atmosphere and surface water (and this latter mainly in lakes). Water in soils or sediments below the water table is described as groundwater, and is used in many areas for wells. While in the liquid and solid phases, the water carries impurities with it, either in solution or in suspension.

Along this path, the impurities may be exchanged with both the atmosphere and the lithosphere. The transfer of materials from water bodies to the atmosphere is limited mainly to volatile substances, which are relatively uncommon as mining residuals. Transfers to the lithosphere are of greater significance. These include materials spread by wave action and periodic flooding of adjacent land, as well as through groundwater movement and contamination of aquifers (Aldous and Smart 1988).

## Liquid Wastes

Wastes in this category range from leached soluble acids and alkalis to leached metals and soluble metallic salts. The most common and serious problem faced by Canadian mines in waste disposal derives from the mining and processing of sulphide ores, which can produce the phenomenon known as acid drainage. Although it does occur in nature, acid production from sulphides is greatly accelerated when the rock is crushed and ground. Acid drainage is described in more detail in Chapter 4, but in essence, it is the result of the chemical reaction that takes place when a metal sulphide combines with oxygen and water, yielding a metal hydroxide precipitate and sulfuric acid. The results are leaching of metals and reduced pH of water that comes into contact with the oxidized surfaces. The metals, particularly the so-called heavy metals such as arsenic,[3] lead, and cadmium, are toxic in some forms (speciation is discussed in Chapter 3) and so must be prevented from leaving the receiving environment. Techniques for accomplishing this are described in Chapters 4 to 11.

The major metal involved in sulphide-oxidation reactions is almost always iron, because pyrite and pyrrhotite occur in association with many other ores and are difficult to separate economically. Iron, therefore, is commonly left behind in the waste rock and mill tailings. The pyrite content of tailings may be greater than 70%, although acid-drainage problems may occur even when the content is as low as 2%–3% (Hawley 1972a, b). There have been difficulties in establishing vegetation on tailings containing as little as 6% pyrite in spite of the addition of massive doses of lime to aid in neutralization (Bell 1974). Some ores, such as those high in carbonate, have a natural buffering effect and help raise pH without the need for alkali treatments. Chapter 4 discusses rehabilitation in some detail.

In addition to sulphide oxidation, the waste material itself may contain soluble acids or alkalis that will affect the pH level of drainage waters, which are therefore treated before discharge to the environment. Reclamation practices in which waste piles are covered (e.g., by soil and vegetation) tend to reduce wind and water erosion, sulphide oxidation, and acid leaching.

It is possible to eliminate many of the liquid discharges to the hydrosphere by impounding or burying solid waste, although some seepage from waste dumps is inevitable. Process water can be, to a large extent, recycled. Mine drainage is the most difficult of the hydrospheric residuals to manage, especially if, as is usually the case, it is contaminated by acids and high concentrations of dissolved metals. It usually requires containment and treatment over a period of time before being returned to the environment.

In addition to the direct physical effects of mining waste disposal, there are a number of other hydrospheric effects that may have an even greater impact on the biosphere (Larkin 1974; Goyette 1975). The two most important of these are the toxicity and the closely related acid-generating capability of many mining wastes. Toxicity is due primarily to the presence of residual metals and process chemicals in the tailings slurry. After deposition in water bodies, wastes may continue to oxidize, increasing acidity and metal dissolution. Paradoxically, as described in Chapter 4, it has been found that one of the most effective ways of controlling the chemical reactivity of sulphide wastes may be to deposit them under deep water, an environment which lacks the oxygen that is an essential element in the process. One of the close relations between acidity and metal release is that, for most sulphides, solubility (and thus metal release) increases dramatically as pH decreases. In addition, both bioaccumulation by benthic organisms and sediment diagenesis may substantially increase the metal concentrations available to other marine organisms (Reynoldson 1987; Salomons et al. 1987; Campbell et al. 1988; Pedersen and Losher 1988).

Other non-toxic substances may have secondary effects of similar magnitude. For example, acidity may induce the dissolution of toxic materials previously inert in the sediment. Highly oxidizable substances may also deplete the oxygen content of the water to a point where organisms suffocate.

Some residuals may add nutrient value to aquatic ecosystems and result in eutrophication. Although this is not usually a major problem with mining wastes, it may become more prominent in the future as more processing plants turn to ancillary fertilizer production to reduce their sulphur residuals.

The disposal of organic materials from mining operations is usually limited to human wastes and a relatively small quantity of organic process chemicals and petroleum wastes. The decomposition of these materials after discharge may reduce dissolved oxygen content of a water body to some degree.

### Emissions from Metallurgical Processes

Smelting and further refining generally produce mixtures of gases, liquids, and solids that may be emitted to the atmosphere. The liquids and solids range in size from several millimeters in diameter to less than 0.05 micrometers and exhibit a wide variety of shapes (Varney and McCormac 1971).

Theoretical fallout velocities of particles can be calculated by Stokes' Law. Larger solids and liquids tend to fall out quickly under the action of gravity. Particles and droplets small enough to remain in suspension for a long period of time are called aerosols; a diameter of 1 mm is often used as the upper limit of aerosol size, which corresponds to a fall velocity of 0.01 $cm \cdot s^{-1}$, or about 10 $m \cdot d^{-1}$ (Humphreys 1964; Fennelly 1976). The dividing line between the groups, however, is not sharp because shape and density also affect fallout velocity.

The main types of atmospheric emissions produced by mining activities are the oxides of sulphur, nitrogen, and carbon, radioactive particles, metals, and particulates. The emissions from ore crushing and grinding operations consist mainly of particles greater in size than 1 mm. It is likely that most of these will fall out fairly quickly, so that they will have only local effects. Smaller particles, such as some of those emitted by smelting, will have such a low fallout velocity that they remain in the atmosphere for days, weeks, or even years, providing that they are not involved in chemical reactions.

The flow path of an emission from its source, through the atmosphere, to a sink where it has a biological effect in illustrated in Figure 2.7. The main factors influencing each stage of the pathway are indicated on the diagram.

Quite apart from any pollution control measures, direct emissions from mining activities to the atmosphere may be classified as either controlled or uncontrolled. Controlled, or stack, emissions are usually channelled from several sources to a single point, processed to certain extent, and released through an

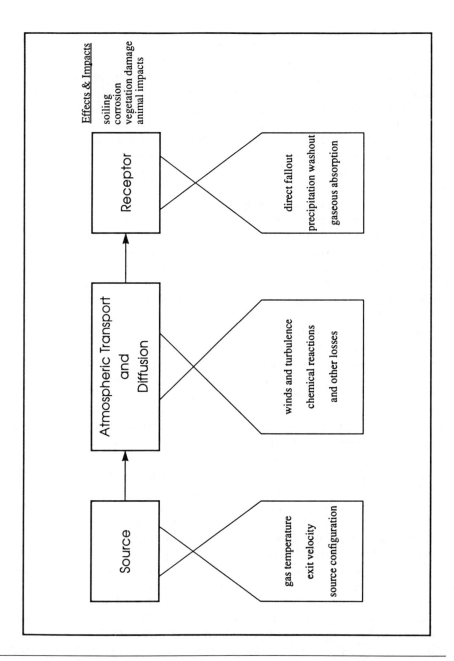

**Figure 2.7** The movement of atmospheric residuals from source to receptor (sink), showing the main factors affecting the emission, transport, and uptake, and the resulting effects and impacts.

elevated stack. Uncontrolled, or fugitive, emissions come from widely dispersed sources such as ore handling, wind erosion of surface ore and waste piles, and ventilation systems and are released at or near the surface.

The atmosphere is characterized by a wide range of scales of motion, ranging from molecular diffusion to large convective eddies. Atmospheric dispersion varies with many factors, including geographical location, weather systems, topography, season of the year, and time of day. The most labile residuals are those that are airborne; the gases and aerosols can travel great distances and spread over the entire globe. They may be absorbed directly by plants or animals through pores or respiration, or be deposited on them by direct fallout or in precipitation. A fraction of the emissions remain in the atmosphere, resulting in increases in trace-gas concentrations.

Unless there is a temperature inversion or strong lateral wind, stack emissions rise because of source momentum and buoyancy before being entrained by the prevailing wind. The movement of a plume depends on a number of factors, including atmospheric temperature and stability, the height of emission, and the initial temperature, velocity and diameter of the plume (Lyons and Scott 1990). The usual method of estimating plume rise is based on the theoretical analysis of Briggs (1975).

After being emitted from the smokestack or other source into the atmosphere, the plume and its particulate and gas components travel downwind, experiencing dilution effects due to stretching (if the wind speed exceeds the stack exit velocity) and to expansion in the vertical and crosswind directions because of "diffusion." A typical dilution rate for a 200 m stack may be one in $10^4$ at a distance of 20 km from the source (Howells 1990). The main effect of stack height is on the local scale. At a distance of 1,000 km, a difference of less than 10% has been estimated between high and low sources (Howells 1990). Concomitant with the rising and dispersal of a plume are changes in its composition due to chemical reactions and losses to the surface due to fallout (Davidson and Wu 1988) and washout (Barrie and Schemenauer 1989).

The aspects of the Canadian climate most relevant to the spread of atmospheric pollution are wind speed, wind direction, and atmospheric stability (Hare and Thomas 1979). Those regions of the country that are located downwind from major sources of air pollution, and which have a high frequency of poor dispersal conditions, are most susceptible to damaging pollution episodes (Charles 1991).

The three main sinks for atmospheric pollutants are chemical reactions, cloud and precipitation scavenging, and sorption at the Earth's surface. It is through the last two of these mechanisms that pollutants reach organisms. Probably the best-known result of such chemical reactions is the formation of acid rain and snow from the escape of gaseous sulphur dioxide during smelting processes. This topic is discussed more fully in Chapters 3 and 4.

Particles larger than 10 mm are removed from the atmosphere mainly by gravitational settling, while other forces are usually more important for smaller particles. Surface impaction and electrostatic attraction are involved in the settling-out of particulate matter, while for gases, absorption and chemical reactions are the main sinks. At times, surfaces may act as a source rather than a sink, releasing absorbed gases by desorption and accumulated particulates through mechanical resuspension. The rate of loss of materials from the atmosphere to a surface depends, first, on the net rate at which they can be captured or absorbed at the surface[4]—a process that is highly variable and poorly understood—and the rate at which they can be supplied through the atmospheric boundary layer to the surface—which is in the realm of turbulent exchange and is similar to the larger-scale "diffusion" discussed above. These calculations are based on the assumption that the materials are sufficiently small for gravitational settling-out to be neglected (Munn and Bolin 1971).

## Heat, Radioactivity, and Noise

Two effects important in particular situations are the release of waste heat and radioactivity. The main heat emissions from mining are from pyrometallurgical operations into the atmosphere (via smokestacks), the hydrosphere, and, to a much lesser extent, to the lithosphere from slag dumps. However, discharges of mill process water and tailings slurry may result in significant heat transfers when their temperatures are much higher than ambient, as in winter. These heat losses are likely to be rather small, however, in comparison with other sources such as electric power production, which is responsible for over 80% of the waste heat released to the environment (Cook and Biswas 1974). At its worst, the waste heat may lead to persistent fog formation or raise the temperature of water bodies sufficiently to kill some organisms. However, even if this is not the case, it still represents a loss of energy that might be put to better use (Cox 1973; Kenney 1973; Coutant and Talmage 1976).

On a local scale, waste energy in the form of heat (among other things) is well known in the case of urban heat islands. Larger mining operations may produce, or at least contribute to, similar effects in some cases, resulting in modification of the local climate (Lowry 1969). The changes in surface properties affecting the natural energy exchange, as well as the entrainment effects of stack plumes themselves, will likely have some influence on local wind circulations.

Flammable wastes, such as those from coal mining, may be ignited either externally, or by spontaneous combustion, producing noxious gases such as carbon monoxide, as well as changes in the physical properties of the waste heap. In general, the lower-rank coals, such as lignite, are the most susceptible to spontaneous combustion. Pyrite oxidation by moist air can itself produce enough heat to ignite inflammable materials (Hocking 1974).

Among emissions, heat probably ranks second to radioactivity, which may be in the form of gases or particulate matter. The gaseous radioactivity persists until the gases are sorbed at the surface. The particulate matter is removed from the troposphere rather quickly, but may persist up to several years in the upper stratosphere (Reiter 1974).

Finally, noise, or "sound without value" (Magrab 1972), is propagated as a pressure wave through the atmosphere. While most mining-generated noise does not spread more than a few hundred meters from the operation, some, such as blasting, may be heard much further away, particularly under certain atmospheric conditions. By far the major effect is on the workers themselves and, in some cases, on residents living in close proximity to an operation. Wildlife populations are also known to be negatively affected by mining noises.

## Deposition Processes

In order for a residual to have an effect upon an organism, there must be interaction—that is, exchange of matter or energy—between the two. In the case of noise, a pressure wave moves through air, land, and water and may affect a sensitive organism. Heat is a form of energy and may be transferred to an organism by conduction, convection, or radiation. The main mechanism is the transfer of pollutants or residual materials from atmosphere, hydrosphere, or lithosphere to the surface or inside of an organism, where they may be partially or completely absorbed.

Motion within the lithosphere is extremely slow, except when assisted by such phenomena as landslides and earthquakes. Dispersion is generally restricted to diffusion, the efficiency of which depends on the characteristics of the substance being dispersed. Heat flows fairly well through dense rock and less so through loose, dry material. Gases and liquids diffuse most readily through loose material and scarcely at all through solid rock. Solids do not diffuse, but change slowly through weathering processes, and the actions of gravity and of microorganisms do cause them to move.

Residuals in soils and sediments, apart from radioactive substances, are therefore sufficiently immobile that they can only affect organisms on which they are deposited, or which move into contact with them. Plants are affected mainly by hydrospheric transfer from the lithosphere to their roots or atmospheric transfer to their leaves. Some animals may absorb some substances directly through their skins, and others through their digestive or respiratory systems, while feeding in an affected area.

### From Lithosphere to Organism

Most terrestrial plants live partly in the atmosphere and partly in the lithosphere. They derive elements from both spheres and are susceptible to contamination from either one. Lithospheric residuals include those deposited from drainage water, or from fallout or washout from the atmosphere. When dissolved in soil water, the material may be taken up by plant roots. The degree of extraction depends not only on the plant, but on many other factors including, in the case of metals, speciation, temperature, soil exchange capacity, and, of course, soil moisture (Friedland 1989; Shaw 1989; Tyler et al. 1989).

Thus, soil treatments such as liming can reduce the availability of metals to plant roots. Plants growing where soils are naturally rich in metals have developed a tolerance for them. Evidence that this is not achieved by exclusion mechanisms suggests uses of the plants not only in reclamation of mine-waste areas, but also for removal of metals from the soil; this metal sequestering by the plants amounts to a detoxification of the soil (Ernst 1975). The eventual fate of the plants could be that they are recycled as humus, thus returning the metals to the soil, or they are removed as toxic waste. This amounts to a relatively inexpensive way of concentrating toxic materials.

The most common pollutant in the vicinity of potash mines is sodium chloride. This can spread from waste piles and brine ponds via the atmosphere to plant leaves and the soil; it can also seep into surface water and groundwater courses. Similar effects are produced naturally in seaside areas, where, under conditions of high atmospheric humidity and on the windward side of vegetation, salt particles may adhere to plant leaves and be absorbed into leaf tissue (Cassidy 1968).

## From Hydrosphere to Organism

The hydrosphere has the atmosphere as its upper boundary and the lithosphere—in the form of soils, sediments, and rock—as its lower boundary. Its composition depends, to a large degree, upon the material exchanges at these two boundaries. These, in turn, are influenced by the chemical and physical nature of the water-body itself. The effects of pollutants, such as toxic metals, on aquatic organisms are dependent on a wide range of factors, including pH, Eh, the species of metal, presence of iron or manganese oxides, and temperature (Campbell et al. 1988).

Dissolved and suspended substances in well-stirred water bodies are freely available to organisms. Transfer is similar to that in the atmosphere, in that it is governed by concentration gradients, but it is also subject to physicochemical processes at membrane surfaces (Shaw 1992). The motion of most unattached marine life reduces flow resistances in the water.

## From Atmosphere to Organism

Our discussion here involves transfers, directly from the atmosphere, to organisms that live in the air or on land. The processes that deposit residuals upon the Earth's surface are direct fallout or dry deposition of solid or liquid particles, rainout and washout of gases, and gaseous diffusion of solids, liquids, and gases. All of these processes operate between the atmosphere and any component of the Earth's surface: plant, animal, soil, or water surface. The rate of deposition by the various mechanisms depends upon several factors, including the phase of the residual—that is, whether it is solid, liquid, or gas—the particle size and solubility, and the gaseous concentration of the substance at the surface (Kellogg et al. 1972). Larger particles will tend to settle out under the action of gravity. More soluble particles and gases will be readily scavenged by cloud droplets and precipitation, and gases will diffuse readily down a steep concentration gradient.

Although most terrestrial plants receive the bulk of their nutrients through their root systems, they can also exchange a wide variety of materials through their foliage (Bukovac and Witener 1957). The main transfers are transpiration flow from leaves and the flow of carbon dioxide into the leaves, which take place mainly through small pores (stomata) on leaf surfaces. However, there are lesser transfers of both gases and aerosols, mainly through the stomata and roots, which can have either beneficial or harmful effects on the plants. Exceptions are lichens and mosses that absorb directly from the air. Lichens are important in both tundra and boreal forests, where they are forage plants for large animals such as caribou.

In addition to absorption through stomata and cuticle, larger particles may adhere to leaves, become impacted in a waxy coating, or even block stomata. Particles transported by the wind may cause abrasion or other damage to exposed plant parts. These are physical effects that may have significant effects on the growth and productivity of vegetation.

Similar deposition processes also apply to animals, although absorption through the gastrointestinal and respiratory tracts can be more important than absorption through the animal's skin.

In general, the degree of impact of a pollutant upon an organism is related to the product of the pollutant concentration and the length of exposure (i.e., the dosage) (Schmidt 1963). This is only an approximate relationship, and the pollutant concentration must usually be reduced by an amount called the "threshold for injury" value: the tolerable limit that the organism can either discharge or avoid without harmful effects.

## Summary

Measurable properties of ecosystems can be used to compare the state of an area and its organisms before and after mining activity. These properties include energetics, food webs, structure, diversity, and biogeochemical cycling. Populations may be a useful focus of attention and may show evolution of tolerance of a stress.

Classification of terrestrial and aquatic ecosystems allows prediction of some of their properties, which is useful for purposes of environmental impact assessments and in designing environmental controls for the six stages of mining and mineral processing. Most mining activity in Canada is in the boreal forest.

Wastes from mining move through the biosphere in various ways, depending on whether they are discharged to the lithosphere, the hydrosphere or the atmosphere.

Each of these spheres has different transport mechanisms, so organisms that come into contact with contaminants are affected to different degrees.

In the next chapter, specific effects of the mining and mineral processing activities outlined in Chapter 1 are discussed in relation to their effects on the ecosystem components described in Chapter 2.

## Notes

1.  Symbionts living on or in plant roots.

2.  The unsheltered distance over which the wind is able to act.

3.  Arsenic is, in fact, a metalloid.

4.  The sink strength.

# CHAPTER THREE

# ENVIRONMENTAL IMPACT

## Introduction

Chapter 2 listed measurable environmental factors likely to be affected by mining and characterized observable changes in these factors. Using the headings introduced in Chapter 2, this chapter describes how activities involved in mining and mineral processing bring about these changes. More detail about impacts and remediation methods for specific minerals and circumstances is given in the following chapters and in case histories. Such topics as the effects of asbestos fibres and other mineral dusts in mines are not considered, the health of mine workers being beyond the scope of this book.

In order to understand the impacts a mining operation has on the environment and be able to predict and compensate for these impacts, it is essential to understand the biophysical, biogeochemical, and biological environment prior to mining. This implies a need for a baseline study conducted over a long enough period to obtain information that will provide a reliable guide to expected interaction between natural and human activities.

This can be done in advance for new projects, but assessment of the effects of ongoing or abandoned operations is usually made extremely difficult by the lack of numerical data collected before an existing mine began operation. The importance of this gap can hardly be overstated. There is no substitute for site-specific data for several reasons, the most significant being the fact that the area surrounding an orebody may have naturally high concentrations of metals—indeed, elevated values can serve as indicators to prospectors of local mineralization. An orebody that has been exposed to centuries of weathering will contaminate soils and water naturally, even in the absence of any industrial activity like mining. This lack of data can partly be offset by using inferential evidence, such as paleoecological findings, including examination of sediment cores from lake bottoms. Because paleoecological work covers a range that extends from influences over thousands of years to seasonal changes, it can show—and distinguish between—differences due to historical and present environmental practices by mining companies. For examples of the insights from paleoecological research, see the section on biogeochemical cycling below, the section on succession in Chapter 2, and the Sudbury case history in Chapter 5.

In determining environmental effects, the choice of organisms to study is influenced by a number of practical factors and, hence, some are given substantially more attention. For example, more studies are published about mining impacts on fish than on mammals, and among the fish the salmonids are emphasized. It is easier to delimit an aquatic habitat, and lake chemistry is easier to monitor than soil chemistry. Fish are economically important, sports fish are large, and fish kills are obvious. Lastly, terrestrial animals can move from a disturbed site early in its development so that changes in their populations are not apparent.

## Measurement of Energy Flow

Before mining, a piece of land has a characteristic primary productivity that depends on its ecoregion and the use to which it was previously put, in a managed or unmanaged state (see Chapter 2). By damaging or removing vegetation, mining generally lowers the primary productivity of the area until reclamation activities have been successful. Vegetation is destroyed during road-making, establishing of camps, building, making rock or waste piles, and constructing tailings impoundments. Figure 3.1 indicates some of the parameters used to measure ecosystem changes before and after mining. Measurements include population numbers, biomass, necromass or yield, gas volumes, diversity indices, etc.

Phytotoxic air emissions can diminish productivity. Damage to forest vegetation by gold mining, iron mining, sulphide processing (see Sudbury case history), and potash mining (Redmann 1983a) has been reported. Stack emissions are joined by particles eroded from wastes, and metals can reach plants through both aerial fallout and the soil. Some of the physiological and morphological changes which metals cause in plants (including a decrease in productivity) are summarized in Table 3.1. The distance from a source at which metals affect plants differs; near Flin Flon, Manitoba, for example, manganese was detectable about 250 km from a smelter, and copper not more than 60 km away (Franzin et al. 1979). Effects of metals differ, and plant responses and toxic thresholds are specific (Shaw 1989). Sulphur dioxide is among the gaseous wastes to which vegetation is exposed; it is the primary source of acid deposition, and its effect on vegetation is well known. Damage occurs above a threshold value which depends on concentration and duration of exposure (Linzon 1972; Roberts 1984). Exposure for a few hours to a concentration of 25–30 ppb $SO_2$

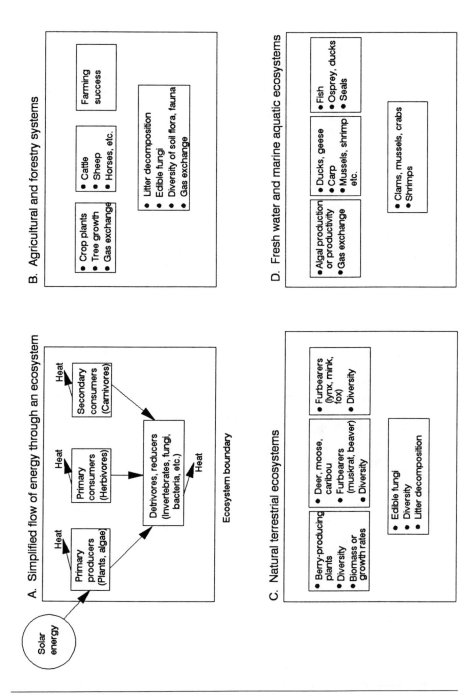

**Figure 3.1** Parameters measured in different types of ecosystems.

**Table 3.1** Symptoms of Metal Toxicity in Plants

| Element | Effect |
|---------|--------|
| Aluminum | Stubby roots, leaf scorch, mottling |
| Boron | Dark foliage; marginal scorch of older leaves at high concentration; stunted, deformed, shortened internodes; creeping forms; heavy pubescence; increased gall production |
| Chromium | Yellow leaves with green veins |
| Cobalt | White dead patches on leaves |
| Copper | Dead patches on lower leaves from tips; purple stems, chlorotic leaves with green veins, stunted roots, creeping sterile forms in some species |
| Iron | Stunted tops, thickened roots; cell division disturbed in algae, resulting cells greatly enlarged |
| Manganese | Chlorotic leaves, stem and petiole lesions, curling and dead areas on leaf margins, distortion of laminae |
| Molybdenum | Stunting, yellow-orange coloration |
| Nickel | White dead patches on leaves, apetalous sterile forms |
| Uranium | Abnormal number of chromosomes in nuclei; unusually shaped fruits; sterile apetalous forms, stalked leaf rosette |
| Zinc | Chlorotic leaves with green veins, white dwarfed forms; dead areas on leaf tips; roots stunted |

Source: Antonovics et al. (1971).

(FPACAQ 1987) is sufficient to damage white pines, for example, which are more sensitive than most crop plants.

In water, there are several ways in which mining decreases photosynthesis by aquatic plants and plankton. These include acidification because of fallout of $SO_2$ or acid drainage, increased pH of waste water, salinization, addition of toxic metals, and inputs of other toxic substances such as ammonia. A comparison of a pristine lake with a neighboring lake receiving uranium mill wastes showed a decrease of primary productivity from 130 to 30–70 mg•m$^{-2}$•d$^{-1}$ carbon (Johnson et al. 1970).

Increased turbidity caused by suspended sediment also decreases primary productivity by blocking out light reaching photosynthetic plants (De Nie 1987). In coal mining areas, for example, increased turbidity can be the major impact on waters receiving wastes. Iron mining and aggregate extraction frequently increase turbidity of waters downstream. Increased turbidity has been reported in marine as well as freshwater sites (Littlepage et al. 1972; Goyette 1978). Secchi disc readings (a measure of water transparency) decrease as turbidity increases, and are a rapid means of monitoring suspended sediment. In recommendations for preservation of aquatic habitats in Ontario, each waterbody is assumed to have a natural level that was acceptable, and so it is suggested that Secchi disc readings should be maintained within 10% of their original values (Ontario 1994).

Increased productivity, rather than the usual decrease, results from local loading of plant nutrients. Areas where this occurs include mining camps with inefficient sewage disposal (Trinh 1981) and tailings sites where fertilizers are leached into water bodies.

## Food Webs

While discussion of the impacts of mines on food webs is hampered by the lack of background data acquired before the beginning of operations, it is obvious that decreased production by plants must affect animals eating them. For example, the reduction of epiphytic lichens noted by Tyler et al. (1989) is likely to affect caribou, which use them for winter food.

Where lakes have become acidified (see the Sudbury case history), changes in food webs have been recorded. For example, when fish that eat insects decline, insect-eating ducks can increase in numbers (Eriksson 1984; McNicol et al. 1987).

Biomagnification of metals such as mercury and lead in food webs occurs in Canada. Top predators poisoned by these metals include humans. In general, the immediate sources of the metals in areas where biomagnification has been reported are not mines, but lead products and other industrial processes (Campbell et al. 1988). Biominification of metals can also occur as animals can excrete, detoxify, or sequester them.

## Structure and Niches

The most obvious impact on terrestrial ecosystems structure is destruction of landforms and vegetation during construction of open-cast operations. Vegetation may also be destroyed indirectly when microclimates are changed. Toxic wastes added to the atmosphere, soil, groundwater, and surface waters may kill vegetation or destroy part of it, such as a forest canopy (Amiro and Courtin 1981). The physical alteration of habitat, including vegetation structure, in turn alters the available sites and food for animals.

Waste storage areas and barrens are discussed in Chapter 4 and examples are provided by the case histories. Some of the largest barrens have been created by the smelting of sulphide ores (Amiro and Courtin 1981) and near asbestos mines, where the barrens resemble those occurring naturally on serpentine rocks (Ernst 1989).

The most significant damage to structure in aquatic ecosystems is done when placer mining causes stream beds to be washed away. The sediment washed downstream then destroys structures there, including refugia and spawning sites (McLeay et al. 1987; Kelly et al. 1988).

Destruction of marine habitat is similarly due to movement of particulate matter. Over 8,000 ha of benthic habitat are reported to have been destroyed by tailings off the coast of British Columbia (Hay and Ellis 1982; Canada Department of Fisheries and Oceans 1991).

Some organisms may benefit from changes of structure. Bighorn sheep, for example, use walls for shelter near coal mines in mountain regions (MacCallum 1989), and disused gravel pits can become productive ponds (Milliken and Mew 1970; Maneval 1975; Browning 1987).

## Communities and Diversity

Lack of data from periods prior to mining generally precludes comparison of diversity indices from before and after a mine's operation. Two indirect approaches can be used to compare diversity, however. The first is to compare matched areas, and the second is to use paleoecological techniques.

In a comparison of different areas, some believed to be affected by a mine are compared with pristine matched areas. It is assumed that all the areas were similar initially; this kind of assumption underlies the vegetation zoning shown in the Sudbury case history. A comparison of aquatic sites is relatively easy, because upstream sites can be presumed to show less impact than those downstream from a mine. An example of this approach was a comparison of diversity in sites from coalfields in western Kentucky; Bosserman and Hill (1986) found that "in severely impacted wetlands, motile fish tended to exist only in unimpacted feeder streams or adjacent wetlands. Similarly, the diversity of macrobenthos was significantly less in severely impacted areas than in unimpacted areas. Water chemistry parameters (pH, $SO_4$, conductivity, Fe, Mn, Na, Mg and K) were related to indices of community structure."

In a paleoecological approach, fossil organisms from dated layers of sediments or peat, such as lake varves, are compared. Changes in algal communities in lakes in the Sudbury basin have been compared by Dixit (1986) showing diversity decreasing until the 1970s, when control of emissions began to reverse the trend.

A problem arises in using diversity indices about whether or not to include introduced exotic species. Exotics include weedy plants, fish such as carp, and many insect pests. After disturbance, the exotics may thrive, while indigenous species are declining. For example, starlings are not disturbed by noise and traffic to the same extent as many native birds. An index of diversity of native species may thus show an opposite trend to an index including all species. Weeds may dominate disturbed ground such as the spoil banks around potash mines and gravel quarries.

## Populations

The population of most interest to people is, of course, our own. Mining affects humans through fallout of gases and solids, and through possible contamination of drinking water and of crops, farm animals, or food obtained in wild areas, such as fish or game.

**Table 3.2**  Some Effects of Selected Metals Species and Arsenic on Human Health

| Substance | Health Effects |
| --- | --- |
| Antimony | Heart disease; skin disorders |
| Arsenic | Cumulative poison at high levels; possibly cancer |
| Cadmium | Heart and artery disease; high blood pressure; bone embrittlement; kidney disease; fibrosis of lungs; possibly cancer |
| Lead | Brain damage; convulsions; behavioral disorders; death |
| Manganese | Nerve damage |
| Mercury (as methyl mercury) | Nerve damage; death |
| Nickel (as nickel carbonyl) | Lung cancer |
| Selenium | Possibly cancer; possibly tooth decay |

Source: Miller (1985).

A major concern for human health is possible contamination of groundwater and thus of wells. Mining-related contaminants reported in groundwater include salt, cyanide (Scott 1987) and radioactive elements (Gilbert 1994).

Metals generally have several possible uptake pathways, including inhalation, ingestion, and absorption through skin. Some of the effects of metals on human health are summarized in Table 3.2, but it should be remembered that some metals are trace elements that are essential in proper human nutrition. Regulatory standards and guidelines for emission of metals in liquid effluents have been established by the federal government and by some provincial governments.

Fallout of wastes includes $SO_2$, CO, $CO_2$, radon, metals, asbestos fibers, etc. Acceptable limits for exposure have been determined for each toxic substance, although provincial, national, and international standards and guidelines frequently differ (see Table 3.3 for the various Canadian jurisdictions). For sulphide dioxide, for example, limits vary from the "desirable limit" for one hour of 0.62 ppm in federal objectives and the objectives or regulations of several provinces to the (federal) maximum "tolerable limit" of 0.33 ppm (Van der Hoven 1985). Excessive exposure can be lethal (Nebel 1987) and relatively low exposure is a health concern. In New Brunswick, for example, the occurrence of asthma cases has been linked to atmospheric $SO_2$ (F/P/TBWG 1994). $SO_2$, which is easily transported over long distances, is a component of most urban smog. Emissions from the mining industry are decreasing, however; in Ontario, for instance, $SO_2$ dropped by half from 1970 to 1991, and mineral processing now accounts for less than half of total anthropogenic emissions (Ontario MOE 1992a).

The limits for exposure include levels for human exposure (e.g., in ambient air or in drinking water) and limits for aquatic life (as indicated by tests for waterfleas and trout generally). Emitters must ensure that loading does not cause concentrations to rise above these levels. Problems arise in that times of exposure vary, so that time of exposure must be considered as well as concentration. For controlling emissions, not only upper levels of concentrations should be considered, but also total loading. If, for example, a lake receives radioactive material,

**Table 3.3** Federal and Provincial Air-Quality Objectives, Criteria and Regulations for a Number of Substances in Canada*

| Substance | Time | Federal | | | British Columbia (O) | Alberta (R) | Saskatchewan (R) | Manitoba (O) | | Ontario (C) | Quebec (R) | New Brunswick (R) | Nova Scotia (O) | | Newfoundland (C) |
|---|---|---|---|---|---|---|---|---|---|---|---|---|---|---|---|
| | | D | A | T | | | | D | A | | | | D | A | |
| particulates | 24 h | | 120 | 400 | 260 | 100 | 120 | | 120 | 120 | 150 | 120 | | 120 | 120 |
| sulpher dioxide | 24 h | 150 | 300 | 800 | 373 | 150 | 150 | 150 | 300 | 275 | 228 | 300 | 150 | 300 | 300 |
| carbon monoxide | 8 h | 6 | 15 | 20 | 15.2 | 6 | 6 | 6 | 15 | 15.7 | 15 | 15 | 6 | 15 | 15 |
| ozone | 24 h | 30 | 50 | | | 50 | 30 | 30 | 50 | | | 30 | 50 | 50 | |
| nitrogen dioxide | 24 h | | 200 | | | 200 | 200 | | 200 | 200 | 207 | 200 | | 200 | 200 |
| hydrogen sulphide | 24 h | | 5 | | 7.1 | | | | | | | | | | |
| hydrogen flouride | 24 h | 0.4 | 0.85 | | | | | | | 0.86 | | | | | 0.9 |
| lead | 24 h | | | | 6 | | | | | 5 | 5 | | | | 5 |

O — objectives  
C — criteria  
R — regulations

D — desirable limit  
A — acceptable limit  
T — tolerable limit

*Concentrations shown are upper limits, for the time periods indicated. All units are µg·m⁻³, except for carbon monoxide, which is mg·m⁻³. Source: van der Hoven (1985).

it may result in a dangerous loading even though concentrations in effluents always meet regulatory standards.

Effects of metal fallout include elevated levels of metals in hair, which are comparable to effects on squirrels' fur and birds' feathers in the Sudbury Basin (Ranta et al. 1978; Rose and Parker 1982; Lepage and Parker 1988). As an example of possible food chain effects—and their potential to affect humans—the kidneys of white-tailed deer sampled within 8 km of a Zn/Cd smelter had elevated Cd (Sileo and Beyer 1985). Elevated levels of metals related to mining have been reported in other human foods such as shellfish.

Partly for this reason and partly because fish populations are intrinsically important, study of impacts on them is nearly as prominent. Spills have caused fish kills, for example, because of cyanide (Leduc et al. 1982). The major impact of mining on fish populations, however, has been through acidification. Most fish populations cannot breed below a pH of 5, and some populations die out about pH 6 (Scheuhammer 1991). Acid spikes often occur in the spring as a result of snow-melt and can affect marine fish coming up river to spawn (Watt et al. 1983). The speciation and solubility of metals are affected by pH, with the metals most likely to affect fish in acidified waters being Hg, Co, and Al (Scheuhammer 1991). Under acid conditions, aluminum is highly toxic, affecting the function of gill tissues.

Aquatic invertebrates, such as those that burrow in mud, are often exposed to high levels of metals. Clams and mussels tend to ingest metals if the mud is contaminated (Luoma and Jenne 1976; Luoma and Brian 1982; Campbell et al. 1988). Off the west coast of Canada, copper is elevated in sediments at sites such as Howe Sound, Rupert Inlet and Britannia Beach, and in invertebrates including oysters (Thompson and McComas 1974). High metal concentrations in sediments and invertebrates, where a mine is the sole source of the metal, occur off the outfalls and docks of Canada's two Arctic mines at Nanisivik on Baffin Island and Polaris on Little Cornwallis Island. Sea urchins and clams have elevated levels of Zn and Pb, but their population density appears unaffected (Canada Department of Fisheries and Oceans 1991).

Like fish and amphibians, many freshwater invertebrates cannot reproduce in acidified waters. Altered invertebrate communities have been reported in acidified waters (McNicol et al. 1987). Metal concentrations are elevated in many invertebrates from areas contaminated by mines, including crayfish with high Cu, Cd, and Ni (Bagatto and Alikhan 1987; Wren and Stephenson 1991). The specific form of the metal makes a difference; speciation is discussed below.

In natural areas, as conditions change because of mining, plant populations may die out locally and other populations move in. Such changes are brought about, for example, by gradual increases in the salt content of the soil or gradual changes in its acidity. In crops and natural populations, more dramatic changes occur because of toxic substances in the air, in soil, and in precipitation. These substances include $SO_2$ and metals. Plant responses differ between contaminants and between species, and thresholds of toxic exposure must therefore be defined for particular species. In contaminated areas, sensitive species such as coniferous trees and lichens disappear earlier and cannot survive as close to a source as, for example, tolerant horsetails and grasses. The most tolerant species are known as accumulators; concentrations of Ni up to 10% of dry weight have been found in some accumulators (Cannon 1960; Ernst 1976; Shaw 1989).

Symptoms of toxicity vary with plant species and with metals, and some are listed in Table 3.1. Generally growth is slowed, parts are stunted, and necrotic spots appear. The root system can be the part most affected (Barcelo and Poshenreider 1990). Metallic ions can enter a plant by following nutrient pathways (e.g., anions following phosphate pathways and cations following calcium pathways). Tolerance has several mechanisms; for example, plants may immobilize metals and keep them away from the most sensitive metabolic systems, e.g., in the roots (Taylor and Crowder 1983; Shaw 1989; Bagatto et al. 1993).

Tolerance of a particular metal, or of other toxic substances such as $SO_2$, has evolved in some populations near mines. These are generally plants with a short period between generations, such as grasses. In Canada, tolerant strains of grass have become widespread, for example, on old roasting beds near Sudbury which have not been used since the 1920s (Rauser and Winterhalder 1985 and the Sudbury case history).

Plant populations in aquatic ecosystems include both macrophytes and plankton. As acidity increases in lakes, diatoms generally decline, and dinoflagellates become dominant (Yan and Miller 1984). Algae have specific niches not only for acidity, but for conductivity, salinity, temperature, etc., so that their populations respond to several effects of mining. Indicators of increasing lake acidity include increased growth of the large green alga, *Mougeotia*, and mosses such as *Sphagnum*.

As wetlands acidify, species of both mosses and flowering plants replace each other; Gorham et al. (1987) described sequences occurring in peatlands undergoing cither anthropogenic or natural acidification.

## Biogeochemical Cycling

Disruption of biogeochemical cycles is perhaps the most obvious environmental effect of mining, since an element or compound is extracted from the ground and moved elsewhere. During the various processes, waste materials are placed in the atmosphere, in water, and in the soil.

As described in Chapter 2, larger particles tend to fall out close to the source of the emissions by direct or "dry" deposition. The gases and aerosols stay aloft longer, spreading out farther from the source and undergoing oxidation and other chemical reactions. Although most of them ultimately reach the surface through precipitation scavenging or "wet" deposition, a fraction remain aloft, adding to the concentrations of these trace substances in the atmosphere.

Fallout of these substances may have several possible effects on soils: salinization, increased soil pH, altered concentrations of nutrients such as potassium, and—most frequently—acidification and addition of metals. Examples of fallout effects on soil are included in the Sudbury case history. Acidification generally causes decreases in the density of bacteria and invertebrates in the soil, resulting in less efficient cycling of nutrients from dead plants and animals and their wastes.

There are a very large number of chemical reactions involving atmospheric emissions attributable to the mining industry. Johnston (1971), for example, listed 31 different reactions involving stratospheric ozone alone likely to occur in the atmosphere. Most of these are influenced by temperature, sunlight, and the concentrations of other gases and aerosols, and many of these responses are imperfectly understood. Raindrops and snowflakes collect gases and particles and deposit them on the surface of the Earth (Shaw and Munn 1971; Stern et al. 1984). This scavenging is also done by cloud particles that may subsequently grow into precipitation and fall to the ground. The relative contributions of cloud-droplet and precipitation scavenging have been estimated from theory and laboratory data to be 5% and 70% for sulphur dioxide and 20% and 5% for sulphate aerosols, respectively (Shaw and Munn 1971). This was for an atmosphere in which the sulphur dioxide concentration decreased sharply with height. In a well-mixed atmosphere, cloud-droplet scavenging predominates (Munn and Bolin 1971). With regard to precipitation scavenging, Beilke and Georgii (1968) found that there was an increase in the rate of scavenging with increasing rainfall intensity, decreasing droplet size (at constant precipitation rate), and increasing pH of the rainwater.

Although a considerable amount is known about precipitation scavenging, there is not yet a completely adequate quantitative treatment of the process (Frohliger and Kane 1975; Venkatram et al. 1988). This is particularly so in the case of snow, where there is disagreement as to whether the flakes collect or shed particles (Shaw and Munn 1971). Data from the Acidic Precipitation in Ontario Study (Chan and Chung 1986) indicated that rain scavenged the sulphate ion more effectively than sulphur dioxide, and that sulphate was scavenged slightly more effectively by snow than rain. However, the scavenging of sulphur dioxide by snow was found to be negligible. Kellogg et al. (1972) estimated that, on a global basis, scavenging processes were responsible for 86% of the transfer of sulphur (in the forms of sulphate and sulphur dioxide) to the Earth's surface, the remainder being attributable to dry deposition and direct diffusion.

Particles larger than about 10 mm are removed from the atmosphere mainly by gravitational settling, while other forces are usually more important for smaller particles. Surface impaction and electrostatic attraction are involved in the settling-out of particulate matter, while absorption and chemical reactions are the main sinks for gases. At times, the surface may act as a source rather than a sink, releasing absorbed gases by desorption and accumulated particulates through mechanical resuspension. The rate of loss of materials from the atmosphere to a surface depends on both the net rate at which they can be captured or absorbed at the surface, that is, the sink strength, and the rate at which they can be supplied through the atmospheric boundary layer to the surface (assuming the materials are small enough so that gravitational settling-out can be neglected) (Munn and Bolin 1971). The former process is a highly variable and poorly understood one. The latter is in the realm of turbulent exchange and is similar to the large-scale "diffusion" discussed above.

A study was made of the dry and wet deposition of sulphur oxides and trace metals between 1978 and 1980 in the region around Inco's Sudbury smelter (Chan et al. 1984). Within 40 km of Sudbury, the smelter was found to contribute 25% and 64%, respectively, of the dry deposition of sulphur and nickel, but less than 20% of other metals. It was found that effects of the smelter on wet deposition could be measured at distances of 40 km from the smelter, although most of the wet deposition in the smelter area was attributable to non-local sources. Larger-sized particles were deposited almost equally by dry and wet fallout, while sub-micrometer particles fell primarily in precipitation. A subsequent study (Tang et al. 1987), covering a larger area (up to 500 km from Sudbury), indicated that the smelter contributed up to 45% of the sulphur fallout by dry deposition, but less than 12% of that by wet deposition.

A survey of atmosphere-to-water dispersion by Whelpdale (1992) concluded that airborne pollutants make a significant contribution to water pollution. Freshwater bodies were found to be more susceptible than marine waters because of their chemical composition and low buffering capacity. Further information on the dispersion of pollutants from the atmosphere to water and land surfaces may be found in Hill (1971), Shaw and Munn (1971), Yamartino (1985), Nicholson (1988), and Sullivan et al. (1988).

Metal additions to soil may at first have a beneficial effect, as many metals are trace elements in which soils can be deficient. This is also true of sulphur, a component of proteins: up to 4 mg·m$^{-3}$ of additional $SO_2$ is generally a harmless addition to soil (Smith 1990). In forest soils, metals may come into contact with the litter layer, composed of needles, leaves, and other organic remains, and react with its humus by adsorption, chelation, or complexation. The mobility of the metals is altered by these processes (Friedland 1989). Movement of metals within soils was discussed by Shaw (1989). Reactions within the soil complex generally mean that measurements of total metal concentration in soils do not reflect the amount available for uptake by organisms (Campbell et al. 1988). Mycorrhizal fungi, which are an integral part of the uptake mechanisms of woody and many other plants, may fail to grow in contaminated soils (Balsberg and Pahlsson 1989), thus further complicating prediction of the uptake of metals and nutrients.

Loss of fertility and changes of structure in soil were described by Hutchinson and Whitby (1974), Freedman and Hutchinson (1980), and Hutton (1984). Such changes are important in forested areas and for farming, as well as in natural areas. Salinization lowers the fertility of soils, particularly in dry areas where soil water moves upwards in summer. More detail on soil fertility is given in Chapter 4 in the context of revegetation.

After being freed from rock by mining, metals may come into contact with organisms as gases, liquids, or solids. Because most metal uptake by organisms occurs in water—from seawater, fresh waters, and the interstitial waters in soils and sediments—it is important to know the speciation of metals in solution. Not all metal compounds are harmful; to understand the environmental impact of an industrial activity and to devise protective and treatment systems, therefore, it is necessary to understand what forms of an element are being produced by the activity.

In soils or sediments, solid particles, adsorbed metals, and dissolved metals are all likely to occur together and to interact. For example, a metal may occur in solid particles as a silicate, may be adsorbed on clays, on humus and on iron or manganese oxides, or may occur in solution (Campbell et al. 1988). The forms of a metal that occur in solution are influenced by the type of water body and its surrounding geology, whether a lake, for example, is dominated by sulphate ions or carbonate ions. Descriptive data on numerous Canadian lakes and rivers are available, including such characteristics as conductivity and concentrations of major ions (N.R.C.C. 1981).

Factors likely to influence partitioning of a metal include its electronic structure, the nature and concentration of solid phases present (such as iron oxides or clays), the nature and concentration of ligands present (such as humus compounds), the concentration of other cations present, the state of oxidation and acidity, and the degree to which equilibrium is attained (Campbell et al. 1988). Under controlled laboratory conditions, equilibrium may be attained and the speciation of a metal can be forecast. For example, Figure 3.2 shows speciation of copper under two sets of experimental conditions and the resulting forms of the metal. In Figure 3.2a, the variables are acidity (pH) and copper concentration with a given concentration ($10^{-3}$ molar) of total carbonate. The resulting forms of copper are the divalent cation, hydroxides, and carbonates. In Figure 3.2b, the variables are oxidation-reduction and acidity. The diagram shows the copper species formed as $CuO$, $Cu_2CO_3(OH)_2$, $Cu_2O$, $Cu$, and $Cu_2S$.

In a situation such as a tailings pond or a stream receiving acid drainage, however, the loading of different constituents may be too dynamic for equilibrium to be reached, so that forecasting the metallic species present is not possible except in vague terms. Within a tailings pond, the oxidation-reduction conditions will vary with depth, and precipitation will affect acidity, causing metal speciation to alter seasonally.

Speciation of a given metal in a water body affected by mining has been determined for many sites; for example, eleven lakes in the Sudbury Basin were analyzed by Nriagu and Gaillard (1984). In the case of lead, the principal species

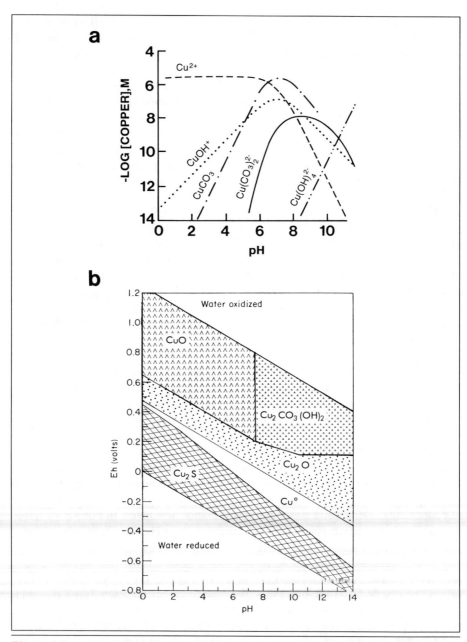

**Figure 3.2** a. Speciation of copper with varied pH and copper concentrations (after Spear and Pierce 1979).

b. Speciation of copper under varied conditions of oxidation-reduction (Eh) and acidity (pH) (after Spear and Pierce 1979).

were the aquo ion $Pb_2^+$, with $Pb(OH)^+$, $PbCO_3$, and $PbSO_4$. For nickel, the divalent cation, the carbonate $NiCO_3$, the bicarbonate $NiHCO_3$, the sulphate $NiSO_4$, and organic nickel were found.

Changes in metal speciation caused by acidity in a water body or sediment are frequently a challenge in managing mining wastes. Between pH 4 and pH 7, changes in speciation are greater in the metals Al, Cu, Hg, and Pb, than in Ag, Cd, Co, Mn, Ni, and Zn (Campbell and Stokes 1985). A pH-dependent biological response was found for Al, Cd, Cu, Hg, and Pb by these authors; in general, increased acidity resulted in less biological response, but the availability of Pb increased.

Even when its speciation in a lake or sediment or soil is known, it is difficult to forecast metal uptake by a particular organism. Uptake not only differs between species of organisms, but it may also be affected by factors such as age. Conditions of oxidation-reduction, acidity, and variables such as the presence of other ions will also affect uptake. Bioavailability of metals in sediments was reviewed by Campbell et al. (1988) and the effects of acidity on uptake and toxicity of Hg, Cd, Pb, and Al to various groups of organisms by Scheuhammer (1991). Both these books are based mainly on Canadian data, and both emphasize the role of organic ligands and the complex covariance of metals and acidity. Metal effects on freshwater organisms, due to mining, were reviewed by Kelly (1988). Kelly's book provides tables showing the pH tolerances of various types of organisms and their sensitivities to metal toxicity.

Because of the difficulties of understanding the complex and dynamic processes of metal uptake under field conditions, the importance of various metal species to an organism is sometimes evaluated by correlations between the metal in the organisms and the metal fractions in the surrounding medium. Correlations with total metal in water or sediment are generally not good, but there may be strong correlation between a particular species of a metal, such as a carbonate, and the metal in the organisms. Such positive correlations indicate that the species of metal concerned is likely to be important for the organisms (Dushenko 1970).

Wetlands, like soils, receive mining wastes from the air, and minerotrophic wetlands also receive them in runoff or from groundwater. Acidification and metal contamination are their major problems. Most wetlands affected by mining, where elevated metals have been measured, are on the Canadian Shield or other acidic rocks; but elevated metal concentrations also occur in wetlands on limestone that have been affected by mine wastes (Greig 1989; St-Cyr 1989). Brown and Kushner (1987) compared thirty wetlands near Timmins, Elliot Lake, Noranda and Sudbury; as Table 3.4 indicates, metal concentrations much greater than background levels occurred in some bogs. For example, Ni and Cu were each over 300 ppm in the Sudbury group, whereas values are usually below 20 ppm. Pb and Zn were also extremely elevated in this group. In the Elliot Lake group, Fe, Al, and K were well above background levels. Oligotrophic raised

**Table 3.4** Mean Metal Concentrations (ppm) in 30 Canadian Shield Wetlands

| Location | Ni | Cu | Pb | Zn | Ca | Mg | Fe | Mn | Al | K |
|---|---|---|---|---|---|---|---|---|---|---|
| Sudbury | 383 | 376 | 86 | 152 | 6,022 | 1,998 | 17,790 | 97 | 24,034 | 5,198 |
| Timmins | 16 | 21 | 14 | 37 | 15,852 | 1,941 | 5,133 | 349 | 10,017 | 2,966 |
| Noranda | 15 | 38 | 15 | 51 | 6,362 | 936 | 4,657 | 226 | 13,974 | 3,418 |
| Elliot Lake | 23 | 42 | 60 | 75 | 4,648 | 1,717 | 32,026 | 367 | 44,308 | 11,866 |

Source: Brown and Kushner (1987).

bogs usually receive all their minerals from the atmosphere and so do not reflect concentrations in local rocks (Moore and Bellamy 1974).

Accompanying metal loading, there is generally replacement of anions of organic acids by sulphates (Moore and Bellamy 1974; Glooschenko and Stevens 1985; Gorham et al. 1987). Berries of blueberry, bearberry, and bog cranberry plants collected near Flin Flon, Manitoba, were analyzed for metals and had elevated arsenic, selenium, mercury, lead, copper, iron, and zinc (Shaw 1982), showing that nutrient cycles have been altered. Changes to wetlands in the Sudbury basin are described in a case history by Gignac and Beckett (1986).

Wetlands may protect water bodies from mine drainage. At Cullaton Lake Gold Mines, Northwest Territories, a wetland blocked a leak of metals and cyanide out of tailings, preventing it from reaching the Kognak River, which flows into Hudson Bay (Diamond and Meech 1984). It is not known how long such immobilization would be effective.

Both natural and artificial wetlands have been used to absorb effluents from coal mines, massive sulphide mines, and other sites such as sewage treatment plants (Gersberg et al. 1984; Giblin 1985). Such wetlands can lose function if overloaded with acids or metals (Wieder et al. 1985); they cannot function if bacterial activity and peat structure is destroyed. Chapter 4 discusses some past and current research on the topic of wetlands as treatment systems.

Cycling in lakes and ponds is altered by aerial fallout, by runoff, and by groundwater. Direct inputs of mine wastes are reported, including spills of acidic effluent, cyanide, and radioactive mining waste (Leduc et al. 1982; Gilbert 1994). Climatic conditions and flow determine the distance from a source to which waterbodies are affected; for example, cyanide was above background levels for 10 km downstream in a river in the Northwest Territories (Scott 1989), and metals in the Clark Fork River in Montana are elevated hundreds of kilometers downstream (Axtmann and Luoma 1987; Moore 1987).

Prediction of the effects of mining wastes on water bodies is difficult because of the many variables of water and sediment chemistry involved, and because of different climatic regimes. Main factors include Eh, pH, the presence of Mn and Fe oxides, and organic ligands (Campbell et al. 1988). Hawley (1972a, b) pointed out that a prerequisite is knowledge of the solubility of wastes under various physicochemical conditions.

Effects on the water bodies include changes in pH, changes in ionic concentration of metals, and additions of toxins. As noted earlier, spring loading may cause particularly damaging spikes of contaminants (Franzin 1984; Sly 1991). Elevated salinity can be a significant change in waters receiving mine effluents; in Island Lake, Saskatchewan, salinity increases were considered to have changed the macrophyte community and caused the loss of some invertebrate species, including leeches (Hynes 1990). The effluents are from the Cluff Lake uranium mine.

Acidification can be due both to acid drainage and to fallout from atmospheric sulphur compounds, to which smelters and mines contribute. Lakes are classified according to their sensitivity to acid loading, depending on their initial chemistry, morphology, etc. (Seip 1986; Jeffries 1990). Some lakes have limestone or marble bedrock, which has a buffering effect that enables them to resist acidification. The part of Canada roughly south of 52°N and east of the Manitoba-Ontario border receives more than 10 $kg \cdot ha^{-1} \cdot yr^{-1}$ of wet sulphate (in the form of acid precipitation). Of the 700,000 lakes in this region, some 10,000 to 40,000 are believed to be at risk of acidification to different degrees. Regional surveys include those of Hammer (1980) for Saskatchewan, Gilbert et al. (1985) for Québec, Clair et al. (1982) for the Maritimes, and Ontario MOE (1990) for Ontario.

The causative relationship between sulphur compounds and acidification has been shown by experimental acidification of a lake in northwestern Ontario between 1976 and 1983 (Schindler et al. 1985). Analyses of sediments from other lakes support the experimental evidence (Dillon and Smith 1984; Dixit 1986; Dixit et al. 1991). The consequences of lake acidification, as well as recovery, are discussed in the Sudbury case history.

An opportunity to distinguish local sources from the general fallout of sulphates and other contaminants was afforded by a shutdown of smelters at Sudbury in 1982-83; sulphate concentrations in Clearwater Lake decreased, resulting in an increase in pH (Dillon et al. 1986; Seip 1986). The local contribution of wet sulphate deposition was calculated to be 15%, and the contribution of dry sulphate (as airborne particulates) up to 20% (Tang et al. 1984; Chan et al. 1984). A combination of aerial fallout and acid drainage had previously caused the pH of local lakes to fall as low as 3.3 (Gorham and Gordon 1960).

Anthropogenic loads settle out in sediments, which may be in ponds, lakes, or the sea. The resulting sediment changes are persistent. The Moira River in eastern Ontario is described in a case history of gold mining and arsenic processing (see Chapter 6). Mining and smelting in the valley ceased in the 1970s, but sediments in Moira Lake, approximately 10 km downstream, are still contaminated with arsenic (Diamond 1990). The river enters Lake Ontario 40 km further downstream, and arsenic and other metals are still elevated in shore sediments near the mouth (Dushenko 1990). Local Lake Ontario marshes have some organisms, such as snails, with elevated copper and arsenic levels (Greig 1989).

In some cases, uranium mining has added radionuclides to lake sediments. At Quirke Lake, Ontario, $^{238}U$ and $^{232}Th$ decay chain radionuclides showed an enrichment of one to three orders of magnitude in surficial sediments relative to background in deeper sediments. Net sediment loading rates included 130–230 $mg \cdot m^{-2} \cdot yr^{-1}$ for uranium and 360–4,000 $Bq \cdot m^{-2} \cdot yr^{-1}$ for other members of the $^{238}U$ decay chain (McKee et al. 1987). Plans for decommissioning waste sites in the region are at present being reviewed. While control of waste sites is generally a provincial responsibility, nuclear-related activity is the domain of the Atomic Energy Control Board, a federal agency.

Disposal of tailings in meromictic lakes depends on maintenance of their vertical water layers, so that metals are retained in a reduced state at the bottom of the lake, separated from upper layers by dense salty water. At the Polaris lead mine in the Arctic, tailings are pumped into Garrow Lake, which is meromictic. This lake has no organisms in the saline layer, which is below 11 to 20 meters in depth. The tailings, therefore, have to be deposited to a depth of more than 26 meters, using pipes placed by divers beneath winter ice. The aim is to have the pipes at a depth of 30 meters. Garrow Lake's discharge creek was dammed in 1990 to prevent possible loss of metals to the ocean (Chadwick 1994).

Marine ecosystems receive aerial fallout similar to that in long-distance transport on land, inputs from rivers, inputs of metals in liquid wastes, and inputs from tailings. The speciation of metals in seawater differs from that in fresh waters. Copper, for example, occurs in seawater largely as suspended colloids; abundant species include $Cu^{2+}$, $CuOH^+$ and $CuCl^+$, depending on the salinity and bicarbonate alkalinity of particular sites (Spear and Pierce 1979).

Distribution patterns of metals in marine sediment are often complex and may not show the neat inverse relation with distance from the source shown by aerially transported material. Seawater is subject to tides and currents, and also may retain discrete bodies of water for some distance; for example, dense acid water with a high iron content can be recognized several kilometers from a titanium oxide factory in the Humber estuary, in England (Mance 1987). A Canadian example of complex distribution of metals occurs in the Arctic waters of Strathcona Sound, off Baffin Island, in effluent from the Nanisivik lead-zinc mine. The patterns indicate either a second orebody under water or possibly an effect of warm effluent being rafted over denser, colder seawater when it is released in the Arctic summer (Thomas and Metikosh 1984).

Tailings can be placed in deep water for permanent disposal. Chapter 4 discusses this practice, with emphasis on subaqueous disposal in lakes and rivers, as a preventive or rehabilitative measure.

Theoretically, marine disposal of tailings in a deep fjord should prevent dispersal (because the material is below wave action) and should prevent oxidation and acid formation from sulphide wastes because the water is sufficiently deep to be reducing (Ellis 1987). In fact, in the 1970s, copper wastes were reported to be released into depths of only 3 m, on a wave-exposed platform. Tailings material was also found to disperse for several kilometers, rather than remaining where it was placed off the British Columbia coast. Resuspension was reported by Hay and Ellis (1982). Some of the west coast sites were in use for several decades—for example, at Britannia Beach near Vancouver. In addition to copper, molybdenum tailings were being put into the Pacific Ocean in the 1970s (Littlepage 1975) at Alice Arm.

Ellis (1982) described four marine disposal sites. In 1986 the only operation listed by Environment Canada as licensed for marine disposal of tailings was at Port Hardy, British Columbia, where Island Copper was producing 48,000 t·d$^{-1}$ of copper tailings. Island Copper, a coastal mine that uses the sea for underwater tailings disposal, is not used as a case in this book because it was the subject of an extremely thorough study by Ellis (1989). It is such an instructive situation, however, that a brief synopsis of the description from Ellis's book is included.

An open pit in a massive porphyry orebody situated beside a fjord at the north of Vancouver Island, the mine started operations in 1971. Its ore is low grade, producing 98 tonnes of tailings for every 2 tonnes of copper concentrate. The decision to place tailings under water in the fjord, Rupert Inlet, was made after an environmental assessment that weighed the risks of broken tailings dams on land against possible damage to marine habitats. The mine is in a very wet forest region, subject to earthquakes and landslides, with numerous streams that are spawning grounds for five species of salmon. By 1989 thick deposits of tailings, with a maximum thickness of 35 m, had formed in about an 8 km length of the inlet; chemical traces were apparent for about another 30 km. The thick or rapidly deposited tailings smothered preexistent benthic habitats, which had been a fine organic sediment.

Monitoring has included the performance of the outfall, in both flow and chemistry, surveys of the seabed and water column, and the pattern of currents. Biological factors measured included primary production, and population dynamics of zooplankton, benthos, invertebrates, and fish. Contaminants have been estimated in sediments and selected organisms. The main concerns have been the rate of discharge of the effluent and its composition. Since the mine began to operate, limits have been imposed on pH, Cu, Mo, Cd, Pb, As, Hg, and cyanide in tailings. The outfall uses a mixture of seawater to ensure that the plume sinks to 50 m or so and travels down the deepest part of the fjord. The silt-textured tailings have, of course, increased turbidity, but it has not continued to rise in the long term. Some changes in turbidity were unforeseen; for example, discolored water eddied up where three branches of the fjord meet. Turbidity has not decreased the primary productivity of phytoplankton, but it has lessened that of some beds of kelp and seagrass.

When the tailings settle, a stable surface can form that is colonized by invertebrates after one to two years. A community of opportunist species gradually develops, and sea anemones are often abundant.

Population sizes of clams, crabs, and several fish species have shown high variance, but the changes were not significantly related to tailings deposition. Some organisms have become contaminated, e.g., mussels by copper. The populations with greatest economic importance, crabs and salmon, are not contaminated, and the crab fishery in Rupert Inlet has increased. Some chemicals used (e.g., a frother) caused fish mortality, but were quickly replaced.

Ellis described careful monitoring protocols at Island Copper, initiated before the mine began operation, good data storage and analysis, and frequent publication of results. He ascribed the success of marine tailings disposal at this mine to thorough planning and to collaboration among mine personnel, government officials, scientists, and members of the public after critical environmental protests.

Groundwater is defined as subsurface water that is below the water table in soils or rocks. A watershed has a complex pattern of groundwater, with recharge and discharge areas, aquifers (strata or formations transmitting water), and aquitards (strata that are less permeable). Methods of measuring groundwater parameters, such as pressure, flow velocity, and density are described by Freeze and Cherry (1979). Groundwater can be affected by any of the processes acting on the soil and is of concern when drinking water comes from wells contaminated by seepage and when the groundwater may pollute water bodies downstream. Mining may directly affect aquifers, such as those in some Albertan coal seams.

Groundwater contains solutes that may diffuse through it or be carried along by advection. The solutes can undergo chemical changes in the groundwater, reacting with substances derived from neighboring rock, such as calcium carbonate, or with each other. The chemical state of a substance in groundwater is generally predictable from Eh-pH diagrams and a knowledge of the availability of major ions such as $Cl^-$ or $Ca^{++}$ (Freeze and Cherry 1979). Groundwater generally contains dissolved gases, including nitrogen, oxygen, carbon dioxide, and methane. The physical state of groundwater may change seasonally; for example, aquifers cooling to freezing point may become aquitards. The patterns of recharge and discharge may alter seasonally or irregularly, with temporary springs and ponds forming or drying up.

Prediction of movement of contaminants in groundwater is generally done by making a computer model based on data from wells and piezometers. It is known that metals generally have different mobilities in groundwater, with Cu, Zn, Ni, and Co moving rapidly, Pb and Ag being less mobile, and Au and Sn even less so (Freeze and Cherry 1979).

Problems with contaminants reaching groundwater can occur at tailings dams. Seepage from tailings occurs because earth and rockfill dams have some permeability. Groundwater is involved in five types of dam failure described by

Freeze and Cherry (1979): geological failure, uplift failure, piping failure, slope failure, and leakage failure. There are a number of reported cases where the damming of streams that have been affected by mining has resulted in metal-rich sediments in reservoirs. Groundwater can be affected by spills such as those caused by the failure of tailings dams; for example, in the 1980s cyanide from gold tailings entered groundwater near Smithers, British Columbia, from a spill at the Baker Mine (Melis et al. 1987).

Groundwater that does not contain contaminants can also be problematic. Open cast pits, for example, may be economically limited by the costs of dewatering, and operations where the rock contains sulphides may incur major expenses excluding water to prevent acidification.

Construction of a conceptual model for seepage migration is exemplified by work at Nordic Main tailings at Elliot Lake (Morin et al. 1988a, b; Morin and Cherry 1988). The uranium tailings are acidic, contain $^{230}$Th and $^{226}$Ra, and generate radon gas, and have been placed over a glaciofluvial sand. Seepage has moved down into the sand and then away laterally; the groundwater moves about 440 m·yr$^{-1}$, but the solutes have moved more slowly, at only 1 m·yr$^{-1}$. Drainage from the tailings enters the Serpent River system, which flows into Lake Huron. The seepage plume was found to have an inner core with reduced pH (as low as 3.2) and high concentrations of iron and sulphates. As this water reaches the edge of the plume, it becomes neutralized; major reactions are thought to be precipitation and dissolution of calcite, siderite, gypsum, and minerals with Al and Fe hydroxides. The computer model was used both to forecast and hindcast. The hindcasting suggesting that the plume had previously been more acidic, that it began to migrate out of the source area in the mid-1960s, and that the source was inside the dam.

On a small scale, a mining operation may drastically alter groundwater flow or direction. A limestone quarry at Westbrook in southeastern Ontario had operated for several years when it flooded so suddenly that workers had to abandon equipment in the pit.

## Succession

Like that of other disturbances, the general effect of mining on succession is to push it back to an early stage. If operations remove part of the vegetation (such as the canopy layer of a forest), succession will tend to replace it when conditions become suitable. If all the vegetation is destroyed (as may happen during road construction or removal of overburden by explorationists), recovery may be slow because some early pioneer species, such as lichens and mosses, are highly sensitive to residual problems such as acidity or metal contamination.

If soil is destroyed or needs to be developed on a new surface (e.g., tailings or rocks), the process is called primary succession and is extremely slow. The establishment of soil processes such as nitrogen fixation and breakdown of detritus depend on the establishment of populations of bacteria, fungi, and invertebrates. Gagnon (1987) described an example of development of soil by natural succession on tailings on northern Québec. Water relations in the developing soil are a major determinant of plant success and may result in zoned vegetation on tailings (McLaughlin and Crowder 1988).

In aquatic systems, acidity or contaminants may destroy plankton and macrophytes; succession depends on the rate at which propagules reach the pond or lake after conditions have become suitable for regrowth. Recovery is discussed in the Sudbury case history.

When vegetation structure is changed, as occurs when a forest is replaced by an early successional stage such as shrubs, the animals using it will be different from those which were there before mining. In some instances, the new vegetation may support more herbivores; for example, moose and white-tailed deer tend to be more numerous in stands of young trees and shrubs than in mature forests.

## Stability and Resilience

It is impossible to compare the resilience of different ecosystems in Canada to mining because adequate data collected before and after disturbances do not exist. The process of environmental assessment that has been conducted in recent years means that there will be more comparative data in future. At present, differences in responses between maritime forests in British Columbia, boreal forest in central Canada or the Maritimes, or sub-Arctic area in the North, for example, can only be surmised. Most current mining operations are in the boreal forest, which is fairly well adapted to disturbance because it is subject to fire and to fluctuating intensities of herbivory by insects.

## Summary

Chapter 2 outlined several types of stress on ecosystems. Examples of all these as they relate to mining and mineral processing are presented in this chapter, ranging from slight perturbations, to lower of rates of productivity, to introduction of exotic species, and to total loss of function.

Environmental effects of mining include some for which there is little evaluation, such as noise disturbance and waste heat. The fallout of particulates is important in urban, agricultural, and wilderness areas and can be more damaging in aquatic systems than on land. Because acidification affects animal populations and plants, soils, the waters of aquatic systems, and their sediments, it is clear that acid drainage is among the greatest challenges to management. Contaminants added to both terrestrial and aquatic ecosystems include metals, radionuclides, and, in some cases, cyanide or other processing materials. Salinization of agricultural land and aquatic systems is a risk in areas where wastes are easily eroded.

The descriptions in this chapter of the effects on the natural environment and the human condition leave no doubt that environmental management needs to be holistic—removing a contaminant from one part of an ecosystem only to place it uncontained or untreated in another does not solve the problems of its existence. Diluting a contaminant only widens the area affected by it. When treating wastes on land, it is always necessary to also consider the potential effects on groundwater and aquatic ecosystems.

Chapter 4 considers methods, techniques, and strategies for preventing environmental damage and for rehabilitating and reclaiming sites that have already been damaged.

# CHAPTER FOUR

# PREVENTION, RECLAMATION AND REHABILITATION

## Introduction

It is widely recognized and accepted by industry and regulatory bodies alike that mining and mineral processing activities create damage to the surrounding land, air, and water that cannot be reversed by nature, but requires the aggressive application of carefully planned restoration programs. Moreover, experience and scientific research are continually identifying new problems and developing methods for avoiding and correcting both potential and historical environmental damage.

Reclamation is now considered an essential element of resource management. It is no longer regarded by either practitioners or regulators as a matter to be dealt with at the end of operations. Rather, it is an ongoing activity throughout the life of the operation to which any mining company in Canada must be committed. Even in those situations where actual reclamation work cannot be started until the close of operations, current activities can generally be carried out in such a way as to accomplish final closure more efficiently and effectively.

Protection is also becoming an integral part of mining and mineral-processing operations. As noted in the Introduction, this takes many forms, such as the establishment of environmental management systems that include environmental audits, policy statements, and guidelines that are communicated to all employees, programs to educate employees about the environmental effects of their activities and how to minimize them, and product stewardship policies.

Proper management of water problems associated with mining requires a baseline study that includes both a preliminary analysis of the geology and hydrology of the area and an assessment of the likely effects of the mining operation (Slagle et al. 1985). Following approval of the mine, an adequate monitoring program must be instituted to assess any changes to the water regimes both on the surface and underground; it also indicates whether and when mitigative action should be taken (Earl 1983). In addition to flow disruption and changes in water quality, a groundwater assessment and monitoring program can help to minimize mine and waste-heap instability due to subsurface seepage, rapid flooding from surface waters or aquifers, excess porewater pressures, increasing rock stresses, corrosion and rotting of materials, and uncomfortable working conditions (Brawner 1986).

While the actual area disturbed by mining may be small relative to the surrounding area or relative to that disturbed by other industries (such as forestry, for example), the impact of mining is, in some ways, more profound. Mining represents a disturbance that is often—evolutionarily speaking—a new experience. When this is the case, the damage can last longer than many human life spans. Observation and experience have demonstrated that natural capacity for restoration of land and water used for mineral extraction and processing usually requires substantial supplementation by remedial activities if an acceptable balance is to be reestablished.

These remedial activities deal with the prevention and amelioration of a broad range of problems, from matters as relatively simple as wind erosion of dust from tailings to complex biochemical reactions between native and introduced substances. This becomes an issue because some substances introduced by mining activities may react with naturally occurring materials— for example, cyanide and arsenic in the processing of gold ores. They involve planning and implementing measures to deal with the problems under widely disparate climatological, geochemical, and biological conditions. As well as the technical considerations, there now exists an understanding that plans must also take into account local cultural norms and social standards. As a result of the *Green Plan* (Environment Canada 1990a), regulatory requirements were gradually being developed or revised to include these issues. Funding for the *Green Plan* was ended in April 1995, and no more of its initiatives are expected to be undertaken.

As described in Chapters 2 and 3, typical mining operations can affect land and aquatic ecosystems and their atmospheric component. Management of these impacts has been regulated in Canada at both the federal and provincial levels for the past two decades, primarily by a system of permits for each component of an operation, based on limits for emissions to air, land, and water. This chapter examines the general issues and techniques in preventing, controlling, and correcting these changes, beginning with land reclamation. While this almost invariably means revegetation, with all the attendant questions of site stability, soil quality, fertility, plant selection, and ongoing care, other forms of reclamation treatment, including some innovative experimental ones, are described briefly as well. A particular emphasis is placed on acid drainage because it is one of the most serious—certainly the most intractable—problem faced in mine reclamation projects.

The chapter also considers the protection and remediation of wetlands and aquatic ecosystems and rehabilitation of abandoned sites, then describes and assesses the development and implications of Canadian environmental regulatory systems.

To a growing extent, regulators are setting the agenda for protection and remediation activities, and issues of risk management are coming to the forefront. These issues, in fact, form the very foundation of environmental impact assessments and underlie corporate decisions. While the need to provide safeguards for human health and the biophysical environment is indisputable, it is also necessary to ensure that those safeguards are not so stringent that their cost cripples the industry's ability to earn the necessary funds for ongoing protection and rehabilitation of decommissioned operations. Such a strategy would be a net loss to society.

With such considerations in mind, it is clear that before protection or rehabilitation are planned, goals should be clearly defined. Is the area to be restored to a previous use, such as agriculture? Is it to be maintained in a semi-natural state? Is it to become self-sustaining wilderness? Are several goals involved, such as creating forest and creating waterfowl habitats? Is the prime consideration human health, for example, through the quality of drinking water? What agencies or local organizations will be involved in future conservation management, and for how long a time?

The case studies at the end of this chapter discuss the approval process being undergone by Consolidated Professor for its Duport gold-mining project in Ontario and by Redland Quarries for its decommissioning of a limestone quarry.

## Terrestrial Ecosystems

Land to be reclaimed after mineral production activities often represents an environment that is hostile to living organisms, a circumstance that has resulted in the development of *land reclamation*, a discipline devoted to stabilizing and revegetating disturbed land (Bradshaw and Chadwick 1980; Cairns 1988; Hossner 1988; Walker et al. 1989). A specialized branch of ecology, *restoration ecology*, recognizes that the ultimate goal of reclamation is the creation of stable and diverse ecological systems resembling local natural ecosystems (Jordan et al. 1990).

Research in this field is now addressing the scientific, technical, and policy issues involved in the restoration of areas affected by mining and mineral processing. The previous chapter described the different impacts to be expected under different circumstances and the various types of land disturbance and solid waste materials. These include the disposal of removed overburden, waste rock from the construction of shafts, tunnels, adits, etc., milling waste, precipitates and other sludges, and tailings. Each type of waste presents its own distinctive difficulties of safe disposal, and each method of disposal has its own idiosyncratic requirements for success. To further complicate the task of rehabilitation, the methods used must be adapted to the conditions specific to the site to be reclaimed. For example, the Polaris Mine in the Northwest Territories must alter its backfilling methods with the season to avoid affecting the thermal equilibrium of the permafrost (Keen 1992).

## Rehabilitating Spoils and Tailings

During a mine's operational stage, irrespective of whether it is an underground or open pit mine, the main impact upon local land surfaces arises from the disposal of waste materials. These wastes may be classified into two broad categories: mine spoils (including topsoil, subsoil, and other strata that must be removed for shafts, tunnels, adits, etc.) and mill tailings. An individual tailings or spoils pile is not necessarily a homogeneous milieu, but may vary considerably within a relatively small area (Murray 1973; Jonescu 1974; Gagnon 1987; McLaughlin and Crowder 1988).

As noted in Chapters 2 and 3, the preferred approach to environmental management uses ecosystems as the basis for evaluation of cumulative impacts, land use alternatives, and management programs and, consequently, of reclamation activities. In analyzing a given situation, the systems approach initially described two decades ago by Wali (1975) appears to be an important advance in reclamation research. By first viewing the disturbed area as a complex system with many interrelated, potentially limiting factors and then expressing these relationships in a systematic, quantitative manner, many of the failures resulting from "trial and error" approaches can be avoided. This will undoubtedly result in savings of time and money.

Lakes and rivers or watersheds are now usually managed as whole ecosystems. For example, contaminant and eutrophication problems in the Great Lakes are analyzed from a systems viewpoint. The Great Lakes have been affected by mining wastes such as taconite tailings and asbestos fines in Lake Superior (Merritt 1994). For Lake Superior, one of the surprises of ecosystem analysis has been the determination that loading of lead and mercury is currently mainly from the atmosphere, with 69% of the lead and 73% of the mercury deposited from the air (IJC 1994).

While surface materials exposed by mining activities frequently appear sterile, they often have the potential to be quite fertile. The raw waste material, lacking in humus and living matter, may provide a site for specialized pioneer species. At the other extreme, it may instead be extremely toxic and entirely unsuitable for plant life. In the former case, colonization by native species may occur quickly, thus minimizing the necessity for reclamation activities. If the site is toxic, two avenues are open. The first is to rely on the natural processes of weathering, leaching and erosion, but the time required for stabilization and revegetation by natural processes could span several human lifetimes, possibly disrupting surrounding ecosystems in the meantime. Government regulations, therefore, generally no longer allow this practice.

The second option is to reclaim the site through an active program; this involves either stabilization and containment by purely physical means, by the use of vegetation or, more usually, by a combination of techniques. The physical containment approach usually has the serious disadvantage of requiring constant surveillance and ongoing inputs to prevent deterioration. An example of some of the risks inherent in this is provided by the case of the Matachewan gold mine, inactive at the time, in northern Ontario. Over a period of a few years, a beaver dam raised water levels in a lake that overflowed into the mine's tailings pond. In 1991, the resulting pressure caused a dam breach, releasing tonnes of tailings to the Montreal River. Drinking water supplies to municipalities along the river were disrupted for several days, but the plume ultimately dissipated (Heffernan 1992).

Such examples point to the importance of good design in protection of the environment. Figure 4.1 shows some approaches to construction of tailings areas. A substantial amount of research is being conducted to improve the design and operation of impoundments for tailings and waste rock, with a particular focus on acid-generating sites because they may need to last for many decades.

## Physical Stabilization of Acid-Generating Wastes

Most metal mines in Canada contain sulphide minerals, either in the ore or in the surrounding waste material. The oxidation of sulphides—particularly iron pyrite—is the major cause of soil contamination by mining wastes and potential for water pollution, through the production of sulphuric acid, which dissolves and releases metals in the wastes. This oxidation process, known as acid drainage, occurs naturally under undisturbed conditions, but occurs slowly due to limited supplies of oxygen or water. Mining and milling processes suddenly expose large quantities of sulphide materials and greatly increase the surface area that is exposed to oxygen, resulting in a sharp increase in the rate of this process. The material is so sensitive and the chemical reaction so fast that changes can be detected from day to day.

The process is also affected by naturally occurring catalysts, the most effective of which by a wide margin are ferric iron and certain microorganisms, particularly *Thiobacillus ferrooxidans* (Peterson and Nielson 1973; Jambor 1994; Sengupta 1994). Tailings and other waste rock impoundments, then, are likely to produce a leachate that is acidic (low pH) and contains high concentrates of metals and sulphate. If uncontrolled, the leachate can "contaminate surrounding ground and surface waters and therefore threaten the aquatic ecosystem" (Yanful et al. 1991).

Recognition of the seriousness of acid drainage from sulphide-bearing mining waste has been relatively recent. As a result, it is likely that mining operations built before the 1980s were not designed to control its effects, and there are many that have operated for decades in ways that are now understood to create the potential for damage from acidic effluents and their solutes. Extensive research activities are now under way to learn more about the process and to develop prevention and abatement techniques for use at new, operating, and abandoned minesites. An example of these activities is the Mine Environment Neutral

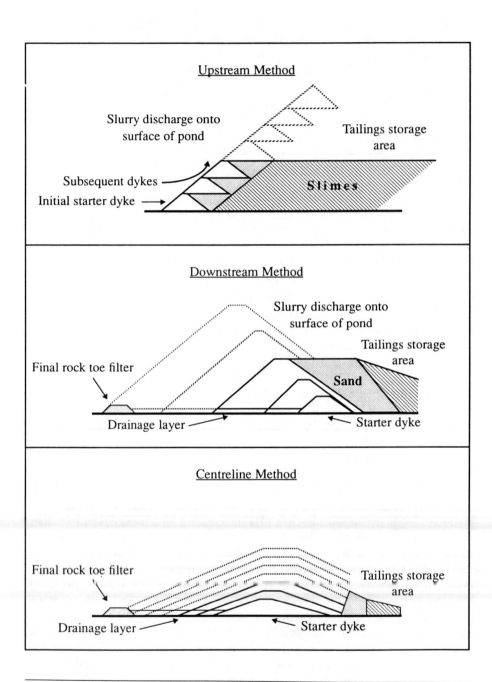

**Figure 4.1** Different methods of constructing tailings impoundments.

Drainage (MEND) Program, which developed out of an initiative started in 1983. It is a cooperative program initiated and financed by the Canadian mining industry, the Government of Canada (through CANMET—the Canada Centre for Mineral and Energy Technology) and the governments of eight provinces. Several Canadian universities and a number of international agencies, including the United States Bureau of Mines and the European Commission, are also involved. After five years of the nine-year program, it has been concluded that more than half of the funds should be appropriated to projects related to prevention and control, the balance being divided between studies related to prediction and projects investigating methods of treatment, monitoring, and technology transfer (CANMET 1993).

Among the most promising approaches to preventing acid generation is the practice of covering the acid-producing material with deep water. This produces an anaerobic environment that is not conducive to the oxidation and bacterial action that cause and accelerate the acid-producing reaction. Excellent results have been obtained by deep-water disposal of tailings in northern lakes; for example, the Polaris Mine deposits its tailings more than 26 meters down in a meromictic lake. The sulphides appear to remain stable, and dissolved metal concentrations are low (Pedersen et al. 1993, 1994). According to MEND's *1993 Annual Report*, water barriers are not suitable for use at "over half of existing tailings sites and most waste rock sites in Canada" because of such factors as the nature of the terrain and the design, placement, and location of the waste material (CANMET 1993). It is also necessary to avoid lakes that have a thermal turnover because waste materials would be forced into upper layers at certain times of the year (Wetzel 1983, Hutchinson 1993). Research into alternative measures involving dry, engineered soil covers will continue, with an emphasis on economic approaches in this area because they are a very expensive option.

Research is being conducted into various approaches to long-term management of tailings. Some of the more promising include flooding impoundments for sulphidic or pyritic uranium tailings at the time of decommissioning (St-Arnaud 1994; Davé and Vivyurka 1994; Pedersen et al. 1994). This approach is discussed in more detail in the section on chemical considerations. Another experimental method (presently being used at the East Sullivan Mine in the Abitibi region of Québec) is the construction of an organic cover—two meters of softwood and hardwood bark—overplanted with grasses (Tremblay 1994). A refinement on the organic-cover approach is Falconbridge's experimentation with compost which, by creating anaerobic conditions, has the effect of halting oxidation of the tailings, thus reversing acid drainage processes and possibly even reducing trace metals (Pierce et al. 1994). Among the materials being tested by Falconbridge is municipal sewage which, it might be noted, does not need to be "clean" as it would have to be if used for agricultural purposes. Other engineered covers include the use of fiber-reinforced, high-volume fly ash shotcrete as a cover for acid-generating waste rock piles, as is being tested for large-scale applications (Bradshaw and Chadwick 1980; Jones and Wong 1994). The MEND program has decided not to pursue further studies on cementitious covers because of the very high cost of such covers (CANMET 1993).

Abatement techniques for new and existing sites include both active collection and treatment methods and passive treatment systems. The latter include passive polishing of effluents in wetlands, either natural or constructed, whereby dissolved metals are absorbed by marsh plants such as cattails (Kalin 1991). This has been used successfully by Falconbridge Limited in Sudbury, as well as in the Appalachian coal region of Pennsylvania (Faulkner and Skousen 1994; McCleary and Kepler 1994). The latter also discuss treatment of acid drainage by means of the construction of anoxic limestone drains. The most common active treatment is by means of adding an alkaline substance, usually lime, to neutralize the acidity. This practice is described in the following section. In some ores, this process occurs naturally: they have a high concentration of calcite or carbonate that self-limits acid production in the tailings or waste rock disposal sites or buffers their acidity. Other treatment strategies seek to limit the oxidation process, either by controlling bacterial access, isolating the sulphide surfaces by coating them with an insoluble precipitate, or by creating an anaerobic environment by means of a protective barrier between the tailings and the atmosphere (Yanful et al. 1991; Blowes et al. 1994). Many of these techniques are still in the early experimental stages, although results are promising.

These facts lead to the conclusion that, at the present time, the best preventive measure is to design new impoundments to take account of the hydrogeological, climatic, and other pertinent features of potential sites, factoring in the geochemical and mineralogical characteristics of the waste to be deposited. A suitable natural site may be available, but it is more likely that it will be necessary to construct an impoundment with an impermeable liner to prevent such potential problems as seepage into groundwater.

Island Copper, a mine at the northern tip of Vancouver Island, British Columbia, provides a very instructive example. Disposal of waste rock at the site has been sufficiently good to win several awards (Ellis 1989). The mine is situated beside a marine fjord, Rupert Inlet, and disposes of huge quantities of tailings under water. In addition to tailings, very large quantities of waste rock are produced and placed on land. The aim of the waste rock disposal is to restore forest habitats with wildlife in this area of very wet coniferous forest. The placement of rock follows a predetermined reclamation plan, followed by conventional bioengineering procedures. Trees are planted without a preliminary cover because they do well in the wet climate. Particular attention has been paid to monitoring freshwater conditions at the minesite, at small lakes, streams, drainage ditches, pit dewatering sites, and seeps from the rock dumps. Some areas with low pH indicate acid water generation and have to be neutralized. This is generally done by combining seepages with neutral or alkaline drainage ditch water. Freshwater streams are of particular importance in rehabilitation of this area, because it has spawning runs of five species of salmon (Ellis 1989).

## Revegetation

Establishment of a mat of vegetation that is both self-sustaining and amenable to predetermined end uses is usually an important element in a rehabilitation program for waste disposal areas. Not only does vegetation stabilize the soil, preventing new acid-generating material from being exposed, but transpirational water movement decreases the amount of water available for deep percolation; in addition, the humus that accumulates at the surface acts as an oxygen sink (Johnston et al. 1975). During the 1960s and 1970s, restoration of tailings impoundments was generally aimed at establishing vegetation on the surface, with grass, shrubs, and trees on the dry parts and aquatic vegetation in ponds. Vegetation was established by controlling pH near the surface with additions of lime or limestone and by adding fertilizers where necessary. In many cases, repeated treatments were necessary before plants were established, particularly for nitrogen (Bradshaw and Chadwick 1980). Even after a grass cover has been established, however, problems of re-acidification can occur (Peters 1984), and vegetation can contain elevated levels of metals (McLaughlin and Crowder 1988). It is also possible for certain hardy vegetation to thrive without producing any improvement in the quality of effluents. At Waite-Amulet in Québec, the trees planted in 1971 are three meters tall, but the pH of waters draining from the toe of the tailings dam has stayed consistently low.

As many attempts at revegetation have demonstrated, this is a complicated task with potential for failure on a number of counts. And even when this practice is employed, it must often be done in conjunction with physical containment— for instance, to control erosion around the perimeter of the revegetated area and hence leaching of metals and other toxic substances into surface and ground waters. As Veldhuizen et al. (1987) observed, "It was not many years ago when a vegetative cover over a tailings area or a rock dump was considered to be a panacea." The realization that even the most carefully thought-out vegetation program was not the solution to acid drainage brought home to the industry and regulators that control of acid-generating sites is both complex and costly.

Success in establishing a self-sustaining vegetative cover on material that has been mechanically and chemically altered to an unnatural condition depends on careful analysis of the physical and chemical characteristics of the new growing medium (Murray 1973; Bradshaw and Chadwick 1980; Lyle 1987). The most frequent problem with mining wastes is the absence of organic and living components. In addition, tailings and other waste may have very different chemical and physical properties from naturally occurring soils and be totally unsuited to plant growth. Treatment is almost always necessary, as is a careful selection of suitable plants at various stages of regrowth. The latter cannot always be predicted, since changes in the texture and composition of the medium are produced by, among other things, interaction of the new plants with the medium. There are four steps to designing a vegetation program: recognition of the factors that are limiting or will limit plant growth; prevention or mitigation of these limiting factors; selection of species and planting techniques that provide the best potential for success; and follow-up programs to assess the long-term success of the operation and, if necessary, additional inputs to ensure long-term stability.

The following sections describe some of the physical and chemical complications to be considered in designing a revegetation program, followed by a discussion of site and soil treatment, and finally a brief discussion of the factors to be considered in plant selection.

### Physical Characteristics

The texture of mining wastes can vary considerably. A spoil embankment may be composed of "particles" ranging from boulders to clay. In tailings, on the other hand, the distribution of particle dimensions may be fairly uniform because the milling process is designed to produce particles in a specific size range. While this homogeneity may be suitable for plant growth, the regular size and firmness of the particles make the surface more susceptible to erosion, compaction by equipment traffic, and all the other problems associated with fine-textured soils. As well as particle size, tailings differ substantially in specific gravity and abrasiveness.

The most suitable soil texture for any site depends, to a large extent, upon the type of vegetation that is desired. The proportion of particles with diameters less than 2 mm (that is, sands, silts, and clays) is most significant to plant growth because small particles, with their greater surface area, contribute most to the water- and nutrient-holding capacity of the substrate. Cereals and forages grow best on finer soils; coarse soil may be adequate for most trees, shrubs, and certain native grasses and forbs. Soil textures directly affect root penetration; fine-textured soils can exhibit great mechanical impedance to root growth, particularly when dry. Water infiltration and rate of movement are faster on coarser soils, with the result that erosion is less of a problem. Depending on local climates, both fine and coarse soils may be disadvantageous. Fine soils tend to be poorly aerated and to hold water near the surface, while coarse textured soils drain more quickly, becoming arid at the surface.

Healthy plant growth is intimately related to the soil property known as aggregation. This refers to the capacity of soil particles to group together into stable structures. A fine-textured, well-aggregated soil is ideal for growth, since it has the desirable properties of both a fine-textured soil—in its capacity for holding water and nutrients—and a coarse-textured soil—in its water permeability, aeration, and resistance to compaction. The binding colloids are derived from organic matter; therefore, a newly created surface generally lacks them. Aggregation will improve with time if suitable soil flora and fauna can be established and maintained.

Soil aggregative properties cannot be greatly improved by artificial means; aggregation depends primarily upon the metabolic activity of soil microflora, microfauna, and roots. On the other hand, compaction, which quite commonly results from moving equipment, is a soil condition that can be substantially improved by mechanical operations. Deep tillage has proved to be the most effective of these, and some success has been derived by combining the tillage with pneumatic injection of low-density organic material into the voids created by the tillage (Sweigard and Saperstein 1991). There are some deeply rooted plants—such as alfalfa, birdsfoot trefoil, and many weedy species—that are effective in loosening subsoil.

The stability of the surface to be revegetated affects the success of plant establishment: erosion or slumping buries plants or prevents them from rooting. Further, as long as a new outer surface is continually being exposed, the processes of weathering and leaching will not be effective in building a soil conducive to plant growth. Wind can have particularly deleterious effects on an unstable surface: plants can be sand-blasted, covered by drifting spoils, or undermined; airborne seeds may become desiccated or transported to undesirable growth areas (Murray 1973; Crowder et al. 1982).

Impervious materials such as concrete, asphalt, rubber, latex, clay, carbonate bonding and alumina/silica gels can be used to seal the surface or restrict water infiltration on waste embankments, thereby reducing acid production and erosion. These treatments are expensive and are subject to chemical and physical breakdown. Attempts to cover sealed surfaces with soil and vegetation have resulted in mechanical damage and root penetration of the seal, and Johnston et al. (1975) note that soils resting on impervious surfaces are inherently unstable.

While it is generally accepted that the most effective long-term soil stabilizer is vegetation, initially establishing plants often requires mechanical treatments or the addition of surface supplements. Placing rock on unstable surfaces can be helpful, but this treatment may seriously limit the accessibility of normal cultivation and seeding equipment. Large quantities of rock should normally be used only when trees or shrubs are the intended future vegetation.

Mulches are widely used as surface stabilizers to increase the cohesion and aggregation of the surface, to provide surface roughness, and/or to improve water filtration and retention. They also moderate extremes of temperature. Hay or straw is an effective mulch, but it should be worked into the soil or otherwise held in place. Murray (1973) noted that this treatment may tend to dry the soil and that mulches such as resin, sodium silicate, latex gelatin, wood chips, bark, limestone chips, and animal litter all worked with varying degrees of success, but were impractical and uneconomical. Certain chemical stabilizers—such as asphalt, latex, or bitumen emulsions—in the proper concentrations, are effective for the temporary prevention of wind and water erosion, while allowing good water filtration.

Snow fencing and other windbreaks reduce wind erosion, but are not effective in preventing water erosion. On sloped areas, solid windbreaks may quickly become covered. Where soil moisture is a limiting factor, the snow trapped by such devices can be particularly beneficial. Because the problems associated with both wind and water erosion of soil are most acute on fine-textured substrates, these difficulties are most often experienced in mill tailings areas, particularly where sludge deposited in settlement basins has been allowed to dry (Thompson and Hutnik 1971). When at or near saturation, fine-textured material may be subject to solifluction or sheet and gully erosion; when dry, it is prone to wind erosion. Freezing and thawing when the moisture content is high can result in severe soil heaving. In mountainous regions, surface instability is usually a greater threat, since spoils dumped on naturally sloping terrain can easily exceed the natural angle of repose (Williams 1973). Slope partly determines ultimate land use: a slope abandoned at 40° may settle to 35°, at which it is satisfactory for forestry, while the maximum slope for improved pasture is usually 15° and for rotation crops, 5° (Bradshaw and Chadwick 1980).

Precautionary measures during embankment construction may head off later toxicity problems, such as salts moving upward and leaching out. Furthermore, since salts tend to move through the soil with water being evaporated, the construction of steep slopes facing south is typically avoided by the industry.

Although adverse soil characteristics are the most common limitation to revegetation of industrial wastes, atmospheric conditions are equally important to survival. Extremes of radiation, temperature, moisture, wind conditions, or air pollution may prove to be limiting factors in any given situation. The capacity of species to tolerate extremes in temperature is quite varied. For terrestrial vascular plants in an active state, the life-supporting range extends from about -5°C to +55°C. However, temperate-zone plants are usually productive only between +5°C and +40°C (Larcher 1983). Mine waste embankments are often constructed in such a way that high surface temperatures are unavoidable. South- and west-facing slopes will be subjected to the highest radiation flux. The color of the substrate will also greatly modify the maximum surface temperature attained.

Water supply is intimately linked to the ability of plants to tolerate high levels of radiant energy and resultant high temperatures (Jones 1983). If the soil water content is high, much of the incoming energy can be dissipated by evapo-ration. The same is true of the leaves themselves; when they are constantly supplied with water, transpiration allows them to remain cooler. In addition, moist soil is better able than dry soil to conduct heat to lower depths and to store heat, thus not only lowering maximum daytime temperatures, but also moderating cool night temperatures. Interception, retention, and distribution of moisture—including both rain and snow—are directly related to topography and soil char-acteristics. Wind is also critical in terms of increased leaf moisture stress, as transpirational water loss is increased.

Rehabilitation must be planned not only in relation to particular types of mining operation, but also in the context of the ecoregion in which operations are placed. The limitations in a sub-Arctic site obviously differ, for example, from those in a wet maritime site.

## Chemical Characteristics

Although physical characteristics are significant in the limitation of plant growth, in many cases it is the chemical environment of mine waste that is the main limiting factor. The problems of infertility, high or low pH, and the presence of toxic elements are interrelated and must be considered together.

The availability of both nutrients and toxic elements changes with pH (Bradshaw and Chadwick 1980; Williamson et al. 1982). Generally, at low pH levels, the availability of nutrients is decreased and the availability of heavy metals is increased because of the greater solubility of metal sulphides (refer to the discussion of metal speciation in Chapter 3). Spoils and tailings are usually low in macronutrients (nutrients that are necessary in large quantities, including nitrogen, phosphorus, potassium, calcium, and magnesium); most have limited availability at low pH levels. Soluble salts are released from rock fragments by strong acidity; the combination can precipitate essential nutrients from the soil solution, thus making them unavailable to plants. The high concentration of dissolved salts may have direct osmotic effects as well.

A pH of approximately 4 appears to be the lower limit for most plant species; below this level, the acidity is directly toxic to roots and attempts to raise pH have only temporary effects. Raising the pH level higher than 5.5 or 6.0, however, will not greatly improve vegetation establishment and may, in fact, hinder the growth of acid-tolerant plants (Murray 1973; Bradshaw and Chadwick 1980). Tailings with a high carbonate content have a good chance of maintaining their vegetation and downstream water quality if carbonate dissolution is sufficient to neutralize pyrite oxidation (Veldhuizen et al. 1987).

There are three generally accepted methods of dealing with acidic substrates: burying, liming, and leaching. Burying extremely acidic or toxic material is most economical if done in conjunction with embankment formation. If only small pockets of phytotoxic materials are present, burial at a later date might be feasible.

On acid-generating wastes, the usual treatment—mainly because it is relatively inexpensive and reasonably effective—is neutralization by application of liming agents such as calcite, hydrated lime, dolomite industrial slag, and fly ash. Ash has the benefit of being hygroscopic and thus enhancing water infiltration. Fuel ash, however, can contain high concentrations of some toxic elements, including boron (Bradshaw and Chadwick 1980). Murray (1973) noted aluminum and manganese present in toxic quantities if excessive liming rates were used. At very high pH, nutrients (e.g., phosphorus) are again unavailable. Liming can also cause other problems if its use is required over long periods of time. For example, it may produce sludges with a low concentration of solids and in sufficient volume that sludges present a storage problem of their own.

The third common approach to raising pH is leaching, which may be implemented in combination with liming. Increasing the pH of the soil solution causes the formation of neutral salts and, thus, the removal of some toxic ions (Beaton

1974); leaching them removes the salts from the surface. The goal is to move acidic salts below the level at which they will interfere with plant growth. Leaching to a depth of about 30 cm should be sufficient for the establishment of most species, but, again, a rising water table or upward water movement by capillarity and evaporation may re-acidify the surface. Flooding to wash away the salts is usually undesirable because it may cause bank instability; however, spraying in such a way that the application rate does not exceed the infiltration rate may be practical. When spraying is conducted over an extended period, in conjunction with mulching and seeding, the establishment of seedlings is very successful (Murray 1973). On the other hand, leaching may substantially increase the acidity and toxicity of water entering the drainage system. Deep leaching, in combination with proper drainage, may help to reduce problems of salinity and/or alkalinity (Beaton 1974). For alkalinity, pH can also be reduced by applying compounds containing sulphur, such as ammonium sulphate fertilizer.

Another approach to controlling acidity is to deal directly with the acidifying bacteria, which oxidize pyrite or ferrous to ferric iron. Bactericides such as surfactants may be added—although this process must be continued—or predators of the bacteria may be introduced. They include the types of organisms that consume bacteria in sewage treatment plants, such as rotifers, ciliates, and zooflagellates (McCready 1987).

From experience and experimentation, it has become clear that, in tailings impoundments, the extent of oxygen diffusion depends on the depth of the water table and on the sulphide content and particle size of the tailings, and that it is independent of the presence of vegetation. In fact, some vegetation could increase the rate of oxidation because marsh and aquatic plants can release oxygen at the root surface and because transpiration can pump water upwards during dry periods.

Chemical treatment may be also aimed at the inactivation or removal of metals. For metal removal, a two-stage process is widely used. First, iron and aluminum are precipitated out at a pH of 5–6. The second stage is intended to remove metals with maximum solubility at a higher pH. The pH can be raised using lime or another alkali to precipitate metals, or sulphide can be added. After metal removal, the effluent pII is lowered again before release (Vachon et al. 1987). A difficulty in adding sulphide is that ions may be lost in the effluent, increasing downstream acidity.

## Soil Fertility

Severe shortages of nitrogen can be expected in most, if not all, mine and mill wastes (Bradshaw and Chadwick 1980). Phosphorus and potassium (Beaton 1974) are generally inadequate as well. The amount and availability of other macro- and micronutrients do not usually present a problem, but should be checked if otherwise unexplainable difficulties arise.

The addition of chemical fertilizers is the most commonly employed method of increasing fertility. Few generalizations may be made with regard to the type or quantity of fertilizer required: conditions vary widely, and species requirements are variable. Improvements in such characteristics as soil stability, texture, pH, and organic matter content all aid in the infiltration of soluble nutrients, their retention, and their availability to plants.

Micronutrients (that is, nutrients required in small quantities) may or may not be present in the wastes, but required quantities are usually so small that deficiency is uncommon. Micronutrients include iron, manganese, zinc, copper, boron, molybdenum, and chlorine.

Soil microflora and microfauna—such as nitrogen-fixing bacteria and mycorrhizal fungi and detritivores—play a vital role in plant nutrition, by binding soil particles and by cycling nutrients through decomposition. While the activity of most bacteria decreases rapidly at pH levels lower than 5, many fungi remain fairly active (Nicholas and Hutnik 1971; Brady 1974). Research is being conducted to develop methods to inoculate severely degraded materials with mycorrhizal fungi (Jasper et al. 1991).

Until the soil's exchange capacity (for bases or ions) is increased and the population of soil microorganisms becomes great enough for effective recycling and augmentation of nutrients, repeated fertilization may be necessary. Massive application of water-soluble fertilizer on coarse-textured wastes may be of only short-term benefit, since most of the nutrients will be leached away. It is better to apply smaller amounts over a longer time period or to use slow-releasing fertilizers. The release of all types of materials can be further modified through the use of chemical additives, such as nitrification inhibitors that affect microbial activity. The time which applied nitrogen can be retained in the soil can be increased by retarding nitrifying bacteria (the soil organisms that change ammonia to nitrate), since ammonia is not subject to leaching and is not as readily taken up by plants (Beaton 1974). The objective is to build up in the system a pool of nitrogen "capital," from which a steady annual increment is available for plant growth, as occurs in natural ecological succession to a stable "climax" ecosystem. The amount of nitrogen capital in soil and productivity of vegetation are closely related.

Biological activity plays an important role in maintaining the fertility of mine wastes. Legumes are commonly planted, since the bacteria associated with their root nodules have the capacity to fix atmospheric nitrogen. When establishing legumes in sterile soil, it is essential that they be inoculated with bacteria. Blue-green algae also can fix atmospheric nitrogen. Since they are a common component of the soil microflora, it has been suggested that they could be valuable in nutrient amelioration of mining spoils (Shubert 1976).

Organic wastes such as sewage sludge and poultry or cattle manure have shown particular promise as fertilizing treatments (Lejcher 1973; Sopper 1992; Pierce et al. 1994). These materials not only supply nutrients, but also improve soil texture when properly worked in; they act as a mulch, as well. The concentration of toxic elements can be increased by sewage application, although this was not a problem in sludge-amended acidic coal mine spoils in Pennsylvania (Lejcher 1973; Sopper 1992). In straw or sawdust—organic mulches that are low in nitrogen—bacterial action may actually induce further nitrogen deficiency and so should be supplemented with nitrogen fertilizer.

Successful colonists of nitrogen-deficient substrates include plants with nitrogen-fixing symbionts (e.g., legumes, alders, and several native shrubs) and mycorrhizal species. The latter group refers to higher plant species, such as most native Canadian trees and shrubs, which have fungi in a symbiotic relationship with their roots. The plants supply carbohydrates to the fungi, which in turn contribute nutrients to the plants. Inoculation of host species with appropriate fungi can improve plant survival and growth (Allen 1991).

Application of topsoil to waste embankments can provide a substantial nutritional benefit and help establish a population of microorganisms (Sopper 1992). Topsoil is effective in slowing oxidation processes in the waste, improves the texture, and may help to prevent later compaction and consolidation (Briggs 1973). In certain cases, the physical and chemical properties of the waste should be partially amended before the soil is applied. On extremely acidic/toxic wastes, initial sealing with topsoil is cheaper in the long term than continued application of lime and fertilizer. While the immediate benefits of topsoil are clear, Curry (1975) has cautioned that leaching will cause a long-term decrease in its fertility.

The problems caused by necessary storage of topsoil during open-cast mining were described by Bradshaw and Chadwick (1980). They include consolidation, loss of nutrients, and change of texture. In British open-cast mines, however, techniques of replacing topsoil have been so refined that replaced sites are not distinguishable after three years.

### Species Selection and Planting Techniques

Since the climatic conditions of an area cannot be controlled, reclamation attempts must center on choosing well-adapted plant species. It should be remembered that individual plants respond not to the mean annual air temperature or to the average precipitation, but rather to the conditions within their own immediate environment. Within spoils or tailings areas, there may be wide deviations of the microclimate and drainage at various locations. Simply choosing species adapted to the general climate may result in plant failure if special microclimatic conditions are not considered. While modification of the general climate is impossible, attempts at microclimate manipulation can be very effective.

The serial replacement of biota that follows a disturbance like one produced by mining is called primary succession; the initial inhabitants are called pioneers. Viewed from an ecological perspective, reclamation can be regarded as a means of hastening succession, so that pioneers are established and a later successional stage achieved quickly.

The amount of time required for natural invasion of mining wastes varies, depending upon the type of substrate and climatic conditions, among other factors. Variation within spoils is also common; some portions may be rapidly invaded, while other parts remain unoccupied for long periods (Jonescu 1974; Crowder et al. 1982; Gagnon 1987).

In a country as large and as diverse as Canada, it is difficult to discuss the plant species that are most suitable for reclamation in more than a very general way. Species lists of potential reclamation plants must be drawn up on a regional basis, in order to take general climatic effects into account, but additional screening programs are also necessary to determine the response of proposed species to special soil conditions. Guides are available for species selection in revegetation of disturbed sites in Alberta (Watson et al. 1980). Such lists are available in Britain (Bradshaw and Chadwick 1980), but a similar information resource has not yet been developed for all ecological zones in North America and worldwide.

The use of commonly available agricultural species is generally favored for several reasons: seed is readily available, cultivation and seeding techniques and nutritional requirements are well known, germination is usually rapid, and initial growth is good. High productivity of introduced forage species often is not sustained, and diversity of such stands is low compared to native vegetation; reclamation laws in some jurisdictions (e.g., Montana) require establishment of permanent, diverse vegetation comprised mainly of native species (Sindelar 1979).

The goal in many reclamation projects is to develop a stable vegetative cover that requires little or no maintenance. For these reasons, the establishment of crop or forage species—which usually require continued inputs of fertilizer, cultivation, and other agricultural techniques—should be only an initial step to prevent erosion and to build organic matter in preparation for more permanent vegetation. A return to regional levels of biodiversity is the aim.

Vegetation can be established in two ways, which can be used separately or together: direct seeding and seedling planting. The choice is dictated by the species used and the site conditions. If both procedures are used simultaneously, care should be taken to reduce competition. Seedlings generally require hand-planting, especially in rough terrain (Winterhalder 1984).

Inadequate or irregular precipitation is probably the most frequent climatic cause of seedling failure. In most cases, irrigation on a large scale is impractical; more feasible are methods aimed at greater interception and longer retention of precipitation. Amendments mentioned previously, such as contour-ploughing, loosening compacted soil, mulching, catching snow, and providing windbreaks, will all improve the moisture regime. The Australian experience of rehabilitating waste dumps and tailings dams at several producing gold mines in the Kalgoorlie-Boulder region has produced some innovative and effective techniques for use where temperatures are high and drought is common (Howard et al. 1991). Water "harvesting" has proved effective in establishing shrubs and trees in the western United States

(Hodder 1973; Aldon 1975): polyethylene sheeting or waxy surface coatings direct runoff toward planted stock, lessening evaporative loss at the same time.

When extreme heat is a limiting factor, it may be necessary to provide shading or to change the reflective characteristics of the surface (Schramm 1966; Howard et al. 1991). In Canada, cool temperatures may be the factor limiting germination and initial seedling growth. If properly oriented, planting furrows or pits can act as radiation traps to maintain warmer temperatures in the seedbed. Loss of seedlings resulting from freezing spells or "radiation frosts" can sometimes be avoided by distributing a mulch material, such as brush, straw, or snow fence, over the surface. In constructing spoil piles, air drainage patterns should be considered so that the creation of low-lying frost pockets is avoided.

In steep or inaccessible sites, hydroseeding is the best technique, spraying seeds up to about 60 meters. The solution containing seeds may have nutrients added to it and also substances to retain moisture or to add organic matter, such as peat, wood fibre, alginates, latex, or oil-based emulsions (Bradshaw and Chadwick 1980).

## Rehabilitation of Abandoned Sites

A significant problem for the mining industry and for governments is to find long-term ways to mitigate environmental effects at abandoned, closed, or inactive mines. Abandoned mines present the risk of environmental problems that include acid drainage and metal and radionuclide contamination. Particularly troublesome from the perspectives of both government and industry are "orphan" sites: abandoned—and contaminated—properties that have reverted to the Crown or that are owned by firms without the financial resources to do necessary reclamation work.

While the Canadian Shield area of Ontario and Québec contains the greatest concentration of abandoned mines, other sites occur across the rest of the country (Campbell and Marshall 1991). Canada lacks an accurate inventory of abandoned mine sites or the quantity of waste stored in abandoned tailings sites. A number of projects have been initiated to remedy this deficiency. With the help of the Ontario Ministry of Northern Development and Mines, Laurentian University in Sudbury has developed an on-line "Mining Environment Database" providing information on topics related to mine reclamation, including abandoned sites (Kelly and Slater 1994). CANMET is also conducting two projects at Elliot Lake, Ontario, to develop databases on active and inactive tailings and waste rock sites and abandoned minesites in Ontario.

Virtually all abandoned mines were operated during a time when environmental standards were less stringent than they are today. Mine owners retain the responsibility for rehabilitation of abandoned sites; provincial and federal governments are responsible for areas that have reverted to the Crown. Regulatory and licensing procedures now in place ensure that mine operators are responsible for the rehabilitation of mines and mills after closure.

The operating life of a mine rarely exceeds fifty to a hundred years, but during this period, semi-permanent landscape features—which can have deleterious long-term environmental effects—may be created. Arkay (1975) has listed the areas of possible concern: land disturbed surficially (by stripping, for example, or by bulldozing), open pits, underground workings, waste rock dumps, tailings areas, and structures such as head frames and mill complexes.

Some of these problems can obviously be solved fairly easily. A general site "clean-up" may be all that is required to improve the aesthetic aspect and eliminate any potential safety hazards posed by abandoned buildings or equipment. Proper maintenance or long-term storage may be necessary.

When the operations involved non-reactive materials, the potential environmental problems are few and often easily corrected. Underground workings may simply require sealing of openings; pollution of groundwater or surface waters is unlikely, but the possibility should not be overlooked. Abandoned pits can be developed into useful recreational or industrial facilities, such as lakes and parks or landfill sites. As an example of the latter, the Redland Quarries case study at the end of this chapter provides a comprehensive description of the technical and policy issues.

## Aquatic Ecosystems

### Wetlands

To an increasing degree, wetlands are being recognized as having high ecological value, both because of their effect on hydrology and because of their plants and animals. Recommendations for pre-mining decision-making include the identification and understanding of on-site and adjacent wetlands—their legal standing varies between provincial jurisdictions. An overview of regulations and current practice in wetland protection and creation was provided by Brooks et al. for the United States at a conference in 1985.

Discussing aggregate mining in the United States, Hart (1992) recommended that wetland functions should be considered when determining practices of clearing, stripping overburden and topsoil, mine dewatering, constructing and operating runways, designing plant processing and stockpiling areas, settling ponds and basins, disposing of waste fines, dredging off-streams, and altering surface drainage. He suggested that the following topics should be addressed in reclamation: post-reclamation land use, topsoil salvaging, grading, surface and subsurface hydrology, revegetation and preparation of soil, fertilizer use, and erosion control. For subsequent monitoring, Hart observed that quarterly hydrologic data would be necessary and that it would be necessary to check the vegetation periodically for up to five years. Monitoring should address function as well as structure in a reclaimed wetland.

Wetlands should be considered when the treatment of tailings and spoils is being planned, as leaching of added fertilizers or of acid drainage may result in excessive levels in their basins. Considerable experimental work has been done to identify the most effective species and their development in wetlands, and their suitability for treatment of mine drainage (Webster et al. 1994).

There is now considerable expertise in rehabilitating or creating wetlands with high diversity. This can be accomplished, for example, by creating varied slopes, indented shorelines, and islands. Their utilization by waterfowl, mammals, and fish has been documented for several areas of coal mining in the United States (Brooks et al. 1986).

## Lakes

Acidified lakes have been found to recover when inputs of acid and metals decrease, after mining stops, or emission control techniques are improved, as well as after chemical treatment. For example, details of the recovery of lakes in the Sudbury Basin, Ontario, are discussed in its case history.

Chemical treatment has been successfully used to raise pH in hundreds of lakes, mostly in Sweden and Canada. A sludge containing calcium carbonate and/or calcium hydroxide is generally spread from boats—in one example raising pH from 4.4 to 6.0. In Sweden, spreading is done from the air for many areas. Use of calcium carbonate depends on the flushing time of the lake and on its neutralizing capacity, and may need to be repeated (Yan and Dillon 1984).

Metal solubility under different pH regimes, rather than metal input, was found to control the concentration of metals in the water column. For example, concentrations of aquo cations of copper, nickel, zinc, iron, manganese and aluminum showed a rapid response to pH rise (Dillon et al. 1986).

Net primary productivity was not affected by raising the pH of Sudbury lakes, but they lost the remarkable clarity typical of acid waters. The decrease in transparency was thought to be independent of the algae and to be due to suspended or dissolved organic matter. Changes in clarity in turn, affect the depth of the epilimnion, which becomes shallower after treatment (Yan and Dillon 1984).

The slow natural recovery in some Sudbury lakes was accompanied by an increase in the species richness of aquatic macrophytes, which numbered only four in 1963, but had risen to 11 by 1980 (N.R.C.C. 1981a). A time-lag is probably characteristic, since Jeffries (1991) reported that chemical recovery was generally not accompanied by a recovery of natural communities. The presence of reservoirs of mobile individuals nearby must affect the rate of recolonization.

## Marine Sites and Meromictic Lakes

Marine disposal of tailings is confined to deep water in current Canadian operations (Ellis 1982, 1992). The aim is to keep sulphide waters in deep anoxic conditions to prevent oxidation. There is, however, evidence of resuspension of tailings (Hay and Ellis 1982). Placing tailings on the sea bottom removes habitat available for benthic organisms, although choice of an anoxic site should preclude most species of coastal areas. The general assumption is that, once dumping is finished, the sea bottom will be recolonized; consequently, no rehabilitation measures are undertaken.

Meromictic lakes are similar to oceanic waters, which remain stratified. In the Canadian Arctic, a meromictic lake is used for disposal of tailings from the Nanisivik zinc-lead mine (Canada Department of Fisheries and Oceans 1991). Management of such a site depends on maintenance of the density regime (Chadwick 1994).

## Regulation and Monitoring

The residuals produced by mining activities, released to land, water and air, comprise a wide range of substances of varying degrees of toxicity to organisms. Most of these substances are not unique to mining, but may also be released to the environment by other human activities, as well as by natural processes (Aneja et al. 1984). The relative importance of anthropogenic sources of atmospheric pollutants is increasing, however, and for most of the trace metals already exceeds that of natural sources (Nriagu 1988). Consideration of ambient and potentially toxic levels of pollutants in the atmosphere and in water bodies has led to the institution of air and water quality objectives by many government and industry bodies (Stern et al. 1984).

Guidelines defining these objectives are of two types: *emission standards* and *ambient air and water quality standards*. The former relate to the concentrations of various air and water pollutants in the immediate vicinity of the source of the emissions. The latter define the maximum allowable concentrations of undesired substances at a specified location in the air or water, regardless of their derivation.

The federal and provincial governments all have many agencies, boards, and commissions to advise them with respect to standards and other matters related to the environment. In addition to legislative requirements, there are a host of informal guidelines for carrying out a wide range of mining activities with minimum disturbance to the environment. A good example is *Environmental Guidelines Pits and Quarries* (MacLaren Plansearch 1982), which addresses pit and quarry development and restoration in the Yukon and Northwest Territories. These *Guidelines* cover the planning, design, operation, and restoration of pits and quarries and deal with such topics as sensitive areas, contouring, visual screening, access roads, erosion control, and permafrost. In its *Guide to Environmental Practice*, The Mining Association of Canada has pledged to "dispose of non-recyclable wastes in an environmentally sound manner" and to "minimize discharges to air and water" (Miller 1990).

Since the biospheric impacts of emissions depend not only on the source concentrations, but also on the dispersive characteristics of the air or water and on long-term loading, it is difficult to determine accurate limits for emissions at source. The problem is to relate values of downstream air, soil, and water quality to source emission rates and dispersal characteristics of the medium. Although dispersion-modelling methods can be used to design a control plan, the results must be verified using site data. This requires monitoring at both the source and sink ends of the dispersal chain.

Regulation in Canada is in a state of flux, with both the federal and provincial governments attempting to develop comprehensive legislative tools to protect and enhance the environment while taking into account the public's expectations for a high standard of living and economic and social progress. With this changing state of affairs, what follows is a very general description of the Canadian regulatory system as it relates to mining and mineral processing.

## Emission Standards

Emissions standards are specified in numerous Canadian federal and provincial regulations, and they vary with jurisdiction as well as for the type of mining operation. Relevant federal legislation includes the *Canadian Environmental Protection Act*, the *Clean Air Act*, and the *Metal Mining Liquid Effluent Regulations and Guidelines* (MMLERs) under the *Fisheries Act*. The MMLERs are the only federal legislation directed specifically at mining. The *Regulations* apply to new, reopened, and expanded metal mines, but not to metal mines existing before 1977 or to gold mines using cyanidation. The *Guidelines* apply to all other metal mines except gold mines using cyanidation. The standards for pH and seven substances (arsenic, copper, lead, nickel, radium-226, zinc, and total suspended solids) are based on "Best Practicable Technology" and are the standards used in most provinces. Environment Canada, which shares responsibility for the MMLERs with the federal Department of Fisheries and Oceans, is currently assessing their adequacy as a means of protecting fisheries.

Most of the provinces rely on the federal regulations and standards, usually applying them under provincial statutes related to environmental protection. Ontario has developed a sector-specific set of regulations and standards under its Municipal/Industrial Strategy for Abatement (MISA) program. The limits are based on the results of a year-long monitoring program carried out individually by every source of water pollution in the province and are based on best available technology. By 1994, legislated standards had been set for five of the nine sectors, including the metal mining and industrial minerals sectors, and they are nearly identical to those under the MMLERs. In Québec, the *Environmental Quality Act* and *Directive 019* of the Ministry of the Environment set the standards, which, again, are very close to the federal standards, as are those employed in British Columbia.

Canada's *Green Plan*, published in 1990, set out a number of plans and requirements for both industry and government, ranging from reporting on and control of toxic substances (e.g., the National Pollutant Release Inventory) to the incorporation of environmental costs into the system of national accounts. As one of these projects, Environment Canada was required to establish a list of Priority Substances used or produced at metal mines and mills, in order to formulate strategies for control of emissions of these substances. Besides numerous organic substances covered under the *Canadian Environmental Protection Act*, these substances included arsenic, cadmium, chromium, lead, mercury, and nickel. Uranium mines and mills were also included.

In the case of many substances requiring control, regulations have been set to limit or to reduce emissions from specific companies. Examples are a Canada-Manitoba agreement requiring the Hudson Bay Mining and Smelting Company to reduce the sulphur dioxide emissions from its Flin Flon smelter by 25%, and the same smelter's airborne particulate emissions by 50%. The Ontario government has been restricting sulphur dioxide emissions by Inco and Falconbridge since 1980.

## Air, Water and Land Quality Standards

Guidelines defining air and water quality standards tend to be rather more uniform across the country than those for source standards, because they are related directly to the effects and impacts of the pollutants on people, other animals, plants, agricultural output, and so on (Purves 1977; Stern et al. 1984). Some plant and animal species are far more sensitive than others to certain pollutants, requiring more stringent standards in areas in which they are found. In addition, consideration must be given to interactive effects (synergism) between different pollutants, and to the processes of bioaccumulation and biomagnification (Minish and Hurley 1991). Standards are generally set differently for aquatic ecosystems, for terrestrial sites, and for air quality.

Canadian water-quality standards are often non-enforceable guidelines. Both federal and provincial legislation tends to deal with surface waters, although groundwater is gaining prominence in planning and regulation (Karvinen and McAllister 1994).

Quality standards should also be applied to land areas and water bodies that are already contaminated. This points to the need to develop criteria for the identification of such areas and to use these criteria to assess contaminated sites in preparation for their rehabilitation. The Subcommittee on Environmental Quality Criteria for Contaminated Sites of the Canadian Council of Ministers of the Environment has developed a set of interim criteria as part of the National Contaminated Sites Remediation Program (Angus Environmental Limited 1991). The criteria, which were adopted from guidelines and criteria presently in use in various Canadian jurisdictions and deal mainly with soil quality, are based on consideration of the following factors: ambient concentrations and environmental mobility of substances; relationships between soil and water quality; the health of plants, animals, and humans; land use and aesthetics; and the limits of analytical capabilities.

In some jurisdictions, these guidelines are being applied as limits for permitting purposes, despite the fact that they are considered "interim" because they lack a complete supporting rationale (Reichenbach 1994).

## Monitoring

Critical to the adoption of emission standards, whether enforced by law or not, is an adequate monitoring program. As indicated above, monitoring must be conducted at source, as well as more generally in the atmosphere, hydrosphere, and lithosphere. Monitoring systems have three aspects: measurement methodology, network structure, and data analysis.

According to the *Green Plan*, the sectoral approach to environmental monitoring, which previously prevailed, was to be replaced by cooperative work among federal and provincial jurisdictions. In November 1992, an Ecosystem Monitoring and Research Network for all Canada was approved. Roberts-Pichette (1994) has pointed out that the Network is uneven and too sparse in the southern Arctic and large Shield areas where mining exploration activities are intense.

Satellite remote sensing is becoming increasingly significant as an instrument for environmental monitoring (Ryerson and Cihlar 1990). It has the advantage of being non-invasive and is suitable for all size scales, from its limit of resolution (0.1 km to 1.0 km) to the global scale. It is also relatively inexpensive to use and has the capacity to detect changes over periods of years and decades. For example, vegetation changes in the Sudbury region between 1973 and 1983 were deduced from LANDSAT images, of 100 m resolution, during this period (Allum and Dreisinger 1987). Vegetation gains in some areas, presumably related to revegetation activities and reduced smelter-stack emissions, were found. The use of Advanced Very High Resolution Radiometer (AVHRR) data to monitor the condition of vegetation was described by Brown et al. (1990). Although it has a coarser, 1 km resolution, the data are available daily, in contrast to the 18-day return period for LANDSAT.

Source monitoring is now usually built into the overall design plan of new mining ventures. This usually involves routine on- and near-site measurements of air and water quality. Depending on the type of operation, this may include the monitoring of smelter-stack plumes, dust, groundwater quality, surface water, and marine waters.

Biological monitoring can be an effective approach to monitoring air and water quality. It may be taken to include the following. bioassay, or assessment of the response of organisms to an applied stress; early detection systems, based on the response of sensitive organisms to low values of an applied stress; the extent of the presence of "indicator organisms," meaning those that are relatively tolerant or insensitive to a particular stress; population studies of particular species; community structure; and bioaccumulation (Humphrey et al. 1990).

The choice of organism for use in biological monitoring is dictated to some extent by the objective of the study. Easily observable species are more suitable for field bioassays, while long-lived species are better for bioaccumulation studies (Humphrey et al. 1990). Although bacteria, algae and protozoa have been used as bio-indicators in aquatic ecosystems, benthic macroinvertebrates and fish are regarded as more practical and useful choices. Duncan et al. (1987) used x-ray fluorescence spectrometry (XRF) to examine the use of freshwater molluscs, such as fingernail clams, as biological monitors of metal pollution in northern Canada. They found the method to be a valuable biomonitoring tool, as both soft tissue and shells could be used for bioaccumulation studies of certain elements, and the method was non-destructive because of the small sampling requirements.

Deniseger et al. (1986) compared the structures of periphyton communities in a pristine mountain stream, above and below a metal mining operation in central Vancouver Island. A major difference was noted in species number and community composition. The small difference during the spring was attributed to low water temperatures and the dilution resulting from snow-melt. A study of macroinvertebrate community structure, in a small stream in an area of Alabama that had been strip-mined for coal (Dills and Roger 1974), found that species diversity was most closely related to water pH. Unpolluted sites showed seasonal variations that did not appear at the most polluted sites. Both macroinvertebrate and fish communities were found to be useful indicators for monitoring mining pollution in the Alligator Rivers Region of northern Australia (Humphrey et al. 1990).

## Summary

Mine spoils and tailings may have physical and chemical properties or qualities that complicate the process of rehabilitation. The physical problems include instability of slopes and surface materials, textures inappropriate for establishment and growth of plants, and inhospitable microclimates. Through early planning, it is often possible to avoid these or, at least, make them more amenable to economical and long-lasting solution. Chemical properties that can create difficulties for reclamation include lack of plant nutrients, excessively low or high pH, and the presence of toxic substances, particularly metals. It may be necessary both to pretreat wastes and to provide subsequent follow-up treatments, such as additions of topsoil, liming agents, nutrients, and organic matter. Temporary expedients include covering wastes with various types of mulch; more permanent treatments may require sealing with impervious materials, such as clay.

Vegetation should be selected according to the aims of reclamation, which may include prevention of erosion, restoration of farmland, or establishment of a self-sustaining ecosystem. The last requires a soil with efficient humus processing and nutrient cycling. The choice of suitable plants will depend mainly on the ultimate use of the land and the local climate.

Recognition of the value of wetlands and improved understanding of techniques for mitigating impacts on them and for their use as part of a rehabilitation program are becoming widespread. This is particularly so in aggregate mining and in the control of acid drainage. Effective reclamation processes for lakes generally involve reducing turbidity and controlling pH. Success has been achieved by additions of lime to reduce acidity.

Ongoing problems are those of cleaning up abandoned mining operations and of upgrading older operations to today's standards of pollution control. The technical, economic, and policy implications of these tasks continue to challenge the best minds in both government and industry.

As awareness of the environmental problems resulting from mining and of their extent has grown over the past two decades, knowledge and understanding of processes and techniques for rehabilitation has been growing rapidly. The industry and regulatory agencies now have the ability to detect, control, and restore the effects of many activities that previously were left to natural forces to remedy. Nevertheless, a realistic approach must be taken in calling for increased reclamation activity. Certain types of wastes may present technically insoluble situations. In these cases, it may be wiser to look for means of preventing the problem rather than trying to cure it after the fact. While there are many programs for monitoring residuals emissions and their dispersal through the biosphere, there is still a need for greater standardization and rationalization, as well as better integration into international surveillance networks. It should also be noted that, as advances in instrumentation have made it possible to detect increasingly small amounts of substances, care should be taken not to establish excessively rigorous standards.

The Canadian mining industry has significantly reduced its emissions of particulates, sulphur dioxide, and carbon monoxide to the atmosphere during the past 25 years. As the voluntary actions of many companies—like the MEND program and sponsorship of university research—demonstrate, they are making a genuine effort to go about their activities in a responsible manner, not simply waiting to be forced into reasonable behavior.

## Case Study 1: Consolidated Professor's Duport Project

Shoal Lake is located in the northwestern corner of Ontario, adjoining Lake of the Woods on the west and extending slightly across the border into Manitoba. Although the region is traversed by the main line of the transcontinental railway and the Trans-Canada Highway, it is still quite remote from the rest of the country. The permanent population is relatively low, but a significant increase in the number of cottages has recently prompted the Ontario Ministry of Natural Resources to impose a freeze on further development.

Consolidated Professor Mines is attempting to bring into production a gold property on Cameron and Stevens Islands in the middle of Shoal Lake. The property has a long history, gold having been discovered there in 1897. Development began almost immediately and work was carried out sporadically by a series of different companies. The mine operated by the Duport Mining Company closed in 1936. Matatchewan Consolidated Mines optioned the property in 1950, but no further development had taken place by 1973 when Consolidated Professor acquired a 60% interest and began a drilling plan. Over the next decade, feasibility, engineering, and environmental studies were completed. Consolidated Professor assumed full ownership of the Duport Mine and has undertaken extensive development work.

The orebody is underwater and is reached by a ramp decline from Stevens Island. It is a high-grade deposit (averaging 0.35 ounces per ton) with most of the gold contained in arsenopyrite. Plans for development include deepening the workings on Stevens Island and connecting them with the old workings on Cameron Island.

Plans for processing the ore have progressed through a variety of proposals since the decision was made to reopen the mine. The present plan is that no processing will be done within the Shoal Lake watershed. The mined ore will be ferried in trucks to a barge landing point on the mainland, then taken 8 km by road to the processing plant. The planned ferry route is well away from the main cottage activity, and the ore will travel on private, rather than public, roads. A bubbler system will make year-round traffic possible, eliminating the need for either an ice road or winter accommodations on Stevens Island.

The processing plant will include grinding, flotation plant, pressure leaching, cyanidation, and gold recovery facilities, a service/office complex, power lines, and a sewage treatment facility. Changes are being made to the development program to answer the environmental concerns, primarily the addition of a secondary treatment in the milling stage. The Inco $SO_2$/Air process will be used to destroy the cyanide, with any residual amounts taken care of by natural degradation in the tailings area. The ore will be processed using a standard autoclave, with improvements in the design and operation to produce stable arsenic precipitates. Tailings—waste from the flotation and pressure leaching processes—will be treated in a staged containment area. Water will be discharged from a final polishing pond into Squaw Lake and will meet Ontario, Manitoba, and Canadian receiving water standards, as will water from the underground mine that is discharged into Shoal Lake.

The nearest population center is the city of Kenora, about 30 miles northeast of the site. At the west end of the Lake, First Nations 39 and 40 have reserves with a population of about 500, whose principal sources of revenue used to be commercial fishing and wild rice harvesting. The fishery has been halted by declining stocks, the harvesting is highly variable, and unemployment on the reserves is above 50%. There are about 300 cottages on the lake, many of them owned by residents of Winnipeg, which is about 90 miles west of the minesite. There have been a number of mines in the area over the past century, but none is operating at present.

The company has maintained ongoing consultations with the First Nations and the parties are negotiating cooperation agreements that will cover aboriginal participation in economic and employment opportunities associated with the project, as well as participation in the development of the environmental study.

Reaction from the cottage-owners has ranged from strong support to very strong denunciation. The main factor that makes this proposal controversial is the fact that Shoal Lake is the primary source of drinking water for the City of Winnipeg. The municipality draws about 100 billion liters of water each year through an intake on the Manitoba side of the lake, about 12 miles west of Stevens Island. Interestingly, the aqueduct was built in 1916 under the supervision of a principal of an operating Shoal Lake mine who became mayor of Winnipeg.

The company's application for permits to develop the project early in 1989 was met by opposition on the grounds that drainage from tailings and waste rock at the site could contaminate Winnipeg's drinking water and damage the quality of Shoal Lake for recreational and other purposes. Intense lobbying by residents of Winnipeg and local communities resulted in a meeting of the premiers of Ontario and Manitoba in Quebec City for the express purpose of discussing the environmental implications of the project and the question of jurisdictional responsibility. The pressure was strong enough that, in August 1989, the Ontario Ministry of the Environment designated the project for review under the Environmental Assessment Act, the first private-sector proposal to be so designated.

Reaction to this decision included recommendations by some mining analysts that investors sell Consolidated Professor stock, which had previously been regarded favorably by the markets.

The main reaction, however, was the company's decision in 1990 to withdraw its application for permits before the assessment process began. In the four years since then, the company has been conducting extensive consultation with all stakeholders and plans to apply again for permits, in the hope that the changes made to the development plans will alleviate the concerns of those opposed to the project on environmental grounds.

## Case Study 2: Redland PLC

Redland PLC, based in the United Kingdom, is one of the world's leading producers of construction materials, with manufacturing operations in 35 countries employing over 27,000 people. Redland PLC's Canadian subsidiary, Redland Quarries Inc. of Hamilton, Ontario, operates one of the largest dolomitic limestone quarries in Canada, annually converting millions of tonnes of limestone into more than 50 products used primarily in the construction, road building, steel, agricultural, and glass industries. Redland Quarries and its predecessors have operated in the Flamborough area, about 90 km west of Toronto, since the late 1800s and currently have reserves that will last for the next 40–50 years.

Quarrying began at the edge of the Niagara Escarpment in what is now known as the Brow Quarry. The quarry's reserves were depleted in the 1930s. Due to the increasing demand for landfill space by the nearby steel industry, the company decided in 1978 to rehabilitate the property by landfilling with solid non-hazardous waste, primarily from the steelmakers in Hamilton. By 1986, the relatively small Brow Quarry Landfill was nearing completion and the need for a replacement site was recognized. Redland began a search of all its Canadian properties for a suitable replacement for the Brow Quarry Landfill.

At that time, Redland's main operations in Flamborough consisted of the Processing Area, the South Quarry, and the North Quarry. Most of the stone required to feed the crushers and the lime kilns in the processing area was obtained from the North Quarry because mining in the South Quarry was nearing completion. Section 48 of the *Aggregates Resources Act*, requires all quarry owners in Ontario to perform progressive rehabilitation, and final rehabilitation as reserves are depleted. Quarry owners file site plans with the Ministry of Natural Resources outlining specific details of their mining operations, including methods of rehabilitation. The goal is to rehabilitate the quarry in such a manner that it can be used productively after all mineable reserves are depleted. Quarries can be rehabilitated by allowing them to fill with water, by gently sloping the edges of the quarry with fill or by landfilling. The rehabilitated property may then be used as some form of community recreational area or as a natural area.

Redland's search for a suitable replacement for the Brow Landfill involved detailed hydrogeological studies of potential sites owned by the Company to evaluate their environmental suitability. These studies showed that the South Quarry was an ideal site for a landfill. Further studies of other rehabilitation options indicated that, from Redland's point of view, the most suitable rehabilitation measure would be to turn the South Quarry into a landfill combined with recycling facilities. Once landfilling was completed, the property would then be developed into a recreational facility such as a golf course. The proposed site would receive approximately 26 million tonnes of solid non-hazardous waste at a maximum fill rate of two million tonnes of waste per year. The estimated minimum life of the landfill site would therefore be 13 years.

From an environmental perspective, this is a particularly sensitive area, and landfills are especially sensitive projects. Such projects must undergo a review under the Ontario *Environmental Assessment Act*, which embodies the following principles:

- evaluation of potential environmental effects;
- consideration of alternatives;
- a broad definition of the environment;
- documentation of the assessment;
- public and government consultation review;
- and if warranted, review by an independent tribunal.

A proponent—Redland, in this case—must submit an environmental assessment document covering these elements. The prescribed process evaluates a project not only from a technical environmental perspective, but also from planning, social, public consultation, and economic points of view. If the proponent fails to convince the Review Board that the project will enhance the "public good" from all the above perspectives, approval will not be recommended. As designed, it is a rigorous process, but there are ministerial powers to exempt projects, and the minister does not have to accept the Board's recommendations.

The *Environmental Assessment Act* was originally intended for application to public-sector projects, and only recently has it been extended to include the private sector. The South Quarry proposal was the first project of its kind where a private-sector proponent had to complete this process.

The application process was initiated in 1986 with feasibility studies and market studies. The public consultation activities began in 1988 with public meetings, site tours, and community compensation meetings which continued throughout the next three years. A public liaison committee held 15 meetings in 1989 and 1990. Redland submitted in September 1989 a draft environmental assessment document, which ran to over 2,000 pages, to the Ontario Ministry of the Environment and to all provincial ministries and agencies. After many reviews and changes, mostly involving the development of a compensation package and explanation of the evaluation of alternative closure plans, the final report was submitted in November 1990.

The proposal was to locate the landfill below the groundwater table to take advantage of inward hydraulic gradients. This form of natural protection would be combined with the most technologically advanced liner and leachate collection system ever proposed for a landfill of this kind. The engineering of this site is unique and is particularly interesting because access to the leachate collection system will go through underground galleries. This will allow inspection and maintenance of the leachate collection pipes for the life of the site and beyond. The estimated capital cost of the project is over $150 million.

Part of the environmental assessment document submitted by Redland included a proposal to compensate approximately 80 local property-owners for potential impacts associated with living near the proposed landfill. This compensation includes protection of property values. A trust fund payable to the Town of

Flamborough as the host community is proposed to provide approximately $2 million per year (close to 40% of the Town's current revenue), to be used at the Town's own discretion. Redland has also proposed to pay a royalty of $4 million per year to the Region of Hamilton-Wentworth for the purpose of furthering their waste diversion and recycling activities. The final decision on the amount of compensation will be included in the Board's ruling.

Under current Ministry of the Environment and Energy legislation, proposals for landfill projects by private-sector proponents must include some form of financial assurance. The purpose of this financial assurance is to cover all the costs of emergency close-out, contingencies, and any long-term monitoring and maintenance requirements.

As the first private landfill proposal to be affected by these guidelines, Redland Quarries Inc. is pioneering in the design of such a financial assurance plan. Developing the financial assurance package required engineering plans for monitoring, maintenance, and the ultimate closure of the landfill, as well as financial analyses of the cost of implementing those plans.

Redland Quarries Inc. has proposed a fund of $60 million to maintain and close the site and to provide for all contingencies for approximately 300 years. The design engineers estimated the contaminating life of the site to be 100–200 years. However, to be conservative, Redland decided to use 300 years for the calculation of financial assurance. The funds would be contributed by an immediate letter of credit for $10 million and the balance a sinking fund over the estimated 13-year life of the landfill. The real annual yield on the fund was conservatively estimated at 3%. The firm has made a commitment that it will not remove any surplus that the fund might generate.

The provincial government's review of the final report (conducted by the Ministry of the Environment and 23 other ministries and agencies) was published in August 1992. A preliminary hearing was held in December 1992 to determine the parties to appear as intervenors at the main hearing and was followed by two further hearings relating to intervenor funding in early 1993. Under the *Intervenor Funding Project Act*, the proponent must provide funding for eligible opponents of a project. Requests for funding were made by the Region, Greensville Against Serious Pollution (GASP), the Town of Flamborough, and various individuals. The intervenor funding panel awarded GASP $384,000.

The main hearing began on 31 May 1993 and concluded on 8 July 1994, after 147 days of hearings. The hearing was held before a Joint Board comprised of one representative from the Ontario Municipal Board and two representatives of the Environmental Assessment Board.

The Joint Board turned down the application on 13 March 1995, and the proponent filed an appeal on 2 June 1995. All cost awards will be paid by Redland Quarries Inc. A positive or negative decision can be appealed to the Provincial Cabinet. By the end of 1994, Redland Quarries Inc., had invested over eight years and more than $12 million in the proposal.

# CHAPTER FIVE

# SULPHIDE ORES

## Introduction

Sulphide orebodies are a major source of several common metals used in tonnage quantities, including copper, nickel, and especially lead and zinc. They also contain a variety of other metals and metalloids that are generally present in minor proportions and are produced as by-products when it is economically or environmentally advantageous to do so. As Table 5.1 indicates, Canada is a major producer of many of these metals, and one of the world's major producers of minerals from sulphide ores.

The sulphide ores are treated as a separate group here because they have similar environmental implications that are quite distinctive when compared with those of other minerals.

The object of mining and processing sulphide ores is to produce refined metals and alloys for commercial use. This requires complex processes and, since the ore contains only a small proportion of valuable metal, inevitably results in the production of a substantial quantity of waste.

Practices for processing these ores have the potential to allow a small proportion of these metals to escape to the environment. Certain of these practices, such as heap roasting of ores, were particularly damaging and have not been used in Canada for decades. Even the processes that replaced them, regarded at the time as breakthroughs, are themselves being superseded by new technologies based on developments in mineral extraction and processing and their environmental effects. As indicated in Chapters 3 and 4, most of these effects are relatively local and so are relatively easy to manage. Others, however, are considered serious enough that major efforts have been made to reduce such discharges to insignificant levels. Processors of sulphide ores now have two primary problems:

**Table 5.1** Production of Leading Sulphide Minerals, 1991

| | Unit | Canada | | World | Canada | |
|---|---|---|---|---|---|---|
| | | ($ millions) | ('000) | ('000) | Share (%) | Rank |
| Cadmium | t | 13.4 | 1,963 | 19,463 | 10.1 | 2 |
| Cobalt | t | 131.4 | 2,223 | 21,786 | 10.2 | 4 |
| Copper | kt | 2,137.0 | 769 | 9,259 | 8.3 | 4 |
| Gold | t | 2,134.6 | 161 | 2,165 | 7.4 | 5 |
| Lead | kt | 247.3 | 344 | 2,980 | 11.5 | 3 |
| Molybdenum | t | 52.3 | 9,405 | 111,532 | 8.4 | 5 |
| Nickel | kt | 1,502.1 | 186 | 853 | 21.8 | 1 |
| Platinum group | t | 130.2 | 11.9 | 193.4 | 6.2 | 3 |
| Silver | t | 178.7 | 1,214 | 13,825 | 8.8 | 5 |
| Zinc | kt | 1,791.2 | 1,325 | 7,231 | 18.3 | 1 |

Source: (Godin 1994).

1. oxidation of waste rock and concentrator tailings in impoundments, creating the potential for acid drainage containing dissolved iron and other base metal salts; and

2. emissions of sulphur dioxide, fine particulate material as dust, fumes from smelter point sources, and fugitive emissions. The resulting ground-level concentrations from the dispersed gas, dust and fume are sufficient to raise concerns about possible long-term health and environmental effects, primarily related to acid deposition.

Solid wastes that are generally relatively inert, such as slag, are piled or impounded. Sulphide-bearing wastes are increasingly stored under water to prevent oxidation. Iron sulphide in the form of pyrite or pyrrhotite, if not stored under water, will oxidize to generate sulphuric acid and dissolve iron and base metals that can cause environmental damage. Other potentially hazardous wastes, notably oils, greases, and chemical solutions, are commonly recycled either internally or externally. The aim is to minimize waste production and to ensure safe handling of all waste both on and off the property.

The treatment of sulphide ores requires either the rejection of iron sulphide minerals for impoundment or the oxidation of sulphur in sulphides, or a combination of the two. The final products of this oxidation include sulphuric acid, liquid sulphur dioxide, ammonium sulphate, jarosite, and gypsum, which are marketed or impounded. Where the sulphur is not fixed in one of the foregoing forms, it is discharged to the atmosphere as sulphur dioxide or to waters as dissolved sulphate. Environmental concerns include the escape of sulphur in the form of free acid and metal sulphates from waste sulphide impoundments and the conversion of sulphur dioxide in the atmosphere to sulphuric acid, giving rise to acid deposition. These matters are discussed in more detail in Chapters 3 and 4.

## History

The discovery of sulphide ores in Canada can be traced back as far as 1744, with the publication of maps showing the existence of silver-lead ores on Lake Temiskaming, Québec. However, it was not until 1883, and the discovery of copper-nickel ores near Sudbury, Ontario, that large-scale mining of sulphides began to contribute to the Canadian economy. The discovery, in 1892, of two separation processes—the Orford nickel-copper and the Mond copper-nickel—was a boon to the copper-nickel mines of the Sudbury area. Sudbury was the center of sulphide mining (and remains to this day a major contributor) until the discovery of silver-lead-zinc mines near Ainsworth and Slocan, British Columbia, in 1894. Perhaps the single greatest expansion in the sulphide mining industry came in 1916 with the incorporation of the International Nickel Company of Canada (later to become Inco Ltd.) and the discovery of the Falconbridge nickel deposits near Sudbury (which led to the formation of Falconbridge Nickel Mines Inc. in 1926).

The history of sulphide mining includes an unfortunate legacy of environmental damage, as described in the case studies at the end of this chapter. The Moira River area and the Sudbury Basin, both in Ontario, were seriously affected by mining practices no longer used, and the treatment and recovery of the areas are very instructive.

## Orebodies

Sulphide deposits are associated with the Precambrian rock of the Canadian Shield and the more recent Western Cordillera. The main regions for sulphide ore mining and processing are therefore located in New Brunswick, Québec, Ontario, Manitoba, and British Columbia (Figures 5.1 and 5.2).

## Extraction and Processing of Ore

As a group, the major sulphide ores account for about one-third of the total quantity of ores of all types mined in Canada in a typical year (Godin, various years). In 1992, approximately 70% (Godin 1993) of sulphide ore production was obtained from 15 open pit mines and the remainder from more than 50 underground mines (EMR 1993a). This varies not only from year to year with the economic cycle, but also regionally and by commodity, with low-grade, high-tonnage copper ores mined by open-pit methods in British Columbia, copper-nickel in Ontario (as well as some in Manitoba), copper-zinc in northern Ontario, and relatively little in Québec and New Brunswick. The amount of mine production from open-pit operations, while fairly stable at present, has increased in recent decades. The result is growing generation of waste rock.

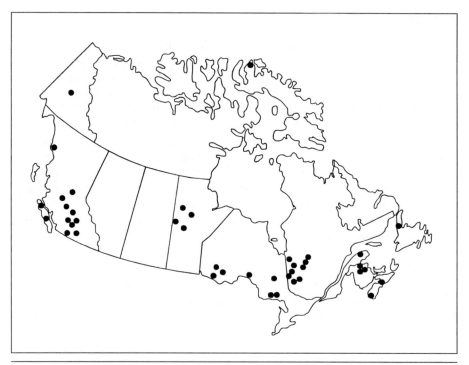

**Figure 5.1** Sulphide-ore base metal mines in Canada, 1993. Source: Godin (1994).

Sulphide ores account for slightly more than half of the tailings produced in Canadian mineral processing and a somewhat higher proportion of the waste rock. Sulphide mines produce about one-tenth of the total overburden removed.

A related recent trend has been declining ore grades. This is due in part to the depletion of accessible richer ores. Improved technology also plays a part, in that low-grade deposits, previously ignored, may now be mined economically, usually by open-pit methods. All sulphide ores are extracted, crushed, and ground as described in Chapter 1, but the sulphides require a variety of techniques to separate the ores, to reject uneconomic fractions, and to produce a high-grade concentrate (10% to 50% paymetal).

Even after the removal of iron sulphides by physical means (such as flotation and magnetic separation), the concentrates still contain substantial amounts of sulphur and iron. Although it is not usually economic to extract iron from polymetallic sulphides, several processes have been developed and operated on a commercial scale. Examples include the Cominco Sullivan steel plant, the Inco iron ore recovery plant and the Falconbridge pyrrhotite plants and nickel-iron refinery. None of these is currently economically viable, necessitating impoundment

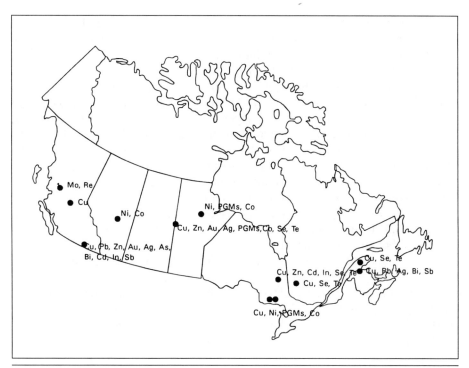

**Figure 5.2** Sulphide-ore base metal smelters in Canada, 1993. Sources: Godin (1994); Natural Resources Canada (1994).

of the rejected iron sulphides either in isolated "cells" or together with the other rejected minerals.

The further processing of sulphide ore concentrates involves the separation of the metals from iron and sulphur and from each other. This usually entails processing by pyrometallurgical or hydrometallurgical processes that result in the formation of iron slags, leach residues or precipitates, and sulphur dioxide gas, as well as a minor amount of elemental sulphur. The solids are impounded, and the sulphur dioxide gas is either converted to sulphuric acid, ammonium sulphate fertilizer, or liquid sulphur dioxide for market, or discharged to the atmosphere. Sulphur containment poses the greatest challenge of sulphide ore mining and processing. Most of the sulphur, in the form of sulphur dioxide, is captured and converted to sulphuric acid, rather than being released to the atmosphere as in the early days of mineral processing. Most of the iron is converted to slag or a leach residue or jarosite precipitate. These may be impounded alone, mixed with mill tailings or, in the case of slag, accumulated in a pile.

Ores are commonly concentrated close to the mine sites to reduce shipping costs. Where mines are located in the same vicinity, several are often served by a single mill. Concentrates may be shipped long distances for processing. Indeed, some concentrate is shipped across the country or even abroad for further processing. Examples are the shipping of nickel concentrate from a mine in Flin Flon, Manitoba, to a hydrometallurgical operation in Alberta; copper concentrate from mines in British Columbia is shipped to mills in Japan and Korea (EMR 1992). Many factors influence the location of a metallurgical plant, including the cost of energy, the proximity of sulphuric acid markets, and the availability of a skilled labor pool. Environmental concerns that must be addressed in the shipping of sulphide concentrates are the potential for wind erosion of the material, dust explosions, and spontaneous combustion.

A materials-flow diagram for sulphide ore mining and processing is shown in Figure 5.3. It indicates the main solid, liquid, and gaseous residuals of the extraction, beneficiation and further processing stages of sulphide ore mining. While the first two stages are similar for all sulphide ore minerals, the stages of further processing are different, at least for the four major metals. This diagram represents generalized mine and processing facilities. Individual operations may differ. The volume and precise composition of emissions will depend on the characteristics of the ore mined and the processing techniques used.

Many minor metals in the ore are also separated as an extension to the processing of the metals of primary economic importance. As a general observation on this matter, it should be noted that the composition of ores varies greatly, and the minor elements may be present in amounts ranging upwards from parts per million. The decision as to whether or not to extract these minor elements may depend on their economic value. It is often necessary to remove them in order to produce the required product quality, for example, greater than 99.9% in the case of lead or 99.99% for zinc. The resulting refinery by-product may have some economic value, but that value may be lower than the cost of removal. Detailed information on the procedures used for extracting them may be hard to find, as the methods are often unique to a single operation.

## Major Sulphide Metals

Nickel, copper, lead, and zinc represent the majority of metal production from sulphide ores, by a wide margin. Because of the wide range of the density and price of the minerals, as well as the fact that production varies from year to year, there is no simple measure that can be used to demonstrate the relative significance of the different minerals. Further, there is no necessary correlation between a mineral's contribution to the quantity of minerals produced in a given year and its contribution to the value of production in the same year. Table I.1 shows the percentage contribution of leading minerals to the total value of Canadian mineral production for the period 1987–93 and Table I.2 shows the

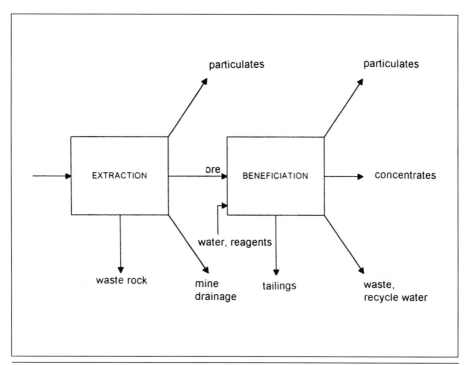

**Figure 5.3** Generalized materials-flow diagram for sulphide-ore mining and milling.

quantity, in thousands of tonnes, for the same period. As an indication of their relative significance, it might be noted that most of the minor minerals recovered from sulphide ores are produced in amounts too small to be included in these tables. Gold, silver, and iron are exceptions, but most production of these metals is derived from primary ores.

### Nickel

Most Canadian nickel is found in nickel-copper ores. Copper, on the other hand, is also found in ores without significant quantities of nickel. In Canada and other northern countries, nickel production is from sulphide ores. Nickel also comes from oxide deposits in the tropics and sub-tropics, the so-called laterite belt (Environment Canada 1987). The proportion of production from each type of ore varies widely with the costs of energy and labor, but in recent years, somewhat more than half has come from sulphides. Ores grading less than 1.0% nickel are not usually economic to mine; grades of 5% are considered very good.

Canada is the second-ranking world producer of nickel, with most production coming from the Sudbury Basin in Ontario and from the Thompson area of northern Manitoba (Figure 5.1). Due to the nature of the ores mined in Sudbury—

the source of most Canadian nickel—the two producers there, Inco and Falconbridge, process the concentrate to recover both copper and nickel.

The mined ore is usually concentrated by flotation to a grade of approximately 10% nickel before being fed to a roaster to drive off some of the sulphur and to oxidize the contained iron. Multiple-hearth or fluid-bed roasters are then used to further drive off sulphur and oxidize iron, producing a calcine. Most of the iron, and some of the minor elements, oxidize to form a slag, which is skimmed off periodically. The calcine is fed to a furnace (reverberatory, electric, or flash-smelting) along with a siliceous flux, which combines much of the remaining iron (Environment Canada 1987). Until recently, Sudbury's nickel/copper ores always required separate smelting for recovery of the two metals. Inco has been flash-smelting copper for some time, but the installation of new processes in 1993 allows for the flash-smelting of bulk nickel/copper concentrates.

The smelter produces a 20% nickel matte, along with residual slag and particulate and gaseous emissions to the atmosphere. The final conversion step removes more iron and sulphur. It also raises the grade of the matte to approximately 75% nickel. There are several facilities, both in Canada and abroad, that refine the converted matte to high-purity nickel. Discussion of these operations is beyond the scope of this work.

Slag production varies from seven to ten times the quantity of nickel produced. Depending on the type of smelting furnace, it contains 38-45% iron, 30-36% silica, 25% magnesium oxide, 0-2% lime, 0.1-0.2% nickel, and about 0.1% each of cobalt and copper (Environment Canada 1987). Under most environmental regulations, slag is considered inert.

All smelting operations produce emissions containing sulphur dioxide, as well as particulate matter containing such elements as arsenic, cadmium, cobalt, copper, mercury, and nickel. Most of the particulate matter is collected by cyclones or in electrostatic precipitators. A relatively small but significant proportion still escapes to the environment. Particulate matter must be removed from emissions before they can be treated in a sulphuric-acid plant. The concentration of sulphur dioxide emitted from multiple-hearth roasters and reverberatory furnaces is usually too low to treat economically. Sulphur dioxide that derives from fluid-bed roasters, from other smelting furnaces and from most converters, however, is of sufficiently high concentration to merit treatment. In addition to the major atmospheric emissions, which are either vented through a stack or treated in an acid plant, all the stages of the process also produce a wide range of fugitive emissions that are much more difficult to control (Joyner 1985). All Canadian operations have systems for recovery of dust and fume, usually using electrostatic precipitators that recover at least 98% of the dust and at least 90% of the fume. Further improvements are expected.

The ores in Sudbury contain sulphur in a ratio of eight parts to one part of nickel. The amount of residual sulphur can be diminished considerably by separating out the pyrrhotite fraction of the ore (see Chapter 1) and processing it separately for its metal values. This practice is now used at all of the Canadian smelters, which also capture a substantial proportion of the off-gas sulphur dioxide content and convert it to acid. This is discussed more fully later in this chapter.

A small proportion, approximately 10%–15%, of Canadian nickel is produced by hydrometallurgy, using an ammonia-leach process (Environment Canada 1987). Concentrate is directly leached at elevated temperature and pressure, producing elemental sulphur and nickel powder; copper and cobalt are also removed. Ammonium sulphate is then produced and finds a ready market as an agricultural fertilizer. At present, the process is limited to concentrates with low contents of copper and precious metals.

### Copper

Although copper deposits are widespread across Canada, relatively few are of sufficiently high concentration to allow profitable exploitation. In 1992, there were 29 operating mines (Figure 5.1). A single heap-leaching operation, located at McLeese Lake, British Columbia, uses solvent extraction and electrowinning to produce copper cathode from waste rock because it is more economical to recover copper than to simply treat the acid drainage. It is doubtful that the process would be undertaken in the expectation of profit. All of Canada's other operations beneficiate the ore to raise its copper content to 15%–35%. Aside from the Sudbury nickel-copper ores, the major ore mineral is usually chalcopyrite ($CuFeS_2$); therefore, typical concentrates also contain 30% sulphur, 25% iron, and minor amounts of lead, molybdenum, nickel, and zinc (Joyner 1985).

In Canada, as is illustrated in Figure 5.2, copper concentrate is smelted at one of six pyrometallurgical facilities (EMR 1993a). The pyrometallurgical processing of copper (see Chapter 1) includes roasting, smelting, and converting before final refining. The roasting step involves the autogenous heating of the concentrate and a siliceous flux to prepare a calcine charge for the smelter. This process drives off 20%–50% of the sulphur in the form of sulphur dioxide, along with small amounts of antimony, arsenic, lead, and other elements. The calcine is heated to melting point, together with more flux, to produce a molten mixture of cuprous and ferrous sulphides called matte, which averages 45% copper. The matte is fed to a converter to eliminate most of the remaining sulphur and iron. Again, flux is added. As with all the stages, stack emissions are produced, at this stage consisting primarily of residual particulates and sulphur dioxide. Technological developments are rapidly making this sort of general description outdated; for example, Noranda's continuous reactor essentially combines smelting and converting in a single reactor.

The principal constituents of the particulate emissions are copper and iron oxides, although oxides of other contained elements—such as arsenic, antimony, cadmium, lead, mercury and zinc—are also present. During the conversion stage, most of the precious metals remain with the copper, as do cobalt, nickel, selenium, and tellurium. Iron and zinc end up in the slag; if not recovered by electrostatic precipitators or similar technique, antimony, arsenic, bismuth, cadmium, germanium, lead, mercury, and tin escape with the stack emissions. The sulphur dioxide is dealt with as described above in the discussion of nickel.

## Lead

Lead is the most highly recycled base metal in North America at 55% (Miller 1991), primarily because its main end-products—motor and other vehicle batteries—can be readily collected for recycling by either secondary or primary lead smelters. This high degree of recycling, as well as reduced use as a fuel additive, has limited demand for increased primary production, which has declined over the past decade (Table I.2).

Mineable lead deposits are found in almost every province and territory of Canada, most commonly in association with zinc, copper, iron, and silver in widely varying proportions. Commercial-grade ores are typically 3%–8% lead, although polymetallic ores may grade considerably lower. Beneficiation, usually by flotation, raises the ore grade to 55%–70% for further processing.

In 1993, as Figure 5.2 illustrates, only two primary lead smelters operated in Canada, producing lead bullion, which was refined by either electrolytic or pyrometallurgical means (Keating and Wright 1994). The smelting process begins with a sintering operation to remove approximately 85% of the contained sulphur, converting the lead sulphide to lead oxide (Joyner 1985). Silica and limestone fluxes are also added during this step. The next step is reduction of the lead oxide to metallic lead in a blast furnace. This stage requires the addition of coke, which serves as both a fuel and a reductant, and large volumes of air that may be enriched with oxygen. Both plants recover sulphur dioxide and produce sulphate, phosphate, and other substances, all of which may be sold as fertilizers.

The main process residuals are gaseous and particulate atmospheric emissions, as well as slag containing silica and trace amounts of arsenic, antimony, and metallic sulphides. Depending on how the slag is cooled, it may be chemically inert. Rapid cooling produces a glassy structure, but slow cooling produces slag of a crystalline structure that may be leachable. Particulate emissions, which are produced mainly by the sintering operation, contain lead, zinc, cadmium, and arsenic. Both particulates and sulphur dioxide emissions can be controlled fairly efficiently through the use of wet and dry scrubbers, and technical improvements in control processes are being developed and adopted rapidly.

## Zinc

Canada is the world's leading producer of zinc, with 20 mines (Figure 5.1) and four metallurgical works (Figure 5.2) in 1992. Of all the non-ferrous metals, zinc is the most abundant. Sphalerite, zinc's most common ore, is two-thirds zinc and one-third sulphur, by weight. It is usually found in association with lead and copper ores and with trace amounts of cadmium, gallium, germanium, indium, and thallium (Joyner 1985). Commercial-value ores are typically 3%–10% zinc; however, much lower grades can be managed in the case of polymetallic ores. Beneficiation, most commonly performed using flotation, raises the mined ore grade to about 50% for further processing.

Canada's four smelting facilities all use a combination process that involves pyrometallurgy, hydrometallurgy, and electrometallurgy. The concentrate is first roasted to oxidize most of the sulphur, which is released from the concentrate, along with a quantity of particulate matter, as sulphur dioxide gas. The resulting oxide is then leached with sulphuric acid to dissolve the metal, purified in several stages, and finally deposited on the cathode of an electrolytic cell. The cathode zinc is melted and cast into shapes to be marketed. Impurities, such as arsenic, antimony, cobalt, copper, germanium, nickel, and thallium, are precipitated out during the purification stage in order to prevent their interference with the electrolysis.

The electrolytic process uses a considerable amount of electrical energy, some of which is lost as heat. Although it generates acid mists, the major particulate and gaseous releases are associated with roasting. Sulphur dioxide concentrations are high enough to warrant sulphur recovery. A high-pressure leaching process using an autoclave is in use at three plants (Timmins, Ontario; Trail, British Columbia; and Flin Flon, Manitoba). This process has the advantage of avoiding the formation of gaseous sulphur dioxide, instead fixing the sulphur in the elemental form, which can then be marketed.

## Minor Sulphide Metals

A substantial fraction of trace metals is in solid solution in major ore sulphides. Most are produced incidentally to other minerals, as described earlier in this chapter. Reference to Tables I.1 and I.2 shows the difficulty of placing the production of minor minerals in perspective.

### Antimony

As well as in sulphides, antimony also occurs in various sulphosalt minerals of varying composition, such as tetrahedrite, with silver replacing copper and arsenic replacing antimony. Deposits of antimony are found in the Yukon Territory, British Columbia, Québec, and the Atlantic provinces (Lymburner 1973). The main ore, stibnite, is frequently associated with lead ores and may also contain small amounts of gold and silver. North America's only primary antimony mine, at Lake George, New Brunswick, was closed at the beginning of 1990 because of unfavorable market conditions. As a result, almost all of Canada's production in 1992 came as a by-product of the smelting of lead ores at Trail, British Columbia, and Belledune, New Brunswick (EMR 1993a; NRCan 1994). The Trail operation produces antimonial lead, which contains 10%–21% antimony and 1% arsenic, by reduction of silver refinery dust and slag; Belledune produces antimony slag, which may be further refined on-site or by a secondary lead smelter to produce antimonial lead (EMR 1992).

Roughly half of North America's total consumption of antimony comes from recycled scrap. Its main uses are in batteries, alloys, paints, and solder. It is widespread (in low concentration) throughout the biosphere (Lymburner 1973) and is toxic to organisms. The drinking water standard is less than $0.2 \; \mu g \cdot L^{-1}$ of antimony, and the toxic threshold value for antimony dust is $0.5 \; \mu g \cdot m^{-3}$ of air (Jaworski 1982). Tests for toxicity to animals have generally used potassium antimony tartrate, for both aquatic organisms and mammals. Effects include skin disorders and heart disease.

## Arsenic

Arsenic is a metalloid and is a common minor constituent of many metal ores, especially those of copper, lead, zinc, silver, and gold. It appears in the dusts and residues that result from the roasting of these ores. The form in which it is usually recovered—arsenious trioxide—is the one commonly called "arsenic." Copper arsenate is frequently used as a wood preservative. Arsenic used to be produced as a by-product of the roasting of refractory gold ores in Ontario and the Northwest Territories, a process that has largely been replaced by pressure oxidation.

Arsenic is toxic to plants as well as animals, including humans. As a cause of metal toxicity in domestic animals, it is second only to lead (Jaworski 1980). Most arsenic emissions to the environment come from the processing of sulphide-ores, particularly from the smelting of copper and nickel concentrates and from primary iron and steel production (Jaques 1987). The movement of arsenic in the environment is very complex. It has several oxidation states and behaves differently than metals. The section below relating to the environmental effects of the extraction and processing of sulphide ores provides more information about arsenic's potential to cause environmental damage.

## Bismuth

Canada's reserves of bismuth are among the highest in the world. Bismuth derives from the sulphide mineral bismuthinite, which is associated with lead ores and tends to remain with them until they are smelted. It is usually recovered as a by-product of the metallurgical processing of lead. Since bismuth and lead have similar physical and chemical properties, bismuth may also substitute for lead in galena. Bismuth is becoming a substitute for lead in fishing sinkers because it is less toxic.

In 1992, bismuth was produced in Canada at the metallurgical facilities at Trail, British Columbia, and Belledune, New Brunswick (EMR 1993a). The Trail electrolytic operation produces refined bismuth metal from silver refinery slags.

The Belledune pyrometallurgical operation produces lead-bismuth alloys (6% bismuth) as a by-product of lead refining.

### Cadmium

Cadmium is a rare element that is always found in association with zinc. It occurs most commonly in solid solution in zinc minerals (sphalerite and marmatite). Cadmium is highly toxic to humans; it is easily ingested, and acts synergistically with other toxic metals. It is also toxic to plants and animals and is readily absorbed from the environment (Jaworski 1980). Canadian zinc ores contain between 0.001% and 0.067% of recoverable cadmium. Zinc concentrates contain about ten times these amounts (Lymburner 1974).

Canada is one of the world's leading producers of cadmium. The main use of cadmium is in nickel-cadmium batteries; other uses are in coatings, pigments, plastics stabilizers, and alloys. The mines and metallurgical works involved in cadmium production are the same as those that produce zinc (Figures 5.1 and 5.2): the cadmium is produced as a by-product of the smelting or refining stages of the zinc concentrate. Some of the material is shipped abroad for refining.

### Cobalt

Cobalt is produced in Canada as a co-product of copper-nickel ore processing (Ignatow et al. 1991; Dupuis 1993). It is a vital industrial metal with major uses in aerospace alloys. It also finds use in cutting tools, magnets, ceramics, and chemicals. Canadian producers in Ontario and Alberta generally represent approximately 20% of the world market (NRCan 1994).

### Germanium

Germanium has been produced as a by-product of the lead-zinc metallurgical operations at Trail, British Columbia, but none has been produced since 1990, following the collapse of the metal's world market (Ignatow et al. 1991).

### Gold and Silver

As well as being mined on their own (see Chapter 6), gold and silver are produced as by-products of the smelting and refining of the major sulphide ore metals. Typically, slightly less than 10% of gold production comes from base-metal mining, a level that has held steady during the past decade, although its share of the market has declined gradually (Couturier 1993). On a global basis, approximately 60% of silver production is a by-product of base metal and gold mining (Keating 1992).

Gold and silver are produced at the smelting operations at Trail, British Columbia; Flin Flon, Manitoba; Copper Cliff, Ontario; and Rouyn-Noranda and Murdochville, Québec. They are also produced at a number of refineries across Canada (EMR 1992). The copper-lead-zinc ores at Bathurst, New Brunswick, graded approximately 100 g·t$^{-1}$ silver (Scales 1986). Silver is distributed to the mill products and is recovered by the Belledune lead smelter as silver doré.

### Indium

Canada has indium recovery plants in Trail, British Columbia, and Timmins, Ontario. Each has a production capacity of 30 t·yr$^{-1}$; world consumption is approximately 110 tonnes (Ignatow et al. 1991). The facility at Trail produced indium as a by-product of lead-zinc operations. The facility at Timmins, which began operating in 1990, treats electrostatic-precipitator dust from the copper-smelting operation.

### Iron

As noted in the earlier sections on nickel and copper, most base-metal sulphide ores contain large quantities of iron that are rejected to mill tailings and/or smelter slag and for which no economically viable recovery method has been found. Chapters 3 and 4 also discuss the environmental and economic considerations involving pyrite and pyrrhotite that often accompany nickel and copper in their ores.

### Molybdenum

Molybdenum is mined for itself, as well as being a by-product or co-product of other sulphide ores. Canada is one of the world's main producers (Table 5.1). The most common ore is molybdenite; the mineral is currently mined only in British Columbia. Molybdenum is used both in its pure metal form and, in other forms, as an alloy additive, as a lubricant, and as a component of chemicals used for such diverse purposes as reagents, pigments, glazes and enamels, fertilizers, and electroplating compounds.

Concentrate from the ore, containing about 60% molybdenum, amounts to about 1% of the ore extracted. Multiple-hearth roasters are used to convert most of the molybdenite concentrate to molybdic oxide for marketing, and the remainder is marketed as molybdenum disulphide.

### Platinum Group Metals

Canada supplies 4% of the world's needs of the platinum group of metals, also referred to as PGMs (Godin 1993). The group includes platinum, palladium, rhodium, ruthenium, iridium, and osmium, and the metals are obtained as by-products of nickel-copper refining at Sudbury, Ontario. Operations at Flin Flon and Thompson, Manitoba, also make small contributions (Couturier 1993).

Recycling of used PGMs constitutes an important source of these metals, which are mostly used for automobile catalysts and jewelry. Any environmental effects, apart from those pertaining to sulphide ore mining in general, are likely to be very small.

### Rhenium

Rhenium is a minor by-product of molybdenum mining operations in British Columbia. It is recovered from those molybdenum concentrates that are shipped abroad for processing (Ignatow et al. 1991).

### Selenium and Tellurium

Selenium and tellurium occur in association with copper, from which they are separated during the electrolytic refining process. The copper-nickel ores in the Sudbury region have selenium contents in the range of 20–80 $\mu$g·g$^{-1}$ (Nriagu and Wong 1983). Both metals are recovered from slimes at copper refineries at Copper Cliff, Ontario, and Montreal East, Québec. Selenium is marketed as selenium cake, selenious acid, sodium selenate/selenite, and iron selenide. Tellurium is produced as tellurium dioxide cake, and tellurium powder, stick and lumps, and is used in alloys.

Selenium is an essential trace element for some mammals. In some areas it is naturally deficient; in others it is toxic to grazing animals. For some animals, the range between deficiency and toxicity is very narrow.

## Environmental Effects

### Aquatic Ecosystems

Sulphide ore mining not only produces large amounts of waste, but the characteristics of the waste are such that it is capable of causing major chemical effects, particularly in inland and coastal waterbodies. As noted at the beginning of this chapter, this is principally due to the presence of sulphur and iron in the ore.

Reduced sulphur tends to oxidize upon exposure to air and water, forming sulphuric acid that leaches out the iron and other toxic metallic components of the waste. The sulphur contained in the ore is either discharged as part of the mill tailings or is converted to sulphur dioxide during smelting. The proportion not captured and converted to sulphuric acid is released to the atmosphere. Control mechanisms now make it possible to contain more than 90% of releases. The greatest portion of the iron, because it generally cannot be recovered, ends up in the mill tailings and slags.

For water, the most serious potential consequence of sulphide ore mining is the process referred to as acid drainage. This is discussed more fully in Chapters 3 and 4. In brief, precipitation and groundwater that come into contact with the broken ore, either inside the mine or in waste or tailings dumps, may become acidic. They dissolve iron and other metals before draining away and affecting lakes and streams. Mine drainage from sulphide ore operations amounts to more than 130 Gl per year, which is almost 60% of drainage from all mines in Canada (see Table 1.4). Untreated water has pH levels that vary widely, often as low as 2–3, as well as high concentrations of sulphate, iron, and other metals.

As a general statement, it might be said that acid drainage is a potential problem for mining wherever the rock matrix contains more acid-producing materials (such as pyrite and pyrrhotite) than acid-consuming materials (carbonate or calcite). This is usually the case in the majority of Canadian orebodies, with the exception of the southern Prairie provinces. It must be noted, however, that some acid-consuming minerals are effective only at specific low pH levels. While sulphides are usually associated with non-ferrous base metals, acid drainage problems also arise from sulphides associated with coal, uranium, and precious metal ores. A survey reported by Environment Canada (Campbell and Marshall 1991) found that hundreds of waste-dump sites across Canada have acid-generation potential. Many of these have severe effects on the biota of lakes and rivers.

In addition to acid-drainage effects, many surface and underground mines disrupt water-movement patterns. They also pollute surface and underground watercourses with suspended matter and toxic metals. Arsenic, a minor constituent of sulphide deposits, may occur in high concentration in surface water or groundwaters which come in contact with sulphide tailings. It is, however, usually ubiquitous in nature: natural concentrations of arsenic in the oceans lie between 1 and 8 ppb, with freshwater values showing a wider spread (Penrose 1974). Many organisms can tolerate high arsenic concentrations in their environment and many accumulate it. Accumulators may pose a risk for organisms at a higher trophic level (see Chapter 2 for more information on this matter).

## Atmospheric Effects

Particulate emissions—which are produced throughout the extraction, beneficiation and further-processing stages (Stages III, IV, and V, as described in Chapter 1)—and sulphur dioxide emissions from smelters (Stage V) are the main airborne contaminants. Sulphide ores are also responsible for 5% of the particulates and 93% of the sulphur dioxide emitted to the atmosphere by mining activities in general (Kosteltz and Deslauriers 1990). The particulate emissions are relatively minor compared with those from mining iron ore and the releases from other industries. Smelters are also potential sources of many other toxic substances, including antimony, arsenic, cadmium, copper, mercury, and nickel. The Ontario/Canada Task Force on Sudbury Smelters reported that, between 1973 and 1981, smelters in the Sudbury area emitted 1,800 t of iron, 670 t of copper, and 500 t of nickel to the atmosphere each year. The bulk of it was deposited within a radius of 50 km from the source (EMR 1984).

As with any other aspect of mineral production, statistics and other data are quite specific to a particular operation at a particular point in time. This fact, along with the fact that such information was not collected in a systematic and consistent form, makes it difficult to present an accurate history of the treatment of sulphur-bearing wastes. The following data, therefore, are offered merely as a guide to changes over the period since 1970. Rabbitts et al. (1971) estimated that, in 1970, copper-nickel operations in Canada produced 1,100 kg of sulphur dioxide per tonne of concentrate, of which one-quarter was converted to acid. A survey of non-ferrous smelters across Canada, taken in 1983, found efficiencies

**Table 5.2**  Ratio of Sulphur Dioxide Emissions to Metal Production (both measured in thousands of tonnes)

| Year | HBMS[1] | Inco | | Falconbridge | | Noranda | BMS[2] | Cominco |
|------|---------|----------|---------|---------|------------|---------|--------|---------|
| | | Thompson | Sudbury | Sudbury | Kidd Creek | (Gaspé) | | (Trail) |
| 1970 | 2.24 | 7.71 | 6.02 | 7.28 | | 2.44 | 0.38 | |
| 1971 | 2.73 | 8.23 | 5.59 | 7.00 | | 2.28 | 1.40 | 0.06 |
| 1972 | 2.12 | 8.63 | 5.16 | 5.11 | 0.02 | 2.50 | 0.61 | 0.05 |
| 1973 | 2.02 | 7.64 | 3.83 | 4.43 | 0.02 | 2.93 | 0.71 | 0.06 |
| 1974 | 2.20 | 7.61 | 3.35 | 4.77 | 0.02 | 0.98 | 0.60 | 0.05 |
| 1975 | 2.24 | 7.17 | 3.35 | 3.94 | 0.02 | 1.06 | 0.61 | 0.06 |
| 1976 | 2.11 | 6.18 | 3.44 | 4.47 | 0.02 | 1.01 | 0.49 | 0.05 |
| 1977 | 2.04 | 6.57 | 3.33 | 3.70 | 0.02 | 1.26 | 0.43 | 0.06 |
| 1978 | 2.11 | 6.39 | 3.14 | 3.25 | 0.02 | 1.02 | 0.41 | 0.06 |
| 1979 | 2.13 | 5.61 | 2.96 | 1.70 | 0.01 | 1.37 | 0.33 | 0.07 |
| 1980 | 1.83 | 5.07 | 3.15 | 2.30 | 0.01 | 1.44 | 0.29 | 0.07 |
| 1981 | 1.84 | 5.40 | 3.16 | 1.99 | 0.02 | 1.73 | 0.46 | 0.06 |
| 1982 | 1.98 | 4.62 | 2.85 | 1.89 | 0.02 | 1.33 | 0.38 | 0.07 |
| 1983 | 1.88 | 4.25 | 3.05 | 1.50 | 0.02 | 0.84 | 0.31 | 0.06 |
| 1984 | 1.84 | 4.24 | 2.91 | 1.31 | 0.02 | 0.89 | 0.19 | 0.05 |
| 1985 | 1.80 | 4.26 | 2.95 | 1.29 | 0.02 | 0.98 | 0.32 | 0.05 |
| 1986 | 1.77 | 3.99 | 2.91 | 1.26 | 0.02 | 0.69 | 0.36 | 0.05 |
| 1987 | 2.11 | 3.93 | 2.82 | 1.67 | 0.02 | 0.75 | 0.27 | 0.05 |
| 1988 | 1.95 | 3.94 | 2.84 | 1.25 | 0.01 | 0.87 | 0.34 | 0.04 |
| 1989 | 1.86 | 3.93 | 2.66 | 1.40 | 0.01 | 0.75 | 0.40 | 0.05 |
| 1990 | 1.86 | 4.34 | 2.71 | 1.36 | 0.01 | 0.71 | 0.18 | 0.06 |
| 1991 | 1.92 | 4.59 | 2.37 | 1.19 | 0.01 | 0.79 | 0.11 | 0.04 |
| 1992 | 1.87 | 4.73 | 2.60 | 0.89 | 0.01 | 0.75 | 0.11 | 0.04 |

[1] Hudson Bay Mining and Smelting; [2] Brunswick Mining and Smelting
Source: MacLatchy (1994).

of sulphur dioxide containment ranging from zero to 95% (EMR 1984). The overall mean, weighted on the basis of smelter capacity, was 60%. The same survey showed that approximately 30% of the sulphur contained in the original ore was released to the atmosphere in 1980, compared with 60% a decade earlier.

A study relating sulphur dioxide production to metal production over the period 1970 to 1992 for ten Canadian base metal smelters was conducted by Environment Canada in 1994, on the basis of data and information supplied by the companies (MacLatchy 1994). Table 5.2 sets out the ratio of $SO_2$ emissions to metal produced, both reported in thousands of tonnes, for those smelters that fall within the scope of this book. For most of the smelters, there has been a clear trend toward reduced emissions. Again, results are very much site-specific. For example, Cominco's Trail smelter was replaced in 1989 with a $100-million-dollar QSL lead smelter that was soon closed because of operational problems that could not be solved. It has had to revert to use of its old smelter while building another new one.

Technology has changed—and continues to change—so rapidly that calculations of emissions should now be based on current values only. For example, Inco completed in 1993 a complete rebuilding of its Sudbury processing operations, incorporating—in some instances, developing—different technologies for improving milling and smelting activities, at a cost of $600 million. Originally intended as a means of meeting new legislated limits on sulphur dioxide emissions, the improvements have reduced those levels well beyond those mandated, as well as emissions of carbon dioxide, nitrogen oxides, and particulates, besides reducing energy requirements and improving worker safety (Sopko 1994). As a result, sulphur dioxide emissions at Inco's Sudbury operations are far less than Table 5.2 would indicate.

Alternative processing methods, including pyrometallurgy and pyrrhotite recovery, reduce atmospheric emissions but increase solid sulphur-containing wastes. Although this can also lead to acid drainage problems, containment can be effective. The possibility remains for future sulphur recovery from the waste heaps.

A preferable approach is to remove the sulphur from the ore as a saleable product. This is done in a number of ways. The most common is by means of an acid plant, in which the smelter gases are converted to sulphuric acid and marketed, for use in the pulp and paper industry, among others. Today, smelter gas production supplies over half of the sulphuric acid marketed in Canada. A variant of this method is to convert the gases to liquid sulphur dioxide, which is used by paper mills. Both of these approaches work well if adequate markets exist. In the absence of such markets, the residual sulphur can be converted to a more stable form. One such form is elemental sulphur, which is much easier to store and to ship, but this is difficult and very costly to achieve. Other possibilities are to react the gases: phosphate rock produces phosphatic fertilizers, ammonia produces ammonium sulphate, and lime produces calcium sulphite/sulphate (Reimers and Associates 1980). These products can be marketed, or stored relatively safely in waste piles. In the future, many sulphide wastes are likely to be impounded under water, both to prevent oxidation and to improve recoveries if the wastes are re-mined.

A special case among the components of gaseous emissions from the processing of sulphide ores is mercury. A toxic metal which, like lead, is found in many rocks and soils, mercury has been used by humans for hundreds of years. Because of its crustal abundance, most sulphide minerals contain at least 3 grams of mercury per tonne of ore (Sheffield 1983). As a result of the close association between the mercury and the ore metals, mercury tends to remain in the concentrates and subsequently to be emitted to the atmosphere during smelting. Mercury emission factors for various base-metal recovery operations are 1 to 117 $g \cdot t^{-1}$ for zinc, 16 to 53 $g \cdot t^{-1}$ for copper and 1.6 to 2.2 $g \cdot t^{-1}$ for lead.

## Reclamation

This is an issue not only for existing operations, but also for the many abandoned sites across the country, many of which may be acid-generating. The main objective in the reclamation of sulphide wastes is to reduce or to eliminate acidification and acid drainage. Other objectives include immobilization of toxic substances, prevention of blowing dust, restoration of the fertility and productivity of disturbed areas, and aesthetic considerations. If future prices or techniques are likely to make reworking of the wastes profitable, they may be left without reclamation, but still require containment measures to minimize seepage, wind and water erosion, etc. Since the technology to rework all waste from sulphide operations is not even on the horizon, most operations are proceeding on the basis that rehabilitating waste areas is a matter of routine.

As noted in Chapter 4, rehabilitation of sulphide wastes involves much more than revegetation because that step alone does not stop acid drainage. In conjunction with other remedial activities, however, establishing a suitable mix of vegetation is a vital component of a reclamation program. On many tailings areas, vegetation may develop through natural succession, often beginning with metal-tolerant mosses or horsetails (Gagnon 1987). In many cases, however, natural succession must be accelerated. Problems may include steep or unstable slopes, which require grading. Texture may vary between all rock and no rock, and may require alteration to produce a suitable substrate for plants. Many tailings, however, have the texture of a sandy field soil suitable for plants, but also have a low cation exchange capacity. This can be improved by the addition of organic matter, such as peat, which would also have the effects of reducing oxygen penetration—and hence acid formation—and of providing nutrients for any vegetative cover

Many tailings slurries also contain toxic metals. The availability of such metals varies with pH, Eh, and other components of the tailings, such as iron oxides. The microclimates of large expanses of sulphide tailings may be extreme enough to prevent revegetation (Crowder et al. 1982). If seedling exposure is a problem, a nurse crop of rye, or of another annual, may be used. Initial planting is generally accompanied by fertilization; at Copper Cliff, for example, 5-20-20 fertilizer was used, as well as limestone (Peters 1984). Legumes may be added to the seed mix to maintain nitrogen levels in the soil. Trees may also be used instead of smaller plants: those most likely to be successful are early successional species—such as jack pine, birch, and aspens. Nitrogen-fixing symbionts such as alders are also useful if they can be established. Metal-tolerant mycorrhizae may have to be added to the soil to start trees successfully (Ernst 1985). Chapter 4 discusses these considerations in some detail. The Sudbury case study at the end of this chapter provides further information.

In 1978, a local initiative at Sudbury, designed and carried out jointly by the citizens and industry, led to revegetation of the most conspicuous parts of the barrens—the steep black rock faces produced by years of severe pollution (Winterhalder 1984). Labor-intensive methods were used: bags of lime, fertilizer, and seeds were carried up and spread by hand. Limestone was applied at a rate of 11.8 t·ha$^{-1}$. The fertilizer was a 6-20-20 mixture. Revegetation has been successful in greening the slopes, and by 1991 some areas were reported to be developing normal soil flora (Maxwell 1991). Revegetation activities have been so successful, in fact, that in 1992 the Sudbury region received an award from the United Nations for the results of the remediation program.

Inco's Copper Cliff tailings disposal area near Sudbury includes more than 1,000 ha in various stages of reclamation and revegetation. A study by Peters (1984) of the revegetation of these sulphide tailings describes the practical aspects of plant establishment, but also recognizes that reclamation must be based on sound ecological principles. Peters considered evidence that those ecological processes, which are necessary for sustainability, have been established in the tailings area, including some development of soil horizons; establishment of a seeded grass cover leading to colonization by some native plants (including trees) and representing succession to a "climax" state; successful introduction of legumes, accelerating the development of a natural nitrogen cycle; presence of some natural fauna, including a number of nesting birds on the revegetated tailings (this provides evidence of suitable habitats and of an apparently non-toxic environment for organisms); and acceptability of the restored habitat as a wildlife management area.

The revegetated areas are unsuitable for farming because of metal uptake by plants (Crowder et al. 1982; Peters 1984). Occasional use of the tailings by wildlife presumably does not cause harm (Peters 1984).

Reclamation procedures must be tailored to each individual tailings area. They may even vary *within* a single area. At Copper Cliff, for example, the tailings slurry has an initial pH of 11. Within a few years, this level falls to acidic values in patches. Even after liming, the soil profile shows a decline from pH 6.6 at the surface to 4.2 at the depths of grasses' lower roots (Crowder et al. 1982, McLaughlin and Crowder 1988). It may therefore be necessary to repeat lime treatments periodically and to treat patches of ground individually.

As discussed in Chapter 4, investigations are being conducted into the use of tailings covers to minimize or prevent the oxidation that leads to sulphuric acid generation. This means that, for sulphide tailings, the acid-generating potential must be handled successfully in order to prevent reacidification of revegetated substrates, as well as to reduce the metal content of seepage waters. At Falconbridge, for instance, two general types of covers are being tested: one dry and one wet. Dry covers consist of either a 2-m thick layer of waste rock, with the interstices filled with fine alkaline material or sewage sludge, or a 2-m thick layer of compostable domestic garbage topped with a 30–60 cm stabilization layer (Peters 1989).

It may be desirable to increase the density of the tailings by packing them down, in order to reduce the supplies of water and oxygen to the sulphides (Knapp and Welch 1991). Another means of slowing down such diffusion is the inclusion of layers of clay between layers of tailings. Research in British Columbia shows that covering tailings with impermeable seals and soil layers does not prevent upward migration of acidity and salts. However, an interface of coarse rock can reduce such upward movement (Ziemkiewicz and Gallinger 1989).

Use of the "porous envelope effect" is being tested by Falconbridge at their Fault Lake tailings area in the Sudbury area. The tailings have been deposited in a kettle lake surrounded by glacial outwash sand and gravel, which is very much more permeable to water flow than the tailings themselves. Groundwater flows around the tailings, rather than through them, minimizing metal leaching from the tailings (St-Arnaud et al. 1994).

Water blocks oxygen transfer to the tailings and can be maintained by the construction of an artificial marsh or swamp. Marsh plants will add detritus to the substrate, encouraging the establishment of bacterial populations that can use organic carbon to reduce sulphur compounds (Kalin 1991). The long-term aim is to develop an anoxic layer of peat as a barrier to diffusion.

The risk of recurring acidity and toxic metal solubility suggests that it may be desirable to use metal-tolerant and acid-tolerant plants for revegetation, producing a very simplified ecosystem with poorly developed nutrient-cycling. This practice produces vegetation types similar to those that occur naturally on areas with high metal concentrations, which have evolved over many thousands of years (Bradshaw 1990).

Exemplifying this approach, Hutchinson and Cox (1984) found that locally adapted populations of the grass *Deschampsia caespitosa* from the Sudbury area are better able to germinate and grow on metal-contaminated soil than can other populations of the same grass, or other acid-tolerant plant species. The authors concluded that commercial production of strains of *D. caespitosa* suitable for tailings revegetation can be readily achieved with lower costs than those involved in establishing agricultural varieties, noting, "It is surely best to suit the plants to the site rather than alter the site to suit the plants." At Sudbury, this approach has been carried out, using metal-tolerant grasses (Rauser and Winterhalder 1985).

Underwater tailings disposal in lakes or in the ocean, while controversial because of actual or potential effects on water quality and aquatic organisms (Ziemkiewicz and Gallinger 1989), is a very promising technique. Recent assessments of subaqueous tailings disposal can be found in MEND reports produced by CANMET. Underwater impoundment of high-sulphur base metal tailings has been evaluated in the area of Flin Flon, Manitoba (Hamilton and Fraser 1978). After 32 years, the high iron and sulphur content of underwater tailings had oxidized very little. Vegetation had colonized the area with a mat of organic matter over the surface of the submerged tailings, which provided a habitat for aquatic animals. The authors concluded that underwater tailings disposal has the advantages of low cost, as well as favorable aesthetics, low erosion, and high water quality. A MEND study reported by Fraser and Robertson (1994) evaluated four lakes—two in Manitoba and two in British Columbia—where tailings had been placed for varying periods of time, some as far back as 1943. The results of the study showed that reactivity of the tailings was low and organic deposits form once deposition ceases, shutting them off from oxygen. For salt water, Ellis (1989) offers a thorough analysis of the environmental effects of Island Copper's operations (on the northern tip of Vancouver Island), where tailings have been dumped directly into the ocean since 1971.

## Summary

The chemistry of sulphide ores creates the potential for their mining and processing to be harmful to the environment. In addition to the particulate emissions and land disturbance associated with other mining activities, the major problem with sulphide ores relates to containment of the sulphur itself, both in gaseous emissions of sulphur dioxide and in sulphide residuals in tailings. While sulphur dioxide emissions from smelters have been dramatically reduced in the past few decades, acid drainage continues to be a great challenge to the industry, both through the direct effect of the acidic water itself, and the indirect effect of metal mobilization. The acidity of the tailings and surrounding environment and the high metal concentrations create a number of problems for reclamation attempts. Further research is therefore necessary regarding the prevention of acid generation in sulphide ore tailings, either by devising better containment strategies that prevent oxidation or by reducing the sulphur content of the tailings.

For operating mines already experiencing acid drainage, the problem goes beyond current control and remediation requirements to the question of what to do after closure. Experience indicates that containment of acid-generating wastes may be necessary for centuries. Hazardous effluents will continue to be produced, and the challenge is to prevent their release.

The environmental effects of sulphur emissions and acid drainage include damage to plant tissue as a result of exposure to sulphur dioxide and acid deposition, as well as damage to fish and other aquatic life from acidic and metal-contaminated drainage. Two regional case studies are used here to illustrate these effects: the Moira River Valley, in which the spread of residuals has been primarily through the *hydrosphere*, and the Sudbury Basin, where the *atmosphere* has been the dominant pathway.

## Case Study 3. Deloro and the Moira River

Ontario's first gold rush occurred in 1866 at the southern edge of the Canadian Shield in Hastings County, where native gold and arsenopyrite are found with veins of quartz and carbonate in diorite near the margin of granitic stock. South of the shield rocks, palaeozoic limestones and dolomites outcrop, and overlying both shield and carbonate rocks there are scattered postglacial deposits such as sands.

The region is one of mixed coniferous-deciduous forest, dominated by white pine, sugar maple, and oaks. Because of rough terrain much of the shield area has remained forested, while the limestone and dolomite areas are farmed. As Figure 5.4 illustrates, the northern half of the watershed of the Moira River is in the shield area, with three main branches. The southern half (about 50 km long) runs through limestone and discharges into the Bay of Quinte, an estuary-like portion of eastern Lake Ontario. The valley contains numerous wetlands and near the mouth of the river, in the Bay of Quinte, there are extensive marshes. The Bay of Quinte has long been a center for fisheries.

Iron ore was already being mined in the Moira Valley for several decades before the discovery of gold. Gold mines were clustered at areas such as Eldorado and Deloro, and villages became established. The topic of this case study is Deloro, where about 25 shafts were sunk. Gold proved scarce, but arsenic, a by-product of roasting the ore, was plentiful and was produced from 1873 to 1961. Its main use was in pesticides, until it was superseded by organic compounds. Other metals were processed at Deloro, including copper, silver after 1903, and cobalt (from Cobalt, Ontario) during World War II. Eldorado Nuclear sent uranium to be refined at Deloro. By the 1950s, the main activity at the site was precision casting. The site was abandoned in the 1960s.

The main works at Deloro were on the west bank of the river, and a large tailings area was established on the opposite bank, with a dam right along the river. The tailings are a complex mosaic of different composition, with some extremely acidic portions and some radioactive patches. On both sides of the river, dumps of material containing several forms of arsenic were left. Buildings, such as an old bag house, were abandoned with sacks of arsenic insecticides lying about. Old shafts remained open, unfenced and unmarked, and no records of underground drainage had been made.

In the 1980s small areas of blackened rock with blackened tree stumps could be seen on both sides of the river. Erosion and damage to the rock could have been due to nineteenth-century ore roasting or to fires. Large fires occurred at several periods, for example in 1918–1919. Apart from the tailings and the immediate area

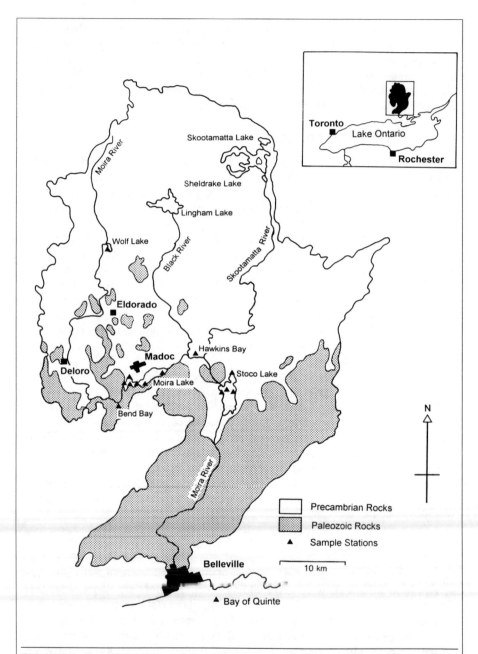

**Figure 5.4** The Moira River Valley of southeastern Ontario, showing the geology of the area and the locations of water-sampling (after Mudroch and Capobianco 1980).

of the mine's buildings, vegetation at Deloro appears healthy and diverse. Proximity to carbonate rocks to the south has probably limited effects from acidic wastes.

Sediment has been affected by the mining in the valley. In Moira Lake, about 10 km downstream from Deloro, sediments are heavily contaminated with arsenic (Diamond 1990). In the Bay of Quinte, dated sediments were shown to have a steady rise in arsenic in the 1860s, and then to have occasional peaks, for example in 1936 (Mudroch and Capobianco 1980). The later peaks are thought to have come from sediment washed out of basins further up the valley, during floods (Sly 1991).

Metals in marshes and sediment in the lower valley of the Moira, and at various distances from the mouth, were estimated by Dushenko (1990) and Greig (1989). High concentrations of metals such as copper, nickel, and cobalt were considered to have been derived from mining activity upstream.

Available records of possible damage to organisms during the century of mining activity include microfossils in Lake Ontario sediments. The Bay of Quinte phytoplankton show considerable changes in community structure, but it is not possible to distinguish an effect of mining apart from the disruptions due to lumbering, farming, and industry (Stoermer et al. 1985: Schelske 1991). The fish community has similarly been unstable but, again, as a result of multiple disturbances, including overfishing of various species (Sly 1991).

Greig (1989) and Dushenko (1990) examined correlations between organisms (plants, snails, muskrats, etc.) at present living in the Bay of Quinte with sediment metals. The only metal that had a negative effect, copper, showed attenuation away from the mouth of the Moira, but also has other sources in the area. Any plants or animals living in the area can be presumed to be adapted to the local conditions.

Underground water, river water, and water in lakes in the Moira Valley were strongly affected. Natural concentrations of arsenic in groundwater are thought to have caused human deaths in the 1930s, from drinking well water. In 1971, cattle drinking river water died from arsenic poisoning. Because of the danger, the Ontario government took over responsibility for the abandoned minesite at Deloro in 1979, considering it was the main source of the metalloid.

At that time, the input of arsenic to the river was approximately 35 kg•d$^{-1}$, resulting in a mean concentration of 330 µg•l$^{-1}$ just downstream of Deloro (Ontario MOE 1991). That concentration was well above the recommended national guidelines for both drinking water and aquatic life, of 50 µg•l$^{-1}$. The drinking water standard for Ontario is lower, at 25 µg•l$^{-1}$. The Ontario Ministry of Environment, therefore, initiated an arsenic treatment program at the Deloro site in 1979 (Ritter 1993). By 1989, daily loads from Deloro in the Moira were reduced from 35 kg to 6 kg (Ontario MOE 1991).

Groundwater carrying arsenic was found to be entering the river at the minesite. To find precisely where, part of the riverbed had to be drained. Pumping stations were installed around the site and an 80 m dyke was constructed parallel to the river. Simultaneously, contaminated buildings were removed or cleaned. Old tailings dams were repaired, since they formed part of the river bank. A large tailings basin on the east side of the river was covered with crushed limestone, to decrease its acid drainage. The limestone had to be placed on the surface of the tailings when they were frozen in winter.

In 1982–1983, the arsenic treatment plant that had been built at the site was improved (Ritter 1993). Drainage water from around the site is pumped into settling tanks and treated with ferric chloride to precipitate arsenic as ferric arsenate. The resulting sludge is sufficiently toxic that it must be contained indefinitely in storage lagoons. The treated water, after 99.5% of its arsenic has been removed, is discharged to the Moira River.

Arsenic concentrations in the lower part of the river valley have been reduced by an order of magnitude since the 1970s, but they still exceed water quality guidelines during the late summer. At present the river is estimated to contribute 2,300 g·d$^{-1}$ to Lake Ontario (Ontario MOE 1990), which is about 76% of the arsenic load to the Bay of Quinte (Diamond and Mackay 1991). By comparison, the Trent River, a bigger system that enters the Bay of Quinte about 15 km west of the Moira, contributes only 13 g·d$^{-1}$ of arsenic to the Bay.

The Bay of Quinte is one of the sites of Remedial Action plans instituted by the International Joint Commission in response to international agreements on water quality in the Great Lakes. The Bay of Quinte plan is now at the implementation stage, and one of its elements is further work to improve the Deloro site. It is estimated that the cost over ten years will be $25 million (Ontario MOE 1992b).

## Case Study 4. The Sudbury Basin

The Sudbury Basin is an example of a region affected by sulphide ore mining, primarily as a result of airborne transport of contaminants produced in the further-processing stage. The Basin is a concave geological formation approximately 60 km by 26 km in size. It consists of a fine-grained micropegmatite granite overlying coarse-grained norite rock. Sulphide orebodies are located in cracks on the underside of the norite. The Basin dips inward at angles between 30° and 70°, exposing the norite at the surface around its rim, where most mining activity is located. Copper-nickel mineralization was discovered in the 1880s during the building of the transcontinental railway, and mining continues to this day. A variety of base and precious metals have been produced, as well as by-products including sulphur compounds. Metallurgical processing has changed from surface roasting, which was used until 1929, to smelters venting all their gases through stacks of increasing height, culminating in Inco's 381 m Inco "superstack". Most recently, this evolution has also extended to the widespread use of scrubbers and acid plants to reduce sulphurous emissions and the 1993 reconstruction noted earlier in the chapter.

The main atmospheric releases during the century of mining operations in the Basin have been sulphur dioxide gas and metallic particulates, with lesser amounts of several other gases and solids. Although no data are available for the early period, it has been estimated that, in the treatment of 25.4 Mt of ore, 10.2 Mt of sulphur dioxide were emitted between 1890 and 1930 from 11 roasting yards

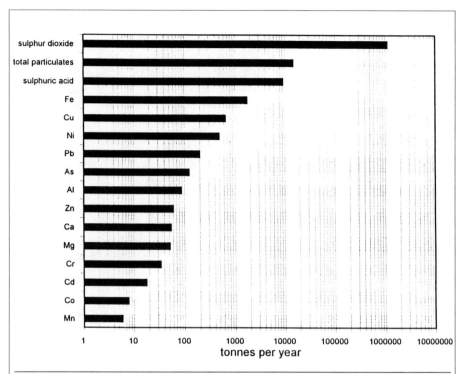

**Figure 5.5** Major pollutant and metal emissions (t/a) in the Sudbury Basin, based on available measurements during the period 1973–1981 (Ozvacic 1982).

(Allum and Dreisinger 1987). This amounted to an annual average emission of 255 kt of sulphur dioxide.

By 1970, emissions of sulphur dioxide had risen to almost 4 Mt·yr$^{-1}$, subsequently declining to 1.5 Mt in 1979 (Campbell and Marshall 1991). Emission factors have been reduced from approximately six tonnes of sulphur dioxide per tonne of metal produced in 1970 to less than one in 1994, as a result of more efficient smelters and increased use of control devices (MacLatchy 1994). The Sudbury Basin now has more than 20 sulphur dioxide monitors, operated either by mining companies or by the Ontario Ministry of Energy and the Environment.

The composition of stack emissions during the late 1970s is depicted in Figure 5.5. The three main metallic components were iron, copper, and nickel. There were also large amounts of lead and arsenic, as well as other toxic elements such as cobalt. Copper and nickel emissions were estimated at approximately 1,100 t·yr$^{-1}$ in the mid-1970s (Scheider et al. 1980).

The impacts of these emissions on soil, vegetation, wetlands, lakes, and animals are presented below, followed by a discussion of the recovery.

## Impacts

### Soils

Sulphur fumes, metal fallout, the loss of forest vegetation due to logging and fires that led to erosion, the fires themselves, and general disturbance all changed soils in the Sudbury Basin. As far as 3 km from the smelters, all soil was destroyed on ridges and steep slopes, creating a desert landscape. Further away, soils were contaminated with copper, nickel, cobalt, and aluminum by the 1970s in quantities sufficient to inhibit growth of crop plants under laboratory conditions. Their phytotoxic effects, however, could be decreased by raising pH levels in the soil (Whitby and Hutchinson 1974). Soil pH values were found, for the most part, to be less than 5.0 in 1978 (Figure 5.6). Near the smelters, acid deposition was sufficient to inhibit the activity of soil organisms; as a result, litter decomposition was abnormally slow (Freedman and Hutchinson 1980a).

Blackened rock, bare of soil and vegetation, resulting from open-hearth roasting and the activities of three smelters (at Copper Cliff, Coniston, and Falconbridge), was unusually extensive in the Sudbury Basin.

### Vegetation

The former extent of forest vegetation is apparent in burnt stumps of large oaks and pines or bare roots standing where soil has been eroded from around them, as well as in living trees that were reduced to the size of bushes. By the 1970s, no tree canopy remained within 3 km of the Copper Cliff smelter. Up to 8 km away, forest survived only in valleys. The exact causes of damage cannot be determined; the possibilities are acidity, elevated soil metals, fire, metal fallout, loss of soil, changed microclimates, or a combination of these factors. Whatever the exact cause, the result is a loss of species diversity not only in the forest but also in the ground vegetation.

In the 1970s, regional norms of vegetation (measured as species richness or cover) were not attained within 30 km of the emission sources. Nine distinct communities were identified in a study conducted in 1978 (Amiro and Courtin 1981), of which six were part of the hemlock/white pine/eastern hardwood regional forest. The remaining three (barren, birch transition, maple transition) were related directly to contamination. The extent of these heavily damaged plant communities is depicted in Figure 5.7.

Conifers are more sensitive to sulphates than hardwoods are. The forest composition was therefore changed by the loss of more conifers. Even a few hours of exposure to 25–30 ppb $SO_4^{2-}$ is sufficient to damage white pine (FPACAQ 1987). White pine was, economically, the most valuable tree in the Basin. Linzon (1971) estimated that production of pine was lost for an area of 1,850 square kilometers.

Plants that have continued to grow under these conditions must be metal-tolerant, at least to some extent. An inverse relationship between metal content and distance from smelters has been shown for several species, such as blueberries (*Vaccinium angustifolium*). Concentrations of copper and nickel in various parts of blueberries,

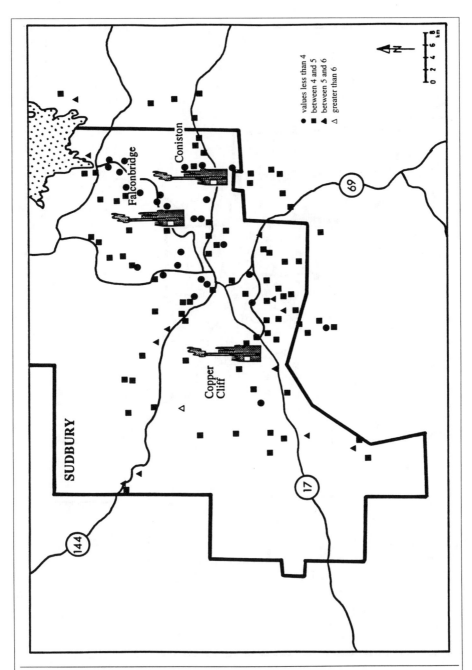

**Figure 5.6** Soil pH measured in the Sudbury area during 1978 (Amiro and Courtin 1981).

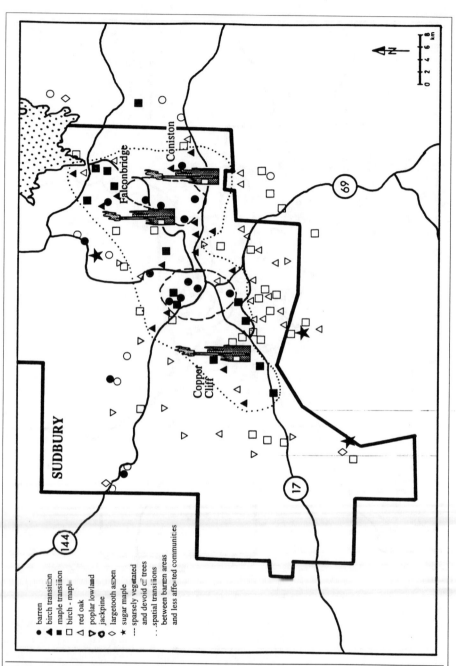

**Figure 5.7** The distribution of plant communities in the Sudbury area as determined in 1978 (Amiro and Courtin 1981).

measured in the Sudbury Basin, are compared in Figure 5.8 with those from a non-polluted site in Newfoundland, which has a similar type of soil.

## Wetlands

In the barren areas, some "fossil" wetland soils can be found buried by sediments that were eroded rapidly a century ago. Like the forests, peatlands within a few kilometers of smelters have depauperate vegetation, often with only leather-leaf shrubs (*Chamaedaphne calyculata*) surviving (Gignac and Beckett 1986). Further than 10 km, a normal complement of vascular plants occurs. However, the full range of *Sphagnum* species is not found within 30 km of the emission sources. The moats of bogs tend to be more species-rich than centers, because nutrients leach into them. In both peats and bog water, copper and nickel are inversely correlated with distance from smelters; such is not the case, however, with manganese, iron, and calcium. Loss of *Sphagnum* has altered the hydrology of some wetlands.

In marshes and areas of open water, species diversity is also low. Often only cattails and reeds (*Typha* spp. and *Phragmites australis*) are sufficiently tolerant to survive close to the smelters. These populations are metal-tolerant (Taylor and Crowder 1983a, b). As in the forest, low pH values, high metal concentrations in soils, and low nutrient levels may all limit wetland plants.

## Lakes

The Sudbury area has a complicated immature drainage system with numerous lakes, possessing a wide range of chemistry and buffering capacity (Semkin and Kramer 1976). Sulphate levels were shown in the 1960s to be elevated near the smelters (Gorham and Gordon 1963). Metal levels in both water and sediment decreased the further they were away from emission sources (Yan and Miller 1982). Within 3 km of smelters, pH levels in lakes ranged from 3.3 to 6.1; beyond 24 km of the smelters, levels ranged from 4.6 to 7.4 (Gorham and Gordon 1963). The diversity of macrophytes and of algae was lower near metal sources. This was also the case for standing crop; algal productivity was found to be limited by nickel and copper rather than by sulphate concentration (Whitby et al. 1976). The plants that grow in the Sudbury-area lakes during this century are obviously metal-tolerant.

Contrary to expectations, insects with aquatic larval stages were found to have low body burdens of metals in most Sudbury sites. They would, however, still be a source of metals to animals eating them, such as birds (Krantzberg and Stokes 1989). Crayfish are fairly abundant in the lakes, and were found to have high body burdens of metals (Bagatto and Alikhan 1987a, b).

Sudbury-area lakes undoubtedly lost fish populations, since a pH level of 4.5 is generally lethal for eggs or young both of fish and of amphibians. Moreover, concentrations of copper greater than 6 $\mu g \cdot l^{-1}$ and of nickel greater than 150 $\mu g \cdot l^{-1}$ are toxic to the early stages of most fish species (Task Force on Water Quality Guidelines 1987). Both salmonids and warm-water fish are affected by the metals, although increasing water hardness does diminish these effects (Mance 1987).

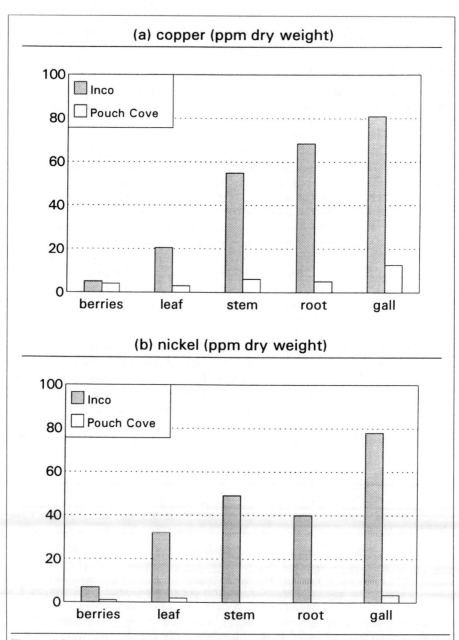

**Figure 5.8** Copper (a) and nickel (b) concentrations in plant and gall tissues of *Vaccinium augustifolium* collected at the Sudbury Basin (Inco) and New-foundland (Pouch Cove) sites. Gall tissues include both plant structure and insect inhabitants. Source: Bagatto (1992).

The distribution pattern of nickel in surface sediments of Sudbury lakes is shown in Figure 5.9, which also shows the historic change in metal concentrations in the sediments. Examples of low pH values were 4.1 at Clearwater Lake in 1973 (Dillon et al. 1986) and 4.05 at Baby Lake, about 50 km downwind of the smelters, in 1972 (Hutchinson and Havas 1986). Indirect effects of smelters include an increase of bioavailable aluminum in lake water, resulting in fish mortality. Food chains are then disrupted; in some cases, ducks may benefit from decreased fish populations, which forage for the same foods.

## Animals

Samples taken in the Sudbury region of the feathers of ruffed grouse, mallards, and black ducks, as well as of squirrel fur and human hair, were all found to contain elevated levels of metals (Ranta et al. 1978; Rose and Parker 1982; Lepage and Parker 1988). Seasonal studies of feathers and pelage indicated that nickel and copper were adsorbed exogenously, presumably from atmospheric fallout.

Herbivores in the area undoubtedly ingest some plants with high metal burdens. Wildlife, however, is fairly abundant. Animals such as moose may have benefited by the replacement of mature forest by scrub.

## Recovery During the 1970s

There was a major reduction of sulphur dioxide emissions in the Sudbury Basin during the 1970s (Figure 5.10), due primarily to improvements in the efficiency and pollution control of smelters in the area. Local deposition was also reduced considerably by the replacement of several shorter stacks with the Inco "superstack" in 1972. It should be noted that, while ambient air quality was improved by this action, it was later realized that the net benefit to the environment was nil. The same amount of acid deposition occurred; it was simply spread over a much greater area.

As indicated in Figure 5.10, emissions were affected by two complete shutdowns of Sudbury smelting operations for economic reasons: the first between the fall of 1978 and spring of 1979 (Scheider et al. 1980), and the second from the summer of 1982 to spring of 1983 (Tang et al. 1987). A study of the 1977–1979 closure found dramatic reductions in local deposition of sulphate and metals (copper and nickel). There was, however, no commensurate effect on the fallout of sulphate 250 km to the southeast. It was concluded that the Sudbury smelters were major contributors to sulphate deposition up to at least 12 km from Sudbury, and to deposition of copper and nickel as far as 50 km. Deposition at greater distances was attributed to long-range transport (Scheider et al. 1980). A study conducted in 1982-83 found that, in central and southern Ontario, 12% of the wet deposition and 20%–47% of the dry deposition of sulphate came from Sudbury smelters (Tang et al. 1987).

The value to the economy of sports fisheries (Minns and Kelso 1986) and of hunting are the main reasons for the numerous studies of recovery in the Sudbury area. These studies have sparked international interest in the process of ecosystem

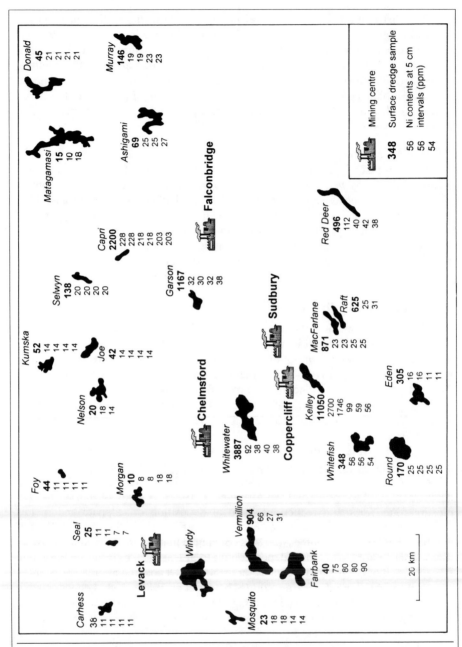

**Figure 5.9** Sediment nickel in lakes throughout the Sudbury Basin in the 1970s. Surface samples are compared with core samples for each lake. Source: Allan (1974).

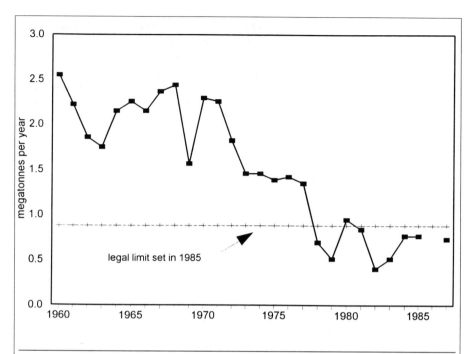

**Figure 5.10**   Sulphur dioxide emissions in the Sudbury Basin between 1960 and 1987 and the legal limit set in 1985 (Dillon et al. 1986; Ontario MOE 1992a).

recovery from acidification (Barth 1987). Since sulphur dioxide emissions began to decline in 1970, dramatic changes have been documented in both land and lake environments and biota.

### Recovery of Lakes

The most striking changes in Sudbury-area lakes during the 1970s have been rising pH (for example, from 4.1 to 6.4 in 13 years); declining sulphate-ion concentrations (for example, from 1,319 to 11.9 mg·l$^{-1}$ in 4 years); and increasing species richness of algae, invertebrates, macrophytes and fish (Beggs and Gunn 1986; Dillon et al. 1986; Hutchinson and Havas 1986; Keller and Pitblado 1986; Keller et al. 1986; MacIsaac et al. 1986; Kelso and Jeffries 1988).

Between the 1960s and mid-1980s, numbers of macrophyte species in some lakes increased from four to eleven, and algae from three to eighteen. Lakes were recolonized by some taxa that had been lacking in the 1960s, including fish and small organisms such as rotifers and crustaceans (Hutchinson and Havas 1986). These improvements were probably related to decreases in metal concentrations in the water—including copper, nickel and cobalt—as great as 90%.

Recovery can be assessed by examining sediment cores for paleoecological evidence, such as changes in algal community structure with lake chemistry. The conditions likely to have coincided with fossil assemblages—in the decades since the 1880s, for example—are inferred from the present niches of diatom species, as defined by the pH levels or metal content of their surroundings (Dixit et al. 1991).

During the early 1980s, trace element loadings of several metals in sediments and fish were compared in 14 lakes across Ontario (Johnson 1987). Loadings of copper and nickel at levels high above background were found only at those lakes within 100 km of Sudbury. Concentrations of some metals—such as lead, cadmium, and mercury—in fish tissue tended to be related to the levels in the sediments. Concentrations of essential trace elements—such as copper and zinc—had no relation to the sediments. Recovery of the Sudbury lakes has been dramatic, primarily because of the sharp reduction in pollutant loading from smelter-gas emissions. This would not be the case where a water body receives acids or metals directly from a local source, as in acid drainage.

### Recovery of Vegetation

Many individual trees that had died back and were surviving only as shoots at ground level began to recover during the 1970s (Freedman and Hutchinson 1980b). Seedings of birch, oak, and aspen were also reported in areas where they had been absent in the 1960s.

More conifers have been lost than hardwoods; experience in Trail, British Columbia, where emission controls were established in 1941, indicates that conifers have recolonized deforested land more slowly than birch and aspen, as a result of a lack of seeds (Archibold 1978).

## Unresolved Issues

Studies such as these described here, as well as simulation models of lake and soil acidification, can explain the main features of acidification and recovery. Lack of an ecosystemic viewpoint is often apparent, so that studies of, for example, tailings, groundwater, and regional particulate fallout are not related. There remain several unresolved issues, including the contributions to local acidification from non-local sources; non-linear and synergistic relationships between emissions, deposition, and sink concentrations; the spatial variability of the soil sulphur pool and the complexity of soil chemical and biological processes; and the timescales of chemical and biological recovery, in response to the various possible pollution control strategies (Howells 1990).

In summary, rapid acidification occurred in many northern-hemisphere lakes and soils in the period between 1940 and the 1980s, which resulted in conditions adverse to many organisms. A general reduction in sulphate emissions during the last two decades has been reflected in reduced sulphate in water. However, acidity of surface water and soil has been slower to decline, because of either continuing nitrate deposition (primarily from fuel combustion) or a slow response time in the system. Although emission controls are essential in the long term, benefits are limited in the short term. This emphasizes the need for remedial measures to enhance the production of base cations from sediments and to reduce aluminum mobilization in soils.

# CHAPTER SIX

# GOLD AND SILVER

## Introduction

There are three sources of gold and silver: primary gold and silver mines, base metal mines and, for gold alone, placer mining operations. In 1992, Canada ranked fifth in world gold production (Couturier 1993). Canadian gold production, as listed in Table 6.1, comes predominantly from auriferous-quartz mines, followed by base metal mines and placer mining operations (Couturier 1993). Figure 6.1 indicates the locations of these operations. The maps of ecozones (Figures 1.3 and 1.4) show that most current operations are in Boreal and Cordilleran zones.

In silver production, Canada also ranked fifth in the world (Keating 1993). Unlike gold, however, nearly all of the silver produced in Canada is derived from the mining of other metals, either as a co-product with copper or gold or as a by-product of base metal mining (Keating 1992). Canadian silver production, as Table 6.2 indicates, comes predominantly from base metal mines.

Because Canadian production of gold and silver as a by-product of base-metal mining is derived from sulphide ores, the methods of extracting and processing those ores are like those discussed in Chapter 5.

## History

### Lode Gold and Silver Mining

In Canada, lode gold was first mined in Nova Scotia in the late 1850s. By 1863, the practice had spread to western Canada, with operations in the Bakerville and Cariboo Gold Quartz districts of British Columbia. Soon afterward, in 1866, lode gold was discovered in the Canadian Shield near Madoc in southeastern Ontario (Boyle 1987). More information about the long-term effects of mining in this region may be found in the Moira River case study in Chapter 5.

**Table 6.1** Canadian Gold Production, 1975, 1980, 1985–1992

|  | Total Production (kg) | Auriferous-Quartz Deposits | Placer Operations | Base Metal Deposits |
| --- | --- | --- | --- | --- |
| 1975 | 51,433 | 73.0% | 0.6% | 26.4% |
| 1980 | 50,620 | 63.1% | 4.0% | 32.9% |
| 1985 | 87,562 | 76.8% | 4.0% | 19.2% |
| 1986 | 102,899 | 80.9% | 2.7% | 16.4% |
| 1987 | 115,818 | 81.8% | 3.5% | 14.8% |
| 1988 | 134,813 | 83.4% | 3.6% | 13.0% |
| 1989 | 159,494 | 86.6% | 3.4% | 10.0% |
| 1990 | 167,373 | 88.0% | 2.4% | 9.6% |
| 1991 | 176,126 | 87.8% | 2.2% | 10.0% |
| 1992 | 160,351 | 88.5% | 2.2% | 9.3% |

Sources: Natural Resources Canada; Statistics Canada.

Because of Canada's long tradition of lode gold mining, its environmental effects date back almost a century and a half. This history makes it very difficult to gauge both the number of abandoned gold mines and their potential and past effects. The National Mineral Inventory indicates that there are at least 859 abandoned gold mining operations across the country (EMR 1989a).

## Placer Mining

The gold rush of 1898 attracted thousands of miners to the Yukon Territory; in the three months following the first major discovery, 500 claims were staked on Klondike creeks (Gilbert 1989), and by 1913, Yukon gold production had risen to a peak of over 11 tonnes (Van Kalsbeek et al. 1991).

Placer mining is still a growing industry in Canada, with the number of claims staked and leases issued increasing every year. The labor-intensive mining methods of the early days gave way to large dredging operations better suited to the lower grades of ore. As the large creek beds became exhausted and richer deposits were found on hillsides, the dredging operations themselves have, in turn, gradually given way to hydraulic mining, in which bulldozers and large quantities of water are used to clear away overburden and separate the gold from its "paydirt."

## Orebodies

The locations of the main primary gold and silver mines that operated in Canada in 1992 are shown in Figure 6.1. In addition to those active mines indicated on the map, a large number of gold mines, located mainly in Ontario and Québec, have suspended operations until market conditions improve.

The ore in most lode gold deposits is native metal, hosted in magnesium- and iron-rich volcanic rocks of the Canadian Shield. Sulphide and/or arsenide minerals such as pyrite and arsenopyrite are often found at these sites. Gold occurs

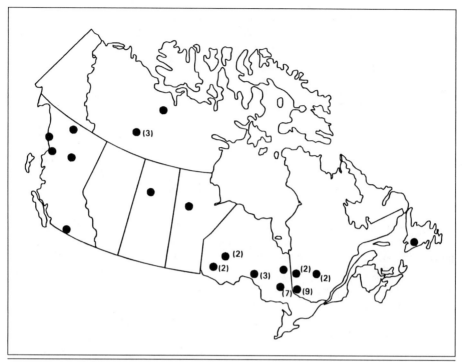

**Figure 6.1** Primary gold mines in Canada, 1993. (Where a location has more than one operation, the number of separate operations is shown in parentheses.) Source: Couturier (1994).

only rarely in chemical combination with other elements, although it is often finely disseminated (in native gold form) in many base metal sulphide ores.

By contrast, silver is more highly reactive than gold and occurs in over fifty different minerals (Thompson 1991); the most common of these are native silver, argentite ($Ag_2S$) and cerargyrite (AgCl). While primary gold mines are numerous in Canada, primary silver mines are much less common. Of Canada's operating lode gold mines, all but one produce silver as well. Gold and silver are recovered as by-products at most of the base metal operations across the country (Jen and McCutcheon 1993).

## Extraction and Processing of Ore

### Lode Gold and Silver Mining

The nature of lode ore deposits tends to favor the use of underground methods—which are more selective—rather than open pit. In 1992, about three quarters of the lode gold mines in Canada operated underground (EMR 1993a).

**Table 6.2**  Canadian Silver Production, 1975, 1980, 1985–1992.

| | Total Production (kg) | Primary Gold and Silver Deposits | Copper-Nickel Deposits | Copper-Gold Deposits | Base Metal Deposits |
|---|---|---|---|---|---|
| 1975 | 1,234,642 | -- | -- | -- | -- |
| 1980 | 1,070,000 | -- | -- | -- | -- |
| 1985 | 1,197,072 | 4.8% | 4.7% | 17.9% | 72.6% |
| 1986 | 1,087,989 | 1.2% | 3.9% | 20.3% | 74.6% |
| 1987 | 1,374,946 | 2.9% | 4.7% | 17.5% | 74.9% |
| 1988 | 1,443,166 | 5.3% | 4.6% | 21.3% | 68.8% |
| 1989 | 1,312,433 | 5.4% | 4.8% | 21.7% | 68.1% |
| 1990 | 1,381,257 | 5.8% | 3.7% | 20.7% | 69.8% |
| 1991 | 1,261,359 | 4.4% | 4.5% | 22.3% | 68.8% |
| 1992 | 1,168,968 | 4.1% | 4.5% | 16.9% | 74.5% |

--No data available. Sources: Natural Resources Canada; Statistics Canada.

The gold and associated silver in auriferous-quartz deposits are usually recovered using one of either the crush and grind, cyanide leach, zinc-precipitation or carbon-in-pulp extraction processes. There may be considerable variation from one mill to another; variation may also exist, occasionally, within a single mill, as necessitated by prevailing economic conditions or environmental-control regulations, or by changes in the ore.

Some operations, particularly in areas where the gold grain size is large, use concentration methods after crushing to recover the bulk of the gold before treating the remainder of the ore by hydrometallurgy. Other mines employ an "all-sliming" process, in which all of the ore is finely ground, leached in cyanide solution, thickened, agitated, and subsequently filtered. The gold is almost always leached from the ore using a cyanide solution, and recovered by filtration and zinc precipitation (known as the Merrill-Crowe method) or by carbon-in-pulp adsorption followed by stripping and electrowinning. A very few operations find it necessary to roast the ore prior to cyanidation in order to free tiny gold particles enclosed in arsenopyrite and, in some cases, pyrite, so that they can be leached.

### Comminution

One example of comminution is provided by Figure 6.2, a simplified process-flow diagram for the Seabee mine (see also the case study at the end of this chapter) that opened in November of 1991 in northern Saskatchewan (Beak Associates Consulting Ltd. 1990). After crushing and primary grinding in a ball mill, gravity concentration is used to remove the larger gold particles, which are sent directly to the refining stage. The remainder of the ore is reground, classified and thickened, and recycled through the gravity concentrator. The underflow from the thickener is sent to the cyanidation vats, along with air and cyanide and with lime to adjust pH.

After 48 hours, the slurry is sent from the thickener to a carbon-in-pulp circuit for gold removal. The dissolved gold adsorbs onto the carbon, which

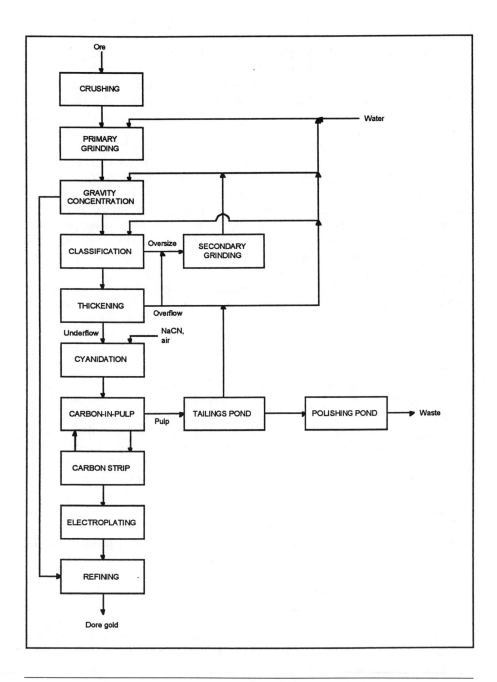

**Figure 6.2** Processing flow diagram for the Seabee gold mine in northern Saskatchewan.

flows counter-currently. After 8 hours, the gold is stripped from the carbon and the eluate electroplated onto steel wool in a heated bath of sodium cyanide and sodium hydroxide. Finally the gold-plated steel wool is refined to gold-silver bullion in a furnace. In addition to the reagents mentioned already, there are requirements for a flocculant for thickening, for fuel oil for heating and for air and water. The mill recovers approximately 92% of the gold in the ore treated.

### Cyanidation

Cyanidation of auriferous ores also recovers silver as a by-product if it is present. Because cyanide is a powerful but non-selective solvent, it brings a considerable number of potentially harmful substances into solution. Cyanide may be present in wastes in several species, depending partly on local geology and climate. In effluents, cyanide may be present as free cyanide ($CN^-$, $HCN$); simple compounds ($NaCN$, $HgCN_2$); cyanide-metal complexes (copper, iron, nickel and zinc); and compounds such as CNS (Mehling and Broughton 1989; Scott 1989a, b). Regulations typically require measurement of total cyanide and weak-acid dissociable (WAD) cyanide.

Cyanide is usually the main toxic residual of lode gold mining. Since it is harmful to a wide variety of organisms, it is the prime candidate for treatment (Scott 1989b). However, a recent study by the Ontario Ministry of the Environment (1992c) indicates that the three most important effluent contaminants in Ontario were ammonia, suspended solids, and copper. Levels of arsenic and free cyanide were found to be extremely low, and inorganic acids were not even discussed because the pH of the effluent varied between 7 and 10.

Cyanide is treated to "destroy" it, generally by converting it into compounds that are insoluble and so cannot be taken up by organisms. Natural degradation has been the most commonly used treatment at most Canadian mills for many years, but it is now being supplemented or supplanted by other methods. For example, of the 50 gold mills using the cyanidation process in 1988, 23 operated chemical-treatment systems as well (Scott 1989a). Some of the methods in use are the Inco $SO_2$/Air process, hydrogen peroxide, Hemlo/Golden Giant precipitation, alkaline chlorination, Homestake biodegradation, and AVR, or acidification-volatilization-regeneration. The first two of these methods are the most widely used in Canada at present. AVR is an example of a process for the recovery (in contrast to the destruction) of cyanide; this method has been used in Australia, and experimentally at one operation in Canada (Ingles and Scott 1987). Some treatments, such as chlorination, convert the cyanide to the less toxic cyanate. Ion exchange is another technique being tested in Canada. Although it is not yet used on a large scale, two pilot-plant operations currently show promising results (Whittle 1992). For example, in a typical system, pre-treatment levels of cyanide ranging from 13 to 1,124 mg·l$^{-1}$ were reduced to 0.1 to 10 mg·l$^{-1}$ afterward.

In the presence of ultraviolet radiation, cyanide volatilizes naturally, but the rate slows dramatically as temperature declines. It takes nine or ten months of retention in a holding pond to reduce cyanide to acceptable levels by means of natural degradation (Scott 1989a, b). As a result, it is often supplemented by chemical methods during cold months in countries with winters as severe as Canada's. The natural degradation of cyanide actually includes a number of distinct processes, each with its own set of controlling parameters (Scott 1989b):

- *volatilization* is most rapid in aqueous systems with low pH and exposure to the air;

- *dissociation* occurs when equilibrium conditions change rapidly;

- *oxidation* is controlled primarily by the availability of oxygen;

- *photo-degradation* requires ultraviolet radiation from the sun; and

- *biodegradation* depends upon both the availability of suitable environmental conditions and the presence of suitable bacteria, algae, and fungi (Mehling and Broughton 1989). It is dependent on respiration by cells and on a suitable temperature—it does not happen in cold winter conditions.

In addition, a number of mechanisms may act to prevent or delay the cyanide breakdown, such as chemical complexation and adsorption onto soil particles.

Konigsmann et al. (1989) describe the effects on natural degradation brought about by a redesign of the tailings pond at the Golden Giant Mine in northern Ontario. Doubling the free water surface and reducing the depth of the retention pond significantly lowered minimum levels of cyanide and metals in the pond solutions. This is consistent with the results of computer modelling of pond design (Simovic and Snodgrass 1989), which were tested at the Lupin Mine in 1987–1988.

Mehling and Broughton (1989) concluded that cyanide species are mobile: they can migrate through groundwater in situations where the underlying soils or tailings are very pervious. Residual cyanide in the tailings could represent a long-term source that might periodically flush through the system as conditions of equilibrium change. The first mine studied was a small gold operation in British Columbia, which began operation in 1980 using both the Merrill-Crowe process and alkaline chlorination to remove cyanide and toxic metals. Five months after its shutdown in 1986, high cyanide levels were found in the tailings and in a creek downstream from the pond. The cyanide in the creek was found to be from leakage through the pervious alluvial deposits, originating at least partly through groundwater flow. Two other small mines, also in British Columbia, operated from 1933 to 1967, depositing tailings on the shore of a nearby lake. Analysis of these tailings in 1986 showed elevated total cyanide levels, despite both the long period of inactivity and a relatively acid environment.

Most work on cyanide toxicity has been on fish. Resistance varies among species; brown bullheads are among the most resistant fish, while salmonids are highly susceptible to cyanide toxicity. Cyanide leaking from tailings in British Columbia, for example, killed approximately 20,000 steelhead smolts in the Coquihalla River (Leduc et al. 1982). Toxicity to freshwater fishes is due mainly to inhibition of cytochrome oxidase, resulting in blocking of electron transport and energy release in cells (Leduc et al. 1982); the functioning of the thyroid gland appears also to be inhibited. The degree of hypoxia and hypothyroidism depends upon the supply of sulphur, iodide, and nitrite in food and water.

Greater use of recycled mill water has the twin benefits of reducing both the requirements for fresh water and the release of cyanide (as well as of other process chemicals). The treatment of mill water for the removal of other substances is similar to that used elsewhere in the mining industry. For example, arsenic can be removed by precipitation with lime (Smecht et al. 1975); another technique is through oxidation to arsenates, followed by precipitation using ferric ion, to form extremely insoluble basic ferric arsenates. In many cases, the wastewater already contains enough iron so that it is necessary to add only a base (Scott 1987).

Alternatives to conventional vat leaching (such as heap and in-situ leaching, as described in Chapter 1), as well as improvements to the vat leaching process itself, may affect the environmental effects of gold mining and milling in future years. Several Canadian gold mines, in addition to mines of copper, uranium, and gold both here and abroad, have been using heap-leaching methods for many years. This process may be effected with run-of-the-mine ore, with waste rock, with comminuted ore, with concentrates, and even with tailings (Werniuk 1987; Hardcastle and Sheikh 1988; Wilder and Dixon 1989). As it is carried out in the open, it appears to pose a greater risk of environmental contamination than does vat leaching, both through leakage and erosion, as well as through winter ice buildup and subsequent thaw. Its lower cost, however, allows the treatment of poorer grades of ore, with the benefit of removing metals that might otherwise pose greater environmental hazards when emitted as waste. The NovaGold case study at the end of this chapter offers an example of another variant on vat leaching that may preempt many of these problems: indoor vat leaching. A number of methods are being developed to render more efficient the leaching of these ores; one of these, for example, is the use of agglomeration (Chamberlin 1986; McClelland 1986).

In-situ leaching has its own benefits and drawbacks (as described in Chapter 1) and is feasible only with porous orebodies not likely to contaminate an aquifer. New methods are being developed to assist in the determination of the former (Brink et al. 1991). Whatever type of leaching is used, however, the fractional recovery and rate of dissolution can be improved, and cyanide use reduced, through the addition of hydrogen peroxide (Lorosch et al. 1991). This peroxide-assisted leach (PAL) process was used first in South Africa and then in Australia and North America. It is in the processing of sulphide ores that the PAL approach has found its main application; in this capacity, increases in gold recovery up to

20% and decreases in cyanide use up to 50% have been achieved. Recent developments include the addition of a surfactant to the carbon-in-leach circuit in order to reduce losses in graphitic ores, and a cyanide recycling process capable of recapturing 99% of the cyanide (Scales 1992).

### Zinc Precipitation

The Parks process is a method more modern than cyanidation for removing silver from lead bullion. Zinc is added to molten lead bullion, alloys with the silver, and rises to the top of the kettle as a dross that is skimmed off. The skimmings are retorted to remove the zinc, and the resulting doré bullion is then separated into its gold and silver components.

### Roasting

The most significant atmospheric emissions from lode gold mining have been associated with the roasting of some types of ores to remove sulphur compounds that would otherwise interfere with the leaching process (Horn 1966). While the practice has been largely discontinued, in 1992 there were at least three Canadian operations, located in Ontario, British Columbia, and the Northwest Territories, which were still roasting their gold ores (EMR 1993a). In a number of cases, the sulphur occurs in the form of arsenopyrites, which produce emissions of arsenic ($As_2O_3$) as well as sulphur oxides when roasted.

High levels of control have been achieved using various control devices such as impingers, electrostatic precipitators and baghouses (Horn 1966; Rabbitts et al. 1971). The lack of an adequate market for these materials means that they have had to be stored in an impounding structure either for eventual recovery or for perpetual isolation. Some of the arsenic trioxide collected in this way has been marketed as a wood preservative (Koren 1992). Other operations have extracted the arsenic through flotation and cycloning (Burckle et al. 1981), alkaline chlorination and ferric-ion precipitation, and, more recently, bioleaching/ bacterial oxidation and pressure oxidation using an autoclave (*Mining Engineering* 1990).

### Mercury

Mercury was widely used to extract native gold from ores both before and after the cyanide process was developed; when mixed with an ore, it forms an amalgam with any gold and silver with which it comes in contact. The amalgam must then be volatilized to liberate the precious metals. Up to 10% of the mercury used in this process is likely to be lost to the atmosphere. Although largely supplanted by cyanidation, amalgamation is still practised at some mills as a supplement to cyanide leaching.

Amalgamation was estimated to have emitted about 37 kg of mercury to the atmosphere in 1978, mainly from operations in Ontario and British Columbia (Sheffield 1983). While these emissions may have important local impacts, they have been considered negligible at the national scale by Jaques (1987). Many of the mines that formerly used mercury-amalgamation to recover gold continue to be potential sources of the toxic metal. Mudroch and Clair (1986) analyzed drainage from an area in Nova Scotia that had been mined for gold between 1860 and 1930. They concluded that, although water concentrations were quite low, there was evidence that arsenic, mercury, lead, and zinc—all originating from past gold-mining activities—are being deposited in lakes downstream and are gradually contaminating a large part of the Shubenacadie River system.

### Tailings

Because grades of gold ores are measured in grams per tonne, the volume of waste rock and tailings is higher than with any other ore: the tonnage of waste rock is almost equal to the amount of ore extracted, and tailings amount to about 60% of the ore. This means that good management of solid wastes, particularly tailings, is a great environmental challenge facing the gold mining industry in Canada. Proper design of a tailings disposal system as part of an overall mine plan can have a major effect on the economic feasibility of the entire operation (Welch and Firlotte 1989).

Surface disturbance related to lode gold mining results mainly from tailings dumps and the relatively few open pit mines. There exist literally hundreds of abandoned sites of gold mining activities distributed across the country, many of them in need of rehabilitation in order to facilitate their natural recovery.

Most of the auriferous-quartz mining sites in Canada (Figure 6.1) are located on the Canadian Shield in major forest resource regions of the boreal and mixed-forest zones. Economic values include wilderness, hunting, fishing, trapping, and food gathering. The main environmental effects of lode gold mining are thus related to the discharge of liquid effluents that are toxic to fish and other aquatic organisms. Toxic emissions that damage vegetation are of less concern than with other minerals.

The seriousness of the effects of effluents makes containment of the liquid component of tailings even more important than that of solids. Although much of this liquid can be returned to the mill for reuse after decantation, there are often situations in which surplus water is released to watercourses. The gross water use of auriferous-quartz mining in Canada was approximately 100 million cubic meters in 1986 (Table 1.4). Two-thirds of this amount was recycled, and almost all of the remainder was ultimately discharged to watercourses, together with about 16 million cubic meters of mine water. Discharges are generally alkaline, with an average pH around 7.5–8.0 (Scott 1987).

Acid-generating potential is a consideration in tailings management. Although acid generation is always a problem in sulphide ore mining, many gold ores contain acid-consuming materials, such as carbonate and calcite, which buffer the effects of acid generation. Knowledge of the acid generation and consumption potentials of the ore, therefore, forms an important part of any plan for the treatment of tailings and effluent. For instance, the ore and waste rock at the Seabee mine (Beak Associates Consulting Ltd. 1990) were found to have acid-consuming potentials well in excess of their acid-generating potentials; draining of acidic water was therefore not expected to be a problem. Less than 60 km away, the Jolu mine ores were found to have high pyrite and pyrrhotite content and an excess of acid-generating over acid-consuming potential sufficient to produce drainage water with pH values between 1.7 and 4.9 (Royex Gold Mining Corporation 1988).

Underground disposal has both advantages and disadvantages, as discussed in Chapter 1. This is particularly true with gold, which occasionally has sufficiently large price fluctuations to justify reworking old tailings for their residual gold values. Reworking tailings is much easier if they are stored on the surface rather than backfilled underground. A number of such recovery projects are currently in operation and still others are planned (Cristovici and Leigh 1986; Werniuk 1987). The greatest environmental concern in underground storage is the possibility of disruption and contamination of groundwater.

The safest way to dispose of tailings appears to be underwater impoundment in natural depressions or lake basins (Scott 1987; Welch and Firlotte 1989). This method considerably reduces ore oxidation and tends to localize environmental effects. In situations where such containment is not feasible, the disposal site should be chosen to have adequate capacity, ground stability, minimal risk of groundwater contamination, and buffering zones between tailings and receiving watercourses.

Groundwater is a concern in the decommissioning of a gold mine: it may be necessary to consider the possible movement of cyanide, arsenic, and metals into groundwater. Sengupta (1994) outlined the stages such decommissioning should take, beginning with a risk assessment consisting of hazard identification, exposure assessment, toxicity assessment (receptor dose-response), and risk characterization. Sengupta's example is from Montana, where it is calculated that when effluent has less than 0.22 $mg \cdot l^{-1}$ cyanide, the levels of Cu, Zn, As, and $NO_3^-$ will have reached compliance. Both migration through surface water and through groundwater are expected in Montana. The exposure assessment includes generation of seepage and transport, calculation of the watershed hydrography, and identification of groundwater location, flow direction, and flow rate. The calculations are based on a storm event 50% greater than the 100-year rainfall event and include the entire watershed. It was concluded that a large rainfall event would cause the groundwater to discharge into the surface system.

The Lupin gold mine, located 400 km northeast of Yellowknife, just 80 km south of the Arctic Circle, was opened in 1982 (Wilson 1989). The area is underlaid by permafrost and has an annual open-water season of only 10 to 12 weeks. The mine/mill site was originally designed for total waste containment; however, a subsequent increase in mill throughput forced a reevaluation of that concept and an investigation into the best way of treating the water that had to be discharged. Annually, the mill produces about 175,000 $m^3$ of solids and

**Table 6.3** Lupin Gold Mine Tailings Treatment System Performance (mg•l$^{-1}$)

| Item | Mill tailings | Pond #1 | Pond #2 | Licence limits Average | Maximum |
|---|---|---|---|---|---|
| Total cyanide | 84 | 7.0 | 0.17 | 1.0 | 2.0 |
| Arsenic | 4.7 | 1.5 | 0.29 | 0.5 | 1.0 |
| Suspended solids | - | 11.1 | 8.5 | 15 | 30 |
| pH | 11 | 8.5 | 7.3 | >6 | >6 |
| Copper | 5.0 | 2.1 | 0.15 | 0.3 | 0.6 |
| Lead | < 0.005 | - | - | 0.05 | 0.1 |
| Nickel | 0.4 | 0.2 | 0.05 | 0.1 | 0.2 |
| Zinc | 20 | 1.1 | 0.11 | 0.5 | 1.0 |

Source: Wilson (1989).

675,000 m$^3$ of liquid tailings; the tailings area covers 750 ha (Wilson 1991). It was recommended that two settling ponds in series be developed for the tailings impoundment area, with a PVC liner to eliminate seepage until permafrost migrated into the structure. It was found that such a treatment facility, using only natural cyanide degradation, could reduce almost all contaminants to acceptable levels. Only ferric-ion treatment proved necessary for arsenic control. Analyses of the major contaminants, before and after treatment, confirm its efficacy (Table 6.3).

The treatment and containment of tailings are important concerns, not only during the life of the mine, but also long after decommissioning. The best long-term solution is likely to be the use of a vegetative cover to provide humus as a substrate for bacteria and to prevent erosion. The "ecological engineering" approach of Kalin et al. (1990) is directed at these ends. It uses wastewater and organic materials to help establish the growth of certain tolerant species of plants long enough to put in place a self-maintaining vegetative cover. A study by Kalin (1987) concluded that this approach could be used satisfactorily for the decommissioning of the Lupin mill tailings after the mine closes. Specifically, the tailings area would be covered with waste rock and/or esker material, sloped to prevent water ponding. The natural colonization of grasses on the esker material would be particularly useful in maintaining water quality and in providing long-term stability. No matter how promising they may seem, however, methods such as these require further long-term testing in a wide range of environments before they may be recommended for general use.

## Placer Mining

In the case of placer gold mining, the methodology of clearing away overburden and separating gold from "paydirt" was outlined in Chapter 1. It is essentially a summertime activity, but the removal of the large quantity of overburden is carried out in the spring, either hydraulically or by bulldozer. The exposed ore-carrying gravel is then allowed to thaw before sluicing operations commence (Van Kalsbeek et al. 1991). Mechanical stripping is most suitable for shallow overburden—that is, overburden that is less than three meters in depth;

it is less practicable for deeper overburden. In these latter cases, powerful streams of water are employed to strip the overburden off the surface and wash it down the valley (DIAND 1986).

No precise data could be found quantifying the overall environmental effects of placer mining. Present operations are many and small. In fact, it is little more than a cottage industry—80% use fewer than six workers and only 5% employ more than nine (LeBarge and Morison 1990)—and, as such, often escapes government inspection. The main environmental effects of placer mining are land disturbance and erosion and the disruption of riverine ecosystems. Subsequent chain effects include stream sedimentation, as well as subsidence and slumping, especially in permafrost areas (French 1976).

In Canada, most surveys of the land disturbance caused by mining tend to overlook placer mining (Marshall 1982, 1983). This is due at least partly to the scarcity of information regarding these numerous, small-scale operations. For example, in his review of waste control progress achieved by the mining industry, Horn (1966) stated: "Placer mining . . . does not pose pollution problems as such. Rather, problems of land use or land reclamation may arise. Due to the remote and, in many cases, intermittent operation of placers, further consideration would appear unnecessary." The estimated 1,740 ha of land disturbed by gold mining in general (Marshall 1982) almost certainly does not include placer mining.

Both direct discharges of water containing suspended solids, and runoff from disturbed land, can have many significant effects on local streams, including increased turbidity and reduced light penetration, channel alteration, and changes in stream-bottom characteristics that may affect water flow within the stream bed (Johnson et al. 1987). The last of these may reduce the normal discharge of a stream to groundwater, lowering the water table and resulting in increased runoff (Kelly et al. 1988). Material discharged to watercourses may contain high concentrations of toxic metals. These metals, however, are part of the suspended material that settles out and becomes sediment; at least in the short term, they pose little threat. In the long term, it is likely that the range of physical, chemical, and biological processes acting on the material over time will cause some problems (Salomons et al. 1987).

The sluicing of paydirt produces large amounts of suspended material; even after treatment, substantial amounts can be discharged to receiving waters. Greater use of water recycling provides a solution to some of the problems associated with suspended material. Johnson et al. (1987) found that recycling did not affect the quality of the water with respect to its process use, and that turbidity did not increase with the number of recycles. This indicates that a naturally occurring coagulation was present and was enhancing the performance of the settling basin. They also found that two settling ponds in series reduced the concentration of suspended solids better than a single pond, both because of a lower surface-overflow rate and because of a reduction in short-circuiting.

A study of the effects of placer mining on sub-Arctic Alaskan streams (Van Nieuwenhuyse and LaPerriere 1986) found that heavily mined streams had levels of turbidity two orders of magnitude greater than unmined streams; the former also showed commensurate reductions in light penetration and primary production by algae. In addition, heavily mined streams manifested a reduction in algal species diversity.

Arctic grayling usually occupy the highest trophic level in the food chains of sub-Arctic Alaskan streams; in heavily mined areas they have been found to disappear completely (McLeay et al. 1987; Kelly et al. 1988). The suspended sediments have effects on different aspects of their biology, including behavior, reproduction and physiology. Sediment deposition had many effects on the grayling, including the removal of refugia and interstitial places in which to lay eggs, the reduction of oxygen availability for eggs and fry, a delay in emergence, and a reduction in the number of fry. Physiological effects of suspended solids on grayling include abnormal gill development, reduction in feeding activity, downstream displacement, and color changes.

Many argue that sedimented streams will purify themselves after a time. This will very likely be the case (Kelly et al. 1988). The question is, exactly how much time is required? In a study of a wide Yukon valley conducted after the cessation of placer mining in the area, reported by Johnson et al. (1987), it was found that water quality restoration took 20 years and habitat restoration anywhere from 30 to 70 years.

Virtually the only pollution-control measure practiced by the placer-mining industry is the reduction of particulate matter in water discharged to surface watercourses. This is achieved through the use of settling ponds alone, a combination of storage areas and settling ponds, or effluent discharge to large areas of gravel or previous tailings.

Some placer mines have used mercury to remove gold by amalgamation. Up to 10% of this mercury may have escaped to the atmosphere (Sheffield 1983); however, no data are available regarding either the losses of mercury or the actual amount used.

## Summary

Lode gold and silver mining have environment effects similar to those of other hard-rock mining operations. Sulphide orebodies present problems as discussed in Chapter 5. The main risk of environmental damage from the extraction and processing of these metals comes from disruptions of streams in placer mining and from waste disposal sites at underground mines. Treatment is usually required for cyanide used in processing and/or arsenic occurring in the ores. The case histories that follow illustrate special situations, but also demonstrate problems facing both the industry and the public.

# Case Study 5. Seabee Mine[1]

The Seabee Mine is an example of a relatively small contemporary gold mine. As a new operation, it was designed to use modern extraction technology and to incorporate techniques to mitigate its effects on the environment. A 400 t•d-1 underground primary-gold mine, it began operation in late 1991. It is located 120 km northeast of LaRonge, Saskatchewan, the Northern Coniferous Ecodistrict of the Northern Boreal Ecoregion, on a small peninsula extending into Laonil Lake. The mill site is a relatively flat area of bedrock covered with a thin veneer of discontinuous glacial till. The operation occupies a total land area of 201 ha.

Native gold, containing less than one percent silver, occurs within quartz-vein material, in association with pyrite, pyrrhotite, and chalcopyrite. At the current rate of extraction of established ore reserves, the mine has a projected life of five years. Table 6.4 contains a summary of the potential environmental effects, including the type, duration, degree of significance, and likelihood of occurrence.

Two-thirds of the mill's daily requirements of 630 m3 of water are taken from Laonil Lake; the remainder is reclaimed from different processes. The proportion of reclaim water used in the mill will be increased as processing requirements allow. Mine water is first recycled underground, with any minor excess discharged to the waste-rock pile. Significant surpluses will be recycled to the mill to reduce freshwater requirements.

The mill tailings are impounded in dewatered East Lake (42 ha), and the water overflow is polished in East Pond (15 ha). Following a retention period of two to three years, excess water will be discharged by way of a natural wetland polishing area to Laonil Lake, which forms part of the Churchill River drainage basin. It is anticipated that natural degradation will be sufficient to lower cyanide levels and metal concentrations (especially those of copper) to meet regulatory release limits. Should difficulties arise, a contingency plan provides for treatment with hydrogen peroxide and ferric ion. As indicated earlier in the chapter, samples of the parent rock were found to contain more than enough acid-consuming materials to neutralize any acid likely to be formed by pyrite oxidation. Although this is not true for all portions of the orebody, the buffering capacity of the overall ore and waste rock is believed to be sufficient to eliminate acid generation as a problem. Scrubbers are used to remove particulate air emissions produced during the crushing process; emissions are being monitored to ensure the scrubbers' efficiency. A comprehensive monitoring plan has also been instituted.

The tailings-impoundment areas are large enough to hold all solid and liquid effluents for a three-year operating period. The areas are well-dyked to protect against contaminating flows caused by leakages. Barring catastrophic dam failure, there are two major sources of potential contamination of surface and groundwater: the treated tailings effluent and mine water. However, the highly fragmented nature of some of the near-surface rock indicates that direct groundwater contamination from the tailings ponds cannot be ruled out. As is the case with most lode

**Table 6.4** Potential Environmental Impacts of the Seabee Gold Mine

| Potential Issue | Degree | Likelihood | Type | Duration | Success |
|---|---|---|---|---|---|
| Treated effluent & mine water | Major | Possible | Direct | Long | High |
| Draining East Lake & using East Pond for polishing | Major | Possible | Direct | Long | High |
| Water diversion & dam construction | Minor | Certain | Direct | Medium | High |
| Sewage disposal | Minor | Possible | Direct | Medium | High |
| Withdrawal of water from Laonil Lake | Minor | Unlikely | Direct | Medium | High |
| Contaminant & spill control | Moderate | Possible | Direct | Medium | High |
| Vegetation clearance | Minor | Certain | Direct | Long | Moderate |
| Rare flora | Major | Unlikely | Direct | Long | Low |
| Wildlife disturbance | Minor | Certain | Direct | Long | Moderate |
| Winter access road & wildlife | Minor | Unlikely | Indirect | Short | High |
| Surface topography | Moderate | Possible | Direct | Medium | Moderate |
| Transport & storage of fuel & chemicals | Major | Unlikely/Possible | Direct | Medium/Long | High |
| Use of explosives | Moderate | Unlikely | Direct | Short | High |
| Atmospheric emissions | Major | Unlikely | Direct | Medium | High |
| Solid waste disposal | Minor | Possible | Direct | Medium | High |
| Traffic | Minor | Possible | Indirect | Medium | High |

Source: Beak Associates Consulting Ltd. (1990).
Notes:
  Degree
    *Minor*: low significance, if small area or population affected or occurrence natural
    *Moderate*: impact possible, but not irreparable
    *Major*: significant, if a large area or population or endangered species is affected
  Likelihood of Occurrence
    *Unlikely*: may occur but highly unlikely on basis of current information
    *Possible*: probability of occurrence not known and may depend on other factors
    *Certain*: certain to occur as a result of mining or milling activities
  Type
    *Direct*: a result of the actual physical presence of the mine or mill
    *Indirect*: secondary impact, related to increased traffic on access roads
  Duration
    *Short*: likely to last only until end of construction period or to occur infrequently
    *Medium*: likely to last the life of the mine and mill
    *Long*: likely to last even beyond the decommissioning stage

gold mines, the most likely contaminants are cyanide, metals, and arsenic, as well as ammonia blasting residues in mine water. Settling-pond discharges exhibited varying pH values; concentrations of metals such as aluminum, barium, and cadmium were well above surface water values. Runoff water from the ore storage area had low pH (3.0) and elevated metal levels.

A number of mitigative measures will be taken if monitoring reveals that effluent water quality does not meet objectives or if seepage from the mine water storage pond affects water quality in Laonil Lake. These include, in the former case, aeration and agitation to enhance natural cyanide degradation, and the possibility of further treatment, such as the addition of hydrogen peroxide. In the latter case, alternative measures include pumping mine water to the tailings pond, directly to the main wetland polishing area, or to an emergency tailings retention area.

The discharge of water drained from East Lake and East Pond to Laonil Lake is not expected to cause any problems, since all three water bodies have similar water qualities. The additions of suspended sediments, toward the end of the drainage period, were expected to produce some local habitat change for littoral and profundal benthos; this would, however, be minimized by reducing the flow rate and leaving a residual amount of water in East Lake. Although the effects on East Lake and East Pond are major, East Pond does not contain fish, and only suckers were found in East Lake.

Since the ores being mined do not require roasting for pyrite removal, the atmospheric residuals from the operation are believed to be of only minor environmental significance. The particulate emissions produced by the extraction, handling, and crushing of the ore are relatively easy to control; since scrubbers have been installed and a monitoring program put in place, this appears to be well managed. Other potential atmospheric residuals include volatile process chemicals and fugitive emissions from the mill, as well as oxidation products from waste rock and tailings. These are not expected to have a significant effect on local flora or fauna. Effects on humans are likely to be restricted to mine and mill workers, who are protected, where appropriate, by ventilation and air-cleaning systems.

The decommissioning plan includes measures to prevent surface erosion by wind or water; to reestablish the area for use by wildlife; and to make the area safe and aesthetically acceptable for human use. This will require grading, covering, and stabilizing disturbed and waste areas to encourage natural revegetation, as well as measures to manage acid generation if it becomes a problem. It is recognized that wildlife may take some years to readjust to the abandoned site after mine closure and decommissioning; the site will therefore be cleared of all structures, debris, and obstructions in order to facilitate this readjustment.

## Case Study 6. NovaGold Resources Inc.

NovaGold Resources Inc. is a small mining company based in Halifax, Nova Scotia. From 1989 to 1992, it used an experimental technology for extracting gold and silver at its Murray Brook Mine near Bathurst, New Brunswick. The process, indoor vat leaching, had been used—apparently unsuccessfully—by other operators at a small number of sites, but was considered worth pursuing because of its appeal with respect to environmental management. For instance, unlike conventional heap leaching, which is carried out in the open on lined pads that may allow some of the cyanide solution to escape into the environment, vat leaching does not discharge effluent, and drainage from tailings is collected and recycled with the process water. Indoor leaching was the preferred choice for Murray Brook because the climate is severe enough that the season during which leaching works efficiently is very short. As well, capital costs had to be kept at reasonable levels for an operation with a short life.

For the type of ore at Murray Brook, a fine disseminated low-grade ore, the process proved successful from a technical point of view, although not from a financial one. Not only were the grade and tonnage lower than had been projected by the feasibility study, but necessary design changes and other modifications to the process increased both capital costs and operating costs.

Nevertheless, the technology was supported by what amounted to a pilot plant at Murray Brook that closed in mid-1992 because the resources were exhausted. The process has been patented in the United States, and a smaller pilot plant has been constructed at the Technical University of Nova Scotia in Halifax to test ores from other properties. It has been determined that the ore at one of NovaGold's new properties in Newfoundland, Pine Cove, is amenable to treatment by this method and development work is under way at that site.

Mining is done in the summer and crushed ore is stockpiled on an impervious till pad underlaid by a drainage bed. The metallurgical process results in 85%–90% recovery of gold and about 15% of silver (a lower rate because the silver is contained in refractory ores for which the process is not suitable). Lime and cement are added to the crushed ore, along with water to produce a 13% solution that is then heated, if necessary, to summertime temperatures. The resulting agglomerates are moved to the leaching vats for curing, a step which takes only eight hours at the higher temperatures. After curing, the agglomerates are sprayed with a dilute cyanide solution with a pH of 11 (to prevent the formation of cyanide gas). Experience has shown that vat leaching allows for better recovery, partly because the spraying can be done more uniformly and without the heap method's problems of treating the edges of the heap.

The gold-bearing solution is drained from the vat and the gold and silver recovered by means of the Merrill-Crowe method (adding zinc dust and lead nitrate to precipitate the metals). The barren solution is held in tanks for reuse and the residue in the vat is washed to recover any remaining pregnant solution, then dried with compressed air and removed to the tailings storage area. The latter is a pad lined with hypalon or clay till. The tailings drain into a lined pond and the captured water is used as process water. At Murray Brook, it was possible to stack the tailings at a slope of 1:2 without major slumping or shear failure. This augurs well for the stability of the ultimate reclamation, since the slope is reduced before revegetation.

By contrast with heap leaching, which takes several months, the leach cycle in the indoor vat process takes less than 24 hours and the whole cycle approximately 72 hours. Not only are recoveries faster and better with indoor vat leaching, but the environmental implications are, on average, better than with either conventional heap leaching or agitated vat leaching. Since process water is consumed, there are no effluents to contend with, yet the method is indicated in situations where water supply is problematic. The Murray Brook site is at the headwaters of a creek that joins a stream three or four kilometers away, but the operations had no impact on the quality of local water.

Murray Brook personnel were responsible for refinements to the vat leaching process which account for its successful application and which help explain better recoveries and much faster results than with heap leaching, while precluding the environmental risks of large, outdoor heaps.

## Notes

1. Detailed information presented here was prepared by Beak Associates Consulting Ltd. and was originally published in 1990 as part of the environmental impact statement submitted by the mine operator to the Environmental Assessment Branch of Saskatchewan's Department of Environment and Public Safety.

# CHAPTER SEVEN

# URANIUM

## Introduction

While much Canadian uranium comes from sulphide ores and thus entails processes that have environmental implications similar to those discussed in Chapter 5, the production of uranium has differences that are significant enough to warrant a separate chapter. Uranium mining is unique in that its purpose is the extraction and concentration of radioactive materials that can cause short-term and long-term damage to the tissues of humans and other organisms. While decisions on opening mines or changing production for other commodities depend largely on commodity market forces, decisions on uranium production cannot depend solely on price. Demand for power-generating capacities and the end uses of uranium products for nuclear power generation and for medical and industrial purposes must also be taken into account. Because waste rock, tailings, and the waste products of military or nuclear fuel use are radioactive, design and operating decisions will have very long-lasting ramifications.

The specialized vocabulary necessary for understanding the effects of uranium mining includes the following terms:

*Radiation* is the process of emitting radiant energy in the form of waves or particles. In traversing material, radiated energy is absorbed. In the case of ionizing radiation (which is the type of radiation associated with the nuclear fuel cycle, including uranium mining), the absorption process consists in the removal of electrons from the atoms, producing ions (thus the term *ionizing radiation*). The two basic quantities in the assessment of radiation levels and effects are the *activity* of the radioactive material and the radiation *dose*.

The *activity* of a radioactive material is the number of nuclear disintegrations per unit of time. The unit for measuring activity is a *becquerel (Bq)*. One becquerel is one disintegration per second.

In this book, the term *dose* is used to mean the energy imparted per unit of biological mass.

The time in which half the atoms of a particular radioactive substance disintegrate is called its *half-life*. Measured half-lives vary from a millionth of a second to billions of years (Table 7.1).

Organisms are exposed to radiation from radioactive isotopes in gaseous, liquid, or solid form. Exposure may be through ingestion, inhalation, contact, or immersion. Types of radiation to which organisms are exposed include alpha-, beta-, and gamma-radiation:

- *Alpha (α) particles* are emitted by elements such as uranium and radium and consist of two protons and two neutrons (a helium nucleus). Because they are large, they do not usually penetrate the surface of organisms. They can be screened by thin material, such as a piece of paper. Alpha particles are usually considered to be a hazard if ingested or inhaled.

**Table 7.1** Table of Selected Istopes

| Isotope | Symbol | Half-life | Mode of Decay |
|---------|--------|-----------|---------------|
| Astatine 218 | $^{218}At$ | $\approx 2$ s | $\alpha, \beta$ |
| Bismuth 210 | $^{210}Bi$ | 5.01 d | $\beta, \alpha$ |
| Bismuth 214 | $^{214}Bi$ | 19.7 min | $\beta, \alpha$ |
| Cobalt 60 | $^{60}Co$ | $5.26 \pm 0.01$ yr | |
| Lead 210 | $^{210}Pb$ | 21 yr | $\beta, \alpha$ |
| Lead 214 | $^{214}Pb$ | 26.8 min | $\beta$ |
| Polonium 210 | $^{210}Po$ | 138.4 d | $\alpha$ |
| Polonium 214 | $^{214}Po$ | 1.64 x 10-4 s | $\alpha$ |
| Polonium 218 | $^{218}Po$ | 3.05 min | $\alpha, \beta$ |
| Radium 226 | $^{226}Ra$ | 1,600 yr | $\alpha$ |
| Radon 222 | $^{222}Rn$ | 3.823 d | $\alpha$ |
| Strontium 90 | $^{90}Sr$ | 28.1 yr | $\beta$ |
| Thallium 206 | $^{206}Tl$ | 4.19 min | $\beta$ |
| Thallium 210 | $^{210}Tl$ | 1.3 min | $\beta$ |
| Thorium 230 | $^{230}Th$ | 8.0 x 104 yr | $\alpha$ |
| Thorium 234 | $^{234}Th$ | 24.1 d | $\beta$ |
| Uranium 234 | $^{234}U$ | 2.47 x 105 yr | |
| Uranium 235 | $^{235}U$ | 7.1 x 108 yr | $\alpha$, S.F. |
| Uranium 238 | $^{238}U$ | 4.51 x 109 yr | $\alpha$, S.F. |
| Xenon 135 | $^{135}Xe$ | 9.2 hr | $\beta$ |

Observed modes of decay: $\beta^-$ = negative beta emission; $\alpha$ = alpha particle decay; and S.F. = spontaneous fission.

Source: CRC Handbook of Chemistry and Physics (58th edition).

- *Beta (β) radiation* consists of electrons and is emitted by atomic nuclei such as strontium 90. It can pass through 2 cm of water and through tissues, but can be screened by a sheet of glass.

- *Gamma (γ) radiation* is strongly penetrating. It is not produced by uranium mining.

Exposure to any kind of ionizing radiation has the potential to damage tissues. Damage may be genetic or somatic, the latter including burns, cataracts, and *inter-alia* lung cancer. In 1992, no Canadian mine or mill worker was reported to have exceeded the permissible radiation dose or exposure; the average dose received by uranium refinery workers was approximately 1.1 mSv (AECB 1992).

According to the 1988 report from the United Nations Special Committee on Effects of Atomic Radiation (UNSCEAR), the total effective radiation dose equivalent received by a human from all natural sources totals about 2 mSv·yr$^{-1}$. Medical exposures constitute the largest anthropogenic contribution to radiation; in some industrialized countries, this source approaches the dose received from natural sources. The legal public radiation limit, set in Canada as well as a number of other countries, is 5 mSv·yr$^{-1}$ from all sources combined (AECB 1990). The operating target for occupational exposure of workers in a nuclear power plant in Canada is 0.05 mSv·yr$^{-1}$. The average dose from nuclear power production, including all steps in the process, amounts to approximately 0.01% of the dosage from natural background sources (UNSCEAR 1988).

## History

The first uranium produced in Canada was the byproduct of a radium mine established at Port Radium, Northwest Territories, in 1933. The demand for uranium increased during and after World War II, when it was required both for weapons production and for use as an energy source. Uranium mining operations were then developed in Saskatchewan, Ontario, and the Northwest Territories.

The quartz-pebble-conglomerate deposits near Elliot Lake in northern Ontario have been mined since the mid-1950s (OWRC 1971). Of the more than one dozen Ontario mines that have operated since then, only four were still mining at the beginning of 1990; two of these closed permanently during the year. One of the remaining mines, owned by Denison Mines Limited, ceased operations in 1992, and the other, operated by Rio Algom Limited, is scheduled to cease operations in 1996.

Uranium production in Saskatchewan has come from two areas in the northern part of the province. The first operations were near Uranium City, in the Beaverlodge district, and were in production from 1953 to 1982. The first of three mines in the Athabasca Basin deposit, at Cluff Lake, Rabbit Lake, and Key Lake, opened in 1975, and these are the only current operations. The Rabbit Lake site closed for two years as a result of poor market conditions, but reopened in August 1991. Development of one extension of that orebody is proceeding and approval is being sought for two others. An extension at Cluff Lake (Dominique-Janine) has also been approved, and several new projects are in the assessment and permitting stages. Average grades at these new sites are considerably higher than at existing operations, ranging from 2.7% at McClean Lake to 7.7% at Cigar Lake, which has some veins grading more than 50% (Whillans 1994).

Ore grades may be expressed in terms of uranium content or, more traditionally, by their equivalent content of triuranium octoxide, $U_3O_8$, which is 85% uranium. Over 60% of the ore was extracted from underground mines in Ontario which, because of their low grade (0.08%–0.09%), produced only 10% of Canada's total uranium product. The remaining 40% of the ores, which came from higher-grade (0.6%–2.1%) mines in Saskatchewan, accounted for 90% of the total output. Processing recovery rates ranged from 94%–95% for operations in Ontario to 96%–99% among the richer Saskatchewan ores.

Uranium mining also took place at two locations in the Northwest Territories: Port Radium and Baker Lake. A high-grade deposit at Kiggavik is in the development stage.

Canada is the western world's largest producer of uranium. Fifteen to twenty percent of Canadian uranium is required for domestic use and the rest is exported. In 1993, the 22 CANDU (an acronym for Canadian Deuterium Uranium) reactors in Canada generated about 15,350 MW of electricity. Seventeen percent of Canada's electricity was nuclear generated; for Ontario, the proportion was 55% and for New Brunswick, 35% (Whillans 1994).

## Extraction and Processing of Ore

Canadian uranium production for a single year, 1992, is shown in Table 7.2, the most recent year for which data are available. It is important to note that production has been extremely variable over the years, at present occupying an interim position as higher-grade deposits were mined out in the mid-1980s, and producers look forward to mining exceptional new deposits around the turn of the century. Figure 7.1 shows the location of current and proposed operations. As Figure 2.3 indicates, most of them lie in the Boreal Shield and Boreal Plains ecozones. The Taiga Shield and Southern Arctic ecozones have also seen some activity.

Uranium occurs most commonly in nature as uranium oxides, of which pitchblende is the one most commonly mined. It is a dark, greasy-looking complex oxide that is often a uranate or uranyl. It occurs, rarely, as a primary constituent of igneous rocks. Uranium also occurs in sulphides, alone or in association with nickel and cobalt sulphides; both types of mineralization have been mined in Canada (EMR 1989b). Ontario ores tend to be high in pyrite (Kalin 1988), most of which becomes waste and is discharged with the tailings. These

**Table 7.2**  Canadian Uranium Production, 1992

| Location and Producer | Mine Type | Processing Method | Processed Ore (kt) | Ore Grade (%) | Recovery (%) | Uranium Output (t) |
|---|---|---|---|---|---|---|
| Athabasca Basin, Saskatchewan | | | | | | |
| Cluff Mining | Open pit, | Acid leaching | 119 | 0.64 | 98 | 742 |
| (Cogema Resources Inc.-100%) | Underground | Solvent extraction | | | | |
| Key Lake Joint Venture | Open pit | Two-stage acid | 263 | 2.10 | 99 | 5,452 |
| (Cameco/Uranerz-67%/33%) | | Leaching, Solvent exchange | | | | |
| Rabbit Lake Joint Venture | Open pit, | Acid Lleaching | | | | |
| (Cameco/Uranerz-67%/33%) | | Solvent extraction | 375 | 0.60 | 96 | 2,160 |
| | | | | | | |
| Elliot Lake, Ontario | | | | | | |
| Denison Mines Limited* | Underground, | Acid leaching, | 235 | 0.08 | 94 | 268 |
| | *In-situ* Leaching | Ion exchange | | | | |
| Rio Algom Limited-Stanleigh | Underground, | Acid leaching, | 914 | 0.09 | 95 | 675 |
| | *In-situ* Leaching | Ion exchange | | | | |
| | | | | | | |
| Totals/Weighted Means | | | 1,906 | 0.50 | 97.6 | 9,297 |

*This mine was permanently closed in March 1993.
Source: Godin (1995).

tailings are likely to be acid-generating. This is discussed later in this chapter as well as in Chapters 3, 4, and 5. In pitchblende, 99% of the uranium is $^{238}$U, an isotope with a slow rate of decay (Table 7.1). The remaining, desirable part is fissile $^{235}$U, which needs to be concentrated and which produces about 450 fission products, most of them radioactive.

Underground extraction has been used at most of the Ontario mines, while the Saskatchewan deposits have been mined mainly by surface methods. Some of the latter currently use a combination of surface and underground methods in order to maximize efficiency. All of Canada's uranium mills use acid-leaching processes to concentrate the mined ores to a magnesium or ammonium diuranate product, called "yellowcake" (Sirois and MacDonald 1983). Uranium in yellowcake is mostly $^{238}$U, with a small proportion of readily fissile $^{235}$U. These concentrates are then either exported or shipped to Canada's only refining and conversion facilities. Refining is done at Blind River and conversion at Port Hope, both in Ontario. The refining process changes the yellowcake into uranium trioxide ($UO_3$). This compound is subsequently converted either to uranium dioxide ($UO_2$) for use in CANDU reactors or to uranium hexafluoride ($UF_6$) for use in foreign light-water reactors (Whillans 1994).

After comminution, classification, and pre-concentration (if necessary), the ore is acid-leached, using atmospheric oxidation (in pachucas or air-lift agitators) and/or pressure oxidation, or bacterial methods. A two-stage acid-leaching process is used, for example, at Saskatchewan's Key Lake mill.

After solids have been removed by filtration or other methods, the pregnant strip solution is concentrated and purified, by means of either ion exchange or solvent extraction. Either magnesia or ammonia is used to precipitate the uranium product. Current operations in Ontario use the ion-exchange method of purification, whereas those in Saskatchewan use solvent extraction. The final step is

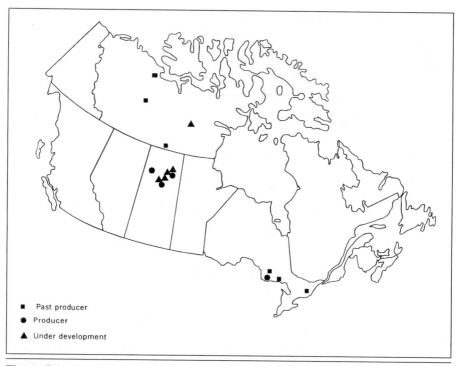

**Figure 7.1** Canadian uranium mines, 1993. Source: Whillans (1994).

drying, which has been performed conventionally, either with direct-fired multiple-hearth roasters or with calciners.

## Environmental Effects

Direct radiation generally falls outside the scope of this book because it is primarily an occupational hazard of uranium mine and mill workers. It may also, however, be a hazard to animals, plants, and humans living in areas of waste rock or tailings. The direct radiation pathway for the public may be of relative importance in some instances. The ingestion and inhalation pathways are not considered to be direct radiation pathways. The major pathway is dependent on the living habits of the critical group.

Like the production of other commodities, the mining and milling of uranium have the potential to produce residuals in the form of waste rock and tailings, liquid effluents, particulate and gaseous emissions to the atmosphere, and organic substances. Some of these can be controlled with relative ease. The direct discharge of particulate matter to the atmosphere, for example, can usually be reduced to acceptable levels by the use of dust collection devices at mines and mills and by the covering of exposed wastes and tailings (Costello et al. 1982; Bruno et al. 1984). Uranium wastes, however, have special complications. They contain radioactive materials that exacerbate the problems of waste rock and tailings disposal, particularly in ores that tend to produce acid drainage. Important radionuclides in the uranium decay chain include radium, thorium, and their decay products (see Feenstra et al. 1985). Some of the most potentially harmful of these are $^{210}$Pb, $^{210}$Po, and $^{230}$Th. Radon is a gas, the immediate radioactive decay byproduct of $^{226}$Ra. It is a hazard not only in mines, but also in parts of the Canadian Shield where houses have been built on rock that liberates it and where it can accumulate in basements. The principal hazard associated with radon is not the gas itself, but the presence of its decay products, or radon progeny. Radon progeny are radioactive solids that can be ingested through breathing and deposited directly within the lungs, where ionization damage results. Windblown dust, containing the entire range of these products, can come from the extraction stage, milling operations, or waste heaps and tailings piles.

Liquid effluents from uranium operations can contain radionuclides, sulphur compounds, metals, and organic toxins. Wastewater streams produced by uranium mining comprise drainage water from underground and surface mines, waste heaps, and tailings piles; surplus mill-process water; miscellaneous washdown water; and precipitation runoff. Mine water is primarily groundwater that has been in contact with the orebody. Unless the ore contains pyrites, it is not acidic, but does contain radionuclides. Mill waters, particularly the raffinate (the pregnant solution after uranium removal), contain both radionuclides and residual acids from the leaching process. Controlling these substances is very important to minimize their radiological and chemical effects (Lendrum 1984). The best treatment method, as regards minimization of residuals and maximization of water recycling capabilities, is usually determined by means of a pilot plant (Al-Hashimi 1989). Other wastewater varies in composition, depending on its source; however, it is usually very low in volume.

Process chemicals used in uranium milling generally include sulphuric acid and sulphates, carbonates, nitric acid, chlorides, ammonia, calcium oxide, magnesia, and sodium hydroxide. In addition, small amounts of potassium permanganate, copper sulphate, manganese dioxide, cyanides, and polyacrylamide flocculants are used. Effluents from mills that use the solvent extraction process may include alkyl-phosphate, amines, alcohols, kerosene, and fuel oil.

## Radionuclides

Since most milling processes are very selective for uranium, progeny radionuclides are present in the wastes in approximately the same concentrations as in the original ore. The acid-leach process leaves 5% of the thorium and 0.1% of the contained radium in the yellowcake (Costello et al. 1982); the remainder is discharged with the solid and liquid wastes (see Table 7.3). It is desirable either

**Table 7.3** Radionuclide Concentrations of Decant-Water Discharge and Local Surface Water at Three Saskatchewan Uranium Mines

| | Cluff Lake | | Key Lake | | Rabbit Lake | | Saskatchewan |
| | Discharge | Background | Discharge | Background | Discharge | Background | Standard |
|---|---|---|---|---|---|---|---|
| Discharge rate $(l \cdot s^{-1})$ | 45 | | 45 | | 80 | | |
| Total U $(\mu g \cdot l^{-1})$ | 341 | 0.42 | $1,000^e$ | 1.00 | 1,022 | 1.000 | 1,000 |
| $^{230}$Th $(Bq \cdot l^{-1})$ | 0.02 | 0.01 | 0.18 | 0.02 | 0.04 | 0.004 | 3.7 |
| $^{226}$Ra $(Bq \cdot l^{-1})$ | 0.056 | 0.005 | $0.37^e$ | 0.02 | 0.30 | 0.004 | 0.37 |
| $^{210}$Po $(Bq \cdot l^{-1})$ | 0.07 | 0.02 | na | na | 0.025 | 0.004 | ns |
| $^{210}$Pb $(Bq \cdot l^{-1})$ | 0.04 | 0.02 | 0.2 | 0.04 | 0.095 | 0.004 | 1.85 |

e = estimated
na = data not available
ns = no standard
Source: IEC Beak Consultants Limited (1986).

to keep radionuclides as solids or, if they have been dissolved, to precipitate them, thus reducing their concentration in surface waters.

Most of the undesired radioactive residuals end up in waste and tailings dumps, where some will continue to release radioactive emissions for thousands of years. The challenge of preventing these emissions from spreading—to air, water, adjoining land, and organisms—is being met through the development of appropriate wastewater treatment methods, tailings management practices, dust collection devices, and alternative processing strategies.

Since approximately 85% of the original radioactive material in the ore is discharged with the mill-tailings slurry (Moffett 1977), it is critical that this material be contained as completely as possible. Fortunately, the solubility of $^{226}$Ra is so low that losses to percolating water are usually minor (Merritt 1971). On the other hand, losses of solid materials by wind and water erosion may well be considerable (IEC Beak Consultants Ltd. 1986). The Ontario Water Resources Commission (1971) recognized that wind erosion of surface tailings, polluted seepage, and physical failure or erosion of dam structures have all been encountered in the past; they concluded, therefore, that proper containment of tailings is perhaps the most important aspect of the long-term control of radiological and chemical pollution from uranium mining. There has been substantial improvement in recent years in the techniques for management of these wastes; according to Knapp (1989), uranium tailings, in fact, are generally better managed than those of other mining sectors.

Analyses of decant and local stream waters from three Saskatchewan mines (Table 7.3) show that, although almost all discharges meet provincial water quality standards, they contain higher radionuclide concentrations than local stream water. Approximately 5 Mt of tailings were produced by a uranium operation in northern Saskatchewan between 1955 and 1964; part of these washed into a shallow bay that opens into Lake Athabasca (Waite et al. 1988).

Periodic runoff from uranium mill tailings continues to this day. A study of the physical, chemical, and biological characteristics of the area, initiated in 1983, found high radionuclide concentrations in the lake sediments. Periodically, as new materials enter the bay, lake water may also show high concentrations of radionuclides.

One problem common to tailings disposal strategies is movement of radionuclides and other contaminants from tailings to the surrounding environment. The potential exists for $^{226}$Ra to be leached from uranium mill tailings, contaminating the surrounding environment. Organic matter and clays play an important role in adsorbing $^{226}$Ra in the soil, primarily because of their high capacities for cation exchange. Organic matter is particularly important because it adsorbs approximately 10 times as much $^{226}$Ra as do clays (Nathwani and Phillips 1979a). As $Ca^{++}$ concentration in soil increases, adsorption of $^{226}$Ra diminishes considerably (Nathwani and Phillips 1979b).

Radon is a decay product of $^{226}$Ra, the decay products of which include $^{210}$Pb and $^{210}$Po. As described earlier, it is a highly radioactive gas. Atmospheric emissions from uranium mines in northern Saskatchewan were assessed by IEC Beak Consultants (1986), who found that radon emissions ranged from 87 to 600 TBq·yr$^{-1}$; roughly 5% of this total came from materials handling, 40% from the extraction process, and the remainder from storage piles of ore prior to processing. Radon emission is not easy to manage. During milling, 90% could be removed by activated charcoal filters. After the completion of mine operations, radon, and dust from tailings must be contained, using a permanent cover. Materials for cover include soil with a thickness in excess of 30 cm, a "wet" covering (i.e., flooding), or asphalt. Moist soil more than 30 cm thick reduces radon emissions to background levels (McCorkell and Silver 1987). Asphalt emulsion covers proved satisfactory in lab and field tests carried out by the U.S. Department of Energy and are currently used abroad. If the wastes are sintered at high temperatures, radon emission is lowered (Dreesen et al. 1984), but sintering wastes in Canada is not economical at present.

Because of their exposure to wind and rain, open pit mines produce more atmospheric particulate emissions than do underground mines. Measurements taken at the three open-pit Saskatchewan operations found particulate emissions ranging from 93–140 t·yr$^{-1}$, almost all of which was attributable to wind erosion of storage and waste piles (IEC Beak Consultants Ltd. 1986). Windblown dust can contain radioactive particles, as well as other toxic substances such as metals.

## Tailings Management

As discussed in Chapters 4 and 5, waste rock and tailings from highly sulphidic ores will continue to produce acidic drainage for many years unless oxidation is prevented. Problems of acidity and of radioactivity are linked: acid formation will lower the pH of the water and lead to the further dissolution of radionuclides, metals, and other toxic substances; even slightly acidic rainwater will contribute to this process (Constable and Snodgrass 1987). The problem is most severe for the unneutralized effluents from abandoned pyritic-tailings areas (OWRC 1971). The continued addition of neutralizing agents will alleviate the problem, but will also increase the levels of dissolved solids in the drainage system to which wastewater is disposed. In these cases, other methods to solve the pH problem must be considered; sulphide removal is an example of one alternative.

After acid neutralization and precipitation, the clarified solution is usually treated with barium chloride to remove radium. The radium co-precipitates with barium sulphate as in the following reaction:

$$BaCl_2 + Ra^{2+} + SO_4^{-2} \Leftrightarrow 2Cl + (Ba,Ra)SO_4$$

Any calcium present will be precipitated as calcium sulphate (or gypsum) and may cause the formation of scales in tanks and pipelines. Because the co-precipitate settles out slowly, the treated water must be impounded long enough to allow the concentration of suspended material to drop to acceptable levels. The system of radium removal at Key Lake is capable of reducing concentrations of $^{226}Ra$ from the 1–5 Bq•l$^{-1}$ range to 0.01–0.15 Bq•l$^{-1}$ (Scott 1987). The same mill discharges effluent that contains approximately 0.25 mg•l$^{-1}$ of arsenic and nickel, and lesser amounts of other metals. A re-design of the Rabbit Lake mill (Hopkins 1987) enabled it to reduce arsenic discharges from 2.0 to 0.5 mg•l$^{-1}$ using a ferric salt addition before lime neutralization. The subsequent BaCl$_2$ treatment lowered the concentration of $^{226}Ra$ to less than 0.1 Bq•l$^{-1}$.

In Canada, recovery by acid-leaching usually involves the use of sulphuric acid. Approximately 20 kg of H$_2$SO$_4$ per tonne of ore is required to achieve uranium dissolution rates above 90% with Elliot Lake ores (Edwards 1992). The presence of arsenic and nickel compounds in some Saskatchewan ores requires their oxidation in order to extract the uranium. This consumes relatively large quantities of the oxidant, which is usually sodium chlorate—in excess of three times the mass of U$_3$O$_8$ produced. There is frequently an economic trade-off between acid and oxidant usage. An ore with a high pyrite content may have the pyrite removed by flotation and the concentrate leached separately using auto-claves (Edwards 1992). This process yields sulphuric acid for use as a leachant and has the potential advantages of enhanced uranium recovery, lower costs, and reduced environmental effects. In the absence of acid-generating substances in the ore, as is the case with many Saskatchewan orebodies, neutralization is required only for those acids used in the leaching process. Treatment of acid wastewater has been mandatory at Ontario uranium mines since 1960.

Examples of tailings-pond wastewater analyses show uranium tailings samples with sulphate concentrations as high as 1,870 mg•l$^{-1}$. Arsenic levels were as high as 0.25 mg•l$^{-1}$, five times the recommended limit for aquatic life and public consumption. Although these residuals are discarded at present, it may eventually be feasible to recover them as valuable by-products (Sirois and MacDonald 1983).

Acid-waste neutralization causes the precipitation of several chemical constituents that are in solution at low pH, such as iron, lead, arsenic, copper, manganese, and magnesium. It also eliminates the direct biological impacts of low pH itself (see Chapter 3). Several alternatives are available for use as neutralizing agents, including lime, limestone, soda ash, and caustic soda. Since each has advantages and disadvantages, the choice must be made on the basis of site-specific conditions (Moffett 1977; Sherwood et al. 1985). In some cases, effluent contamination may be lowered by the use of alternative leachants, such as hydrogen peroxide or Caro's acid (Edwards 1992).

Uranium ores may contain potentially significant amounts not only of radium and thorium, but also of non-radioactive elements such as rare earths, nickel, iron, copper, vanadium, molybdenum, and lead. Arsenic and fluorine are also common contaminants. The Collins Bay uranium ores of northern Saskatchewan are particularly contaminated by arsenic and nickel, requiring special consideration both in mill design and for the handling of mill effluents (Hopkins 1987). The Rabbit Lake mill removes impurities such as zirconium, nickel, and arsenic from the pregnant strip solution, prior to uranium precipitation, and disposes of them with the tailings (Cameco 1990). The development of chemical precipitation techniques for the removal of arsenic and nickel was a major improvement in tailings management technology.

The long-term containment of uranium tailings now includes using natural and synthetic liners for basins and tailings dams; limiting the access of surface water to tailings areas in order to reduce leaching; immobilizing the surface of tailings, primarily by vegetation; reducing the concentrations of radionuclides in wastes before disposal; and adequately considering surficial and sub-surface geology of disposal areas, with respect to long-term waste containment (Haw 1992).

The importance of the long-term stability of tailings impoundments is emphasized by the fact that the half-life of $^{226}$Ra is 1,600 years. Proper planning of tailings-disposal sites has become one of the most important aspects of development plans for new uranium mines—for both the developer and the regulator. By the mid-1970s, these plans typically included site-selection studies, hydrogeological investigations, acid-generation evaluations, detailed assessment of environmental impacts, and the design of systems for seepage collection and control (Knapp 1989). The 1970s also saw the development of high-grade deposits in northern Saskatchewan, bringing with it new challenges in tailings management. Some of the improvements initiated during this period were discharge methods, such as stacking and coning, designed to maximize surface water drainage of the tailings; the use of liners with cutoffs to bedrock or low-permeable soil; grouting of fault zones; chemical treatment of tailings-pond discharges; and the use of groundwater monitoring wells and seepage control systems.

Since the 1970s, several new disposal concepts have been developed. The first concept was a highly "competent" containment system for tailings with a

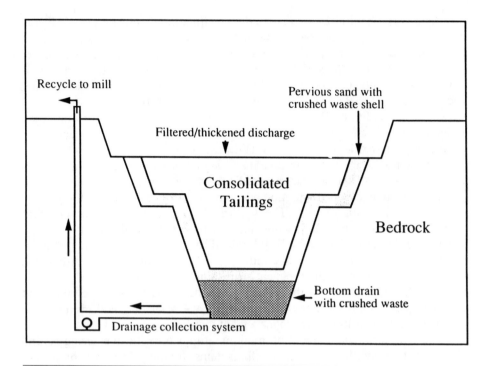

**Figure 7.2** Collins Bay "B" pit disposal concept. Source: Knapp (1989).

large chemical-precipitate fraction that would not drain properly. The solution was to place the tailings in layers in a lined above-ground containment area and to use surface evaporation and an under-drainage system for dewatering (Knapp 1989). The second concept was directed at long-term containment, utilizing below-ground disposal in a mined-out open pit at Rabbit Lake. A pervious layer of crushed rock was placed under the tailings, on top of an impervious barrier, so as to collect drainage through the material and to isolate it from groundwater flow (Cameco 1990). Other improvements have included the development of chemical-precipitation techniques for the removal of arsenic and nickel, and the use of a system of ion-exchange, mechanical thickeners, and sand filters for radium removal (Edwards 1992). In addition, soil covers on uranium mill tailings can reduce above-ground gamma radiation exposure to background levels.

Figure 7.2 depicts an example of a modern disposal concept. It places consolidated tailings inside a pervious shell, which provides a pathway to facilitate drainage collection. Ideally, the pit should be located below ground level and should eventually be covered by a lake.

The future of tailings disposal is likely to see even more emphasis on long-term containment through the use of sub-surface burial, underwater or deep-water disposal (Davé and Vivurka 1994), and the use of covers both to prevent surface erosion and to permit revegetation (Robertson 1986a). In northern areas, burial in permafrost and the application of freezing techniques are likely to be investigated (Knapp 1989). Risk assessment in such areas should include potential climatic changes.

## Effects on Organisms

Higher order animals can ingest or inhale radionuclides directly, and their intake may also be increased by passage of radionuclides through a food chain. At the bottom of food chains are green plants, including algae, mosses, lichens, and higher order plants. Terrestrial plants generally get radionuclides from the air, whereas algae take them from water. Some plants are exposed through their roots and also from the air. Dustfall from a tailings pile in the Uranium City area of northern Saskatchewan was measured using moss bags as sample collectors (Archibold 1985). Because mosses have no roots, their metals contents are derived only from wet or dry fallout. Eight metals were deposited at rates exceeding $100 \text{ mg} \cdot \text{m}^{-2} \cdot \text{d}^{-1}$. Deposition decreased very rapidly with distance from the tailings, because the particles were large—the size of sand. Deposition was lower during rainy periods. It was considered, partly because of the length of the snow-covered season, that the tailings had a low susceptibility to wind erosion.

An example of bioconcentration of $^{226}$Ra is found in low-bush blueberries growing on the Stanrock tailings area near Elliot Lake. They contained 20–290 $\text{mBq} \cdot \text{g}^{-1}$, compared to a background level of 2–5 $\text{mBq} \cdot \text{g}^{-1}$ (Davé et al. 1985). To put this concentration in perspective, a person would have to eat 160 kg of washed blueberries in order to exceed the annual limit recommended by the International Commission on Radiological Protection.

Eating berries is a single-step food chain. Bioaccumulation effects are more likely in a chain with several links. Snowshoe hares are herbivores; concentrations of $^{226}$Ra in their faeces, collected near uranium tailings, averaged 203 $\text{mBq} \cdot \text{g}^{-1}$, implying that they had a bioconcentration factor of 0.03 (Clulow et al. 1986). Three-step food chains of lichens→reindeer→humans have shown transfer of radionuclides (Schultz and Whicker 1974). The transfer of $^{210}$Po and $^{210}$Pb from lichens to caribou to wolves has been estimated in an Arctic tundra region and a sub-Arctic forest region. The tundra site was Baker Lake, close to a uranium exploration camp (Thomas et al. 1994). In the tundra, the main lichen food of the caribou was *Cetraria nivalis* and in the forest region it was *Cladina mitis*. Radioactivity of the two elements differed both between lichen species and between sites. Both elements were more active in caribou bone than wolf bone, while $^{210}$Po activity in liver, kidney, and muscles was similar in wolves and caribou. Wolves retain more $^{210}$Po and less $^{210}$Pb from their diet than do caribou. There was a clear decline in activity of both elements with increasing trophic levels, and therefore biominification. Average concentration ratios from caribou to wolves were 0.40 for $^{210}$Po and 0.13 for $^{210}$Pb. The simplest way of considering the transfer is that for every 1,000 Bq of $^{210}$Pb in dry lichens, 94 Bq get into caribou and 12 Bq into wolves.

Thomas et al. (1994) estimated that caribou take up considerably more $^{210}$Po from lichens than cattle do from their feed. Differences occur not only in ability to take up the element, but in transfers through the gut wall and in the ability to excrete radionuclides. They suggest that the currently used absorption estimates for $^{210}$Po in humans are possibly underestimated at 10%. The transfer from caribou to wolves is 40%, and people living in the Arctic may have a value higher than 10%.

Radioactive elements get into organisms using the transport mechanisms of nutrient ions. For example, $^{226}$Ra can usurp the pathways of calcium, then continue to emit alpha particles within tissues where calcium would ordinarily accumulate, such as bones. Moffett and Tellier (1977) examined radioisotope uptake by studying four grass species that had been growing for years on uranium tailings in the Elliot Lake area. Sedimentation in tailings had produced separate zones of sand and slimes. The slimes contained significantly more $^{226}$Ra, $^{210}$Pb, and $^{210}$Po than did the sands; the mean $^{226}$Ra in tailings was 203 mBq·g$^{-1}$. There was no difference in radionuclides concentrations in grasses growing on the two substrate types. Levels of radionuclide were significantly higher in grasses on tailings compared to those in a control area.

Plants growing on revegetated uranium mill tailings in South Dakota (Rumble and Bjugstad 1986) had a $^{226}$Ra concentration ratio of only 0.026, which suggested that, at least in the short term, some plants there exclude radium and, therefore, that plant uptake might not be a significant pathway for the spread of radionuclides from mill tailings to the adjacent environment. Ibrahim and Whicker (1988) determined concentrations of $^{226}$Ra in soil, tailings, and native vegetation around an active uranium mine/mill in Wyoming. The highest vegetation concentration (1.3 Bq·g$^{-1}$) was found in a grass species growing on exposed weathered tailings. Few differences were found among plant species. The study speculated that the sulphuric-acid leaching process used at this mill sequestered $^{226}$Ra in a sulphate form relatively less soluble than the sulphates of other elements, such as uranium and thorium. The result was probably a reduced availability of radium to plants.

The radium concentration in leaves of plants grown on freshly contaminated tailings substrates may be high, but over time the concentrations drop, partially through decreases in soil and plant radium during leaching by precipitation. The equilibrium concentration of radium in vegetation growing on mill tailings can be calculated from mathematical models incorporating these transfers (Simon and Deming 1986). The concentration of radium in the leaves of mature plants was found to decrease exponentially to a constant value, as the soil-plant system came to equilibrium after experimental additions of radium-enriched solutions.

An "ecological engineering" approach is being used in the Rabbit Lake drainage basin (Kalin 1990), the aim being to immobilize the radium in solid form. Algae, discovered thriving in local water bodies, are cultivated to increase their innate capability to remove radium from surface water and sequester it in bottom sediments (Vandergaast et al. 1988).

Since uranium mining in northern Ontario began in 1953, following the discovery of uranium-bearing ore in the Serpent River Basin, the number of operating mines in the Basin has fluctuated, depending on uranium market conditions. The Ontario Water Resources Commission's 1971 report indicated serious radiological and chemical pollution throughout the Basin. Radium was present in some lakes at concentrations 50 to 200 times normal background levels. The study also showed both high acidity and concentrations of dissolved solids, sulphate, and ammonia. Populations of plankton, fish, and bottom fauna were severely reduced. Although waste disposal practices were improved between 1960 and 1970, levels of $^{226}$Ra continued to exceed acceptable standards in 1975, due in part to contaminated drainage from abandoned tailings and spillage. Low pH values were probably caused by oxidation of the large quantity of sulphides in the tailings. Potentially toxic levels of ammonia continued to be a problem as late as 1975 (Marshall 1982). John (1987) reported that corrective action has been taken, including neutralization of the tailings, more judicious water use, and the precipitation of radium with barium sulphate. New tailings dams were engineered with cores of clay or of plastic; seepage from older facilities was collected and neutralized. John reports that "the Serpent River System is now virtually restored, and the area is being restocked with fish."

For effects of uranium mines on aquatic biota, see the Athabasca Basin case study at the end of this chapter.

## Reclamation of Impoundments

Responsibility for the decommissioning of tailings is the responsibility of the uranium producer. In Canada, the regulation of these sites falls under federal jurisdiction.

The National Technical Planning Group on Uranium Tailings Research, sponsored by CANMET, was formed in 1980 to review and plan future research on tailings management after the shutdown of uranium mines and mills. Their report (EMR 1981) identified the following environmental problems associated with shutdown: surface and groundwater are contaminated by seepage; radium is redissolved from barium/radium/sulphate sludge left in treatment ponds; tailings impoundment dams fail; tailings are removed for outside use; the surfaces of tailings ponds are sources of radiation and radon gas emission; and soil and vegetation are contaminated by windblown radioactive tailings dust.

When surface storage is unavoidable, the exposed surface area should be minimized. Covering tailings or wastes with vegetation can be desirable in some cases, because the roots hold the surface and prevent erosion, the plants improve the microclimate, and the beginnings of natural nutrient cycling are established. In the case of uranium tailings, most experts recommend revegetation (Haw 1982; Robertson 1986). However, there are cautions to be observed, because of potential transfers of radioactive material by vegetation. In the first place, vegetation may bring radioactive substances out of the substrate via the roots, making them available above ground for herbivores. Secondly, plants may accumulate airborne radioactive contaminants, again making them available to herbivores. Thirdly, the leaves may release radon, which is sufficiently soluble to move up the plant with the transpiration stream (Lewis and MacDonell 1990). The most important factor controlling radon release appears to be the leaf area produced per unit area by a plant (its leaf area index). Species now being recommended for planting are therefore those that have shallow roots and small leaves, and which, if possible, are not palatable to herbivores.

The objective of revegetating and reclaiming uranium tailings is both to restore the ecosystem and to ensure that it is stable over the long term. Stability requires the development of ecosystem processes such as the decomposition of organic matter and the cycling of nutrients, both of which are functions performed by soil flora and fauna. On the positive side, at Elliot Lake, McCready (1976) found that the numbers and diversity of soil microorganisms had improved four years after revegetation. On the negative side, however, nodule development on the roots of legumes—which indicates nitrogen-fixing capacity—was poor in plants growing on uranium tailings (Murray and Moffett 1977). Long-term stability of introduced vegetation on the Elliot Lake tailings remains to be proven; however, results today appear to be favorable.

Three types of vegetation have been described on Ontario uranium tailings: native plants in wetlands, native plants in dry areas, and mixed native and cultivated plants in some sites. The wetland plants included cattails, rushes, and sedges (Kalin 1983). Plants in dry areas included trembling aspen and paper birch. At Elliot Lake and Bancroft, 110 species of vascular plants were identified; seeded vegetation occupied 13% of the tailings, colonizing native plants 33%, and a mixture of native and seeded species the rest. Woody plants have mycorrhizal fungal symbionts in or on their roots. One of the problems of establishing woody plants on tailings is often the lack of fungi. On naturally vegetated uranium mill tailings in Ontario, 13 species of macrofungi were found, possibly capable of mycorrhizal association with trees (Kalin and Stokes 1981).

Tailings at Elliot Lake were the site for revegetation experiments by CANMET (Dunn et al. 1972; Murray 1972). Problems included unstable surfaces, infertility, high surface reflectivity, low cation exchange capacity, and low pH. Successful seedings included creeping red fescue, Kentucky bluegrass, redtop, reed canary grass, tall fescue, timothy, red clover, birdsfoot trefoil, and alfalfa. The cost of establishing such crops—and a mixture was recommended—was $6,250 per hectare, in 1977 dollars. CANMET tested coniferous tree species as well as crop plants and found that Scots pine, red pine, and jack pine all had better survival rates after four years than had either white spruce or white cedar (Murray 1978).

At Uranium City, Saskatchewan, vegetation was established to reduce erosion from tailings. Swanson and Abouguendia (1981) considered it more difficult to establish vegetation in this part of Saskatchewan than at Elliot Lake in Ontario. Problems include saline soil, a short growing season, and drought. Forage species would require fertilizers. The Uranium City tailings, however, have some native species growing on them. It has been suggested that species unpalatable to grazers or browsers would be most suitable, in order to limit food chain transfers of radionuclides (Swanson and Abouguendia 1981). Native plants with potential for growing on such sites were listed by Redmann and Frankling (1982).

## Research

Bacterial leaching of uranium ores *in situ* avoids some problems of waste chemistry. *Thiobacillus ferrooxidans* was used as a leaching agent in Ontario for many years (OWRC 1971) and more recently in the United States (McCready 1986a; Edwards 1992). For 20 years, CANMET has had a program in place to research microbial leaching and has isolated a *Penicillium* type of fungus capable of accumulating up to 25% of its biomass in uranium (McCready 1986b). A combination of these fungi and *Thiobacillus ferrooxidans* may be capable of extracting uranium from low-grade ores without creating major problems of waste disposal. However, bacterial leaching of uranium is not economical at present.

The most effective, although not the most economical, way of reducing the environmental effects of uranium mining would be to eliminate all of the potentially contaminating components of the tailings materials. Recent research on the Cluff Lake and Key Lake ores of northern Saskatchewan has evidenced some success in removing residual $^{230}$Th and $^{226}$Ra simultaneously, by using nitric and hydrochloric acids (Muthuswami et al. 1989). Extraction rates greater than 96% were obtained with nitric acid; similar rates were obtained with hydrochloric acid only after the addition of an oxidant such as sodium citrate. Such extraction is closely linked to the economic benefits of recovering by-products such as thorium and radium (Edwards 1992).

## Monitoring

The environmental risks associated with uranium tailings, unless they are adequately contained, will last many centuries. A failure in the containment system itself, or in the associated water-treatment facility, is likely to result in the release of radioactive material to the environment. A theoretical methodology was developed under the National Uranium Tailings Program to estimate the long-term risk associated with uranium mill tailings (Murray et al. 1987). Three questions were asked: What can go wrong? How likely is it to do so? What will the consequences be if it does?

Using this approach, a methodology was developed to calculate the probabilities of occurrence of various radiation dosages as they relate to specific problems; e.g., excessive rainfall, an earth tremor that might lead to tailings-dam failure, or drought stress that might make vegetation more susceptible to damage or provide less protection for animals.

A monitoring operation directed specifically at existing uranium mine wastes (Feenstra et al. 1985) is the aforementioned National Uranium Tailings Program. It identified the principal environmental concerns as surface water runoff from tailings impoundments; seepage from tailings into surface watercourses; exfiltration of porewater from tailings into groundwater; wind erosion from tailings material; and radon exhalation from tailings. Monitoring data include physical, chemical, and radiological characteristics of the tailings solids; chemical and radiological characteristics of drainage and adjacent surface waters; amounts of windblown dust; and concentrations of atmospheric radon. The main objective of the Program is to provide a scientifically credible database to assist regulatory agencies in establishing criteria for the long-term protection of the environment and of human health.

Another monitoring program for uranium mines has been in operation for many years (W & W Radiological and Environmental Consultant Services Inc. 1978). Emphasis has been both on radioactive residuals from operating mines and mills and on decommissioned tailings areas. The program includes air sampling for radon, particulates, and radiation dosage, aquatic sampling for radionuclides in effluents, receiving bodies, and sediments, and biotic sampling in leaves and fish. Reliable instruments to measure many of these parameters, especially radon gas, have become available only in recent years; not all of these parameters are monitored at every site.

Radioactivity in air has been measured routinely at 28 environmental network stations in Canada since the late 1950s (Environment Canada 1991).

Monitoring requirements for the Cluff Lake, Key Lake, and Rabbit Lake uranium mines in Saskatchewan are specified in each mine's "Permit to Operate Sewage or Industrial Effluent Works" (Dirschl et al. 1992). These facilities are licensed by a federal agency, the Atomic Energy Control Board (AECB), and monitoring requirements for radiation are included in their AECB license. Monitoring programs involve periodic measurements of water, sediment, and air quality, aquatic biota, soils and lichens, radon gas, and potable-water-supply systems.

In monitoring air, soil, or water, decisions must be made about how frequently to sample. Alternatively, an integrated sample may be used by analyzing some organism that spends its life in the conditions being monitored. Mosses and lichen are often used to trap air contaminants, because they reflect atmospheric fallout rather than substrate chemistry (Nieboer and Richardson 1981). In northern

**Table 7.4** Accumulation of $^{210}$Pb, $^{210}$Po, $^{226}$Ra and U in Plants from Northern Saskatchewan

| | $^{210}$Pb | $^{210}$Po | $^{226}$Ra | U |
|---|---|---|---|---|
| | | (mBq·g$^{-1}$) | | ($\mu$g·g$^{-1}$) |
| Trees | 0.40 | 0.30 | 3.49 | 0.155 |
| Shrubs | 0.90 | 0.51 | 18.90 | 0.148 |
| Lichens | 4.20 | 2.91 | 4.25 | 1.410 |
| Mosses | 5.40 | 3.67 | 7.67 | 1.570 |

Source: Sheard (1986).

Saskatchewan, for example, as is shown in Table 7.4, lichens and mosses accumulate higher levels of $^{210}$Pb, $^{210}$Po, and U than either shrubs or coniferous trees (Sheard 1986). The zone in which mosses and lichens accumulate radionuclides intensely at Elliot Lake has a radius of 8 km. Their total contaminated zone has a radius of 22 km (Beckett et al. 1982). Tree bark can also be used for monitoring, but is less easy to sample (Brownridge 1985).

Some uranium mines are sufficiently recent in origin to have been licensed after an environmental assessment process. The Cluff Lake project commenced in 1983, after meeting the requirements of Saskatchewan's Environmental Impact Assessment process. It has now operated long enough for valuable research to have been carried out (e.g., Hynes 1990), and comparisons of recent monitoring data with forecasts from the original Environmental Impact Statement are complete (Swanson 1990).

Swanson's observation is that the present database is rich in details for isolated parts of the ecosystem (e.g., water quality), but it lacks total integration. She found that diversity indices for aquatic invertebrates in Cluff Lake were one of the most valuable indicators of change. Data about terrestrial animals and rare plants were insufficient for trends to be determined. Swanson commented on the difficulty of using toxicity tests that are done on species that do not occur in this boreal region. One of the surprising findings of this follow-up work has been that the radionuclides may have been overemphasized in importance, and more mundane contaminants, causing acidity and salination, have been underemphasized.

Decommissioning has the aim of returning the minesite to a condition as close as possible to its original state, including downstream water quality (Dirschl et al. 1992). To ensure this, monitoring has to be conducted over the long term.

## Summary

The amount of land affected by uranium mining in Canada is small. Problems of acid drainage and toxic leachates or other toxic reagents are fairly small as well, at least in comparison to, for example, the sulphide ores of copper and nickel. Uranium mining, however, disturbs or concentrates radioactive minerals, exposing radioactive wastes at the surface where they are in contact with air and water. The main environmental problem faced by uranium mines is, therefore, long-term disposal of radioactive wastes, which may be gaseous, liquid, or solid. Because of their harmful effects on organisms, radioactive wastes will continue to require monitoring, and possibly treatment, until their activity matches background levels. Potential food chain transfers can detract from the value of vegetation in stabilizing wastes.

## Case Study 7. Athabasca Basin, Saskatchewan

The uranium industry has eliminated approximately 100 ha of natural boreal forest in northern Saskatchewan through its mining and milling activities (Dirschl et al. 1992). Considering the size of the Athabasca Basin, the affected area is relatively small, but would be increased substantially if the indirect effects of road construction and associated commercial development in the North were included.

The uranium-producing area in the Athabasca Basin has been the site of 25 mines, which produced large quantities of ore, and six mills (Saskatchewan Environment 1991). Decommissioning is in progress or has been completed for nineteen of these mines. The Beaverlodge mill and tailings are in the monitoring phase, following full decommissioning, and appear to be stabilizing. Mines and mills at Cluff Lake, Key Lake, and Rabbit Lake are still operational.

The Gunnar and Lorado mills tailings were deposited directly into natural basins without pre-treatment or pre-disposal site preparations (Swanson and Abouguendia 1981). Neither stabilization nor control measures have been instituted to date, and tailings material is subject to transport by air, surface water, and groundwater. The tailings affect the terrestrial environment through wind transport off the sites, which continuously increases the area of contamination; bioaccumulation of radionuclides and heavy metals by plants growing on tailings or on adjacent contaminated areas, and subsequent transfer through terrestrial food chains (through moss and caribou, for example); and elevated gamma radiation. Aquatic effects include $^{226}$Ra concentration above permissible levels, reduction in pH near pyritic tailings, increases in sulphates and nitrogen compounds with potential ammonia toxicity at pH greater than 7, elevated levels of toxic metals, especially under acidic conditions, and decreased diversity of fish, macrophytes, and plankton (Swanson and Abouguendia 1981).

The Lorado mill deposited tailings at pH 2 directly into Nero Lake, which is connected to the much larger Beaverlodge Lake. Most of the tailings are underwater, although they also cover an 8 ha area of land. A plume of aluminum precipitate high in uranium was observed where Nero Lake water flows into Beaverlodge Lake (Whiting et al. 1982). Tailings at the Lorado mill site are acidic and moderately saline, with elevated levels of sulphate and low levels of available nitrogen, phosphorus, and potassium (Frankling 1984). Tailings have high concentrations of potentially toxic metals, including lead, nickel, uranium, and vanadium. Areas vegetated naturally by water sedge (*Carex aquatilis*) have higher capacities for holding water and lower salinities than do unvegetated tailings.

Severe acidification, accompanied by high levels of metal and radionuclides, has severely affected Nero Lake. Fish are absent and the abundance and diversity of plankton greatly reduced. The macrophyte community is made up of acid-tolerant mosses. Long-term seepage from Nero Lake may eventually contribute to the pollution of Beaverlodge Lake.

In a 1981 study of Nero Lake, Kalin (1982) observed that pH levels appeared stable. She also noted that the submerged tailings were covered by an aquatic moss carpet, which isolates them from the oxygenating water above and prevents further acidification of the lake. The Nero Lake system appeared to have reached a steady state balance between tailings, moss carpet, and lake water. Kalin cautioned that disturbance of the exposed tailings during any remediation and revegetation could lead to destabilization of the lake system.

In another study, Kalin (1988) determined that $^{226}$Ra in the Gunnar tailings averaged 2,553 mBq·g-$^1$, ranging from less than 37 to 8,806 mBq·g$^{-1}$. Values for $^{210}$Po were similar. Mean uranium concentrations were 135 mg·g$^{-1}$, with a range of 7 to 1,108 mg·g$^{-1}$. Although average values were higher than in control sites away from the tailings, there was significant overlap in the range of values, indicating that radionuclide concentrations in surface tailings are low, despite their location. Radionuclide concentrations were higher in the Uranium City tailings than in those from Elliot Lake, Ontario.

Tailings management at currently active uranium mills in northern Saskatchewan takes a longer-term view than in the past. Current tailings design must ensure that water quality downstream is maintained after decommissioning. Five to ten years of monitoring are usually required to ensure that the objectives are met. Conventional tailings management, in which tailings are deposited in natural depressions confined by dams, has been used in the Cluff Lake and Rabbit Lake regions. The Key Lake mill uses a hybrid facility, which amounts to a pervious-surround system installed above ground. It is intended to virtually immobilize chemical contaminants within the deposit (Dirschl et al. 1992).

Some waterbodies near current uranium operations have been damaged by increased salinity and metal contamination (Dirschl et al. 1992). In particular, sedimentation of watercourses has adversely affected the bottom fauna and rooting zone for aquatic macrophytes. There is evidence of lowering of aquatic macrophyte productivity and species replacement by more salt-tolerant types, but there has been no systematic monitoring of macrophytes or of aquatic biota other than fish.

Northern pike and common whitefish in Wollaston Lake are monitored periodically, but do not show any signs of contamination to levels that would render them unfit for human consumption (Dirschl et al. 1992). The longer-term potential for radiation damage to fish has been evaluated in the Uranium City area where the Gunnar mine released tailings directly into Langley Bay. The lower concentrations in the latter fish were attributed to their periodic excursions to less affected areas of the lake (Waite et al. 1988). Waite et al. (1990) were unable to find physiological differences in contaminated northern pike and whitefish in Langley Bay, by comparison with a control group. Fish from lakes in the Uranium City area affected by an operating uranium mine and mill had radionuclide levels one to two orders of magnitude above those from an uncontaminated lake (Swanson 1983). The primary pathway of radionuclide transfer appeared to be through sediments; bottom-feeding fish had the highest levels of radionuclides. The internal dose rates were low; Swanson and Bernstein (1984) also found no apparent physiological effects on fish from the Beaverlodge area near Uranium City.

Concern about possible effects of elevated levels of radionuclides in fish from Beaverlodge Lake led to a study of lake whitefish, white suckers and lake trout. The fish had elevated levels of U, $^{226}$Ra, $^{210}$Pb, and $^{210}$Po. They were found not to have tumors, lesions, or abnormal levels of parasites. They did, however, have significant differences in blood parameters when compared with the same species in neighboring unaffected lakes. The authors suggested that little is known of long-term effects of alpha and beta radiation in comparison with gamma radiation, and that further research was necessary (Bernstein and Swanson 1989).

Intensive monitoring for sulphur dioxide and radon and its daughters, $^{210}$Pb and $^{210}$Po, is carried out close to the mine/mill complexes in northern Saskatchewan. However, patterns of long-range transport of these contaminants are unknown. Dirschl et al. (1992) recommended the introduction of a region-wide air quality monitoring program, in view of the potential in the area for increased uranium mining.

The long-range transport of radionuclides—as well as the extent to which these substances contaminate human food and affect human health—is at the core of much of the public concern about the uranium mining industry in northern Saskatchewan. A report to Environment Canada and to the Atomic Energy Control Board (IEC Beak Consultants 1986) concluded that the combined regional radiological impact from three operating mines in northern Saskatchewan is negligible. Radiological effects on biota were judged insignificant in all cases. The authors of the report, however, cautioned that the results "do not imply that uranium developments will or will not have significant long-term radiological impact on northern Saskatchewan."

Dirschl et al. (1992), in their review for the Joint Federal/Provincial Panel on Uranium Mining Development in Northern Saskatchewan, noted that the lack of long-term empirical data in radionuclide transfers remains a serious information gap that has yet to be filled. They suggested study of the transfer of radioactive and other chemotoxic contaminants along three food chains: from plankton and bottom fauna to fish to humans; from vegetation to moose and humans; and from lichens to caribou to humans.

In some respects these gaps have been filled. Thomas et al. (1994) have studied the third pathway, using wolves as a substitute for people. The food chain effect was biominification, but the authors concluded that the percentage transfer from caribou to people was probably higher than risks for southern populations eating beef.

The first pathway is paralleled by a study in Saskatchewan by Swanson (1985), who studied sediments, water, insects, forage fish, lake whitefish, and white suckers from Beaverlodge Lake, which is known to receive effluent from the Eldorado Nuclear tailings and drainage from the abandoned Lorado mill. Three uncontaminated neighboring bodies of water were also sampled for total U, $^{226}$Ra, and $^{210}$Pb. The radionuclides were significantly elevated in water, sediment, and biota in Beaverlodge Lake. Again, the food chain effect was biominification, and major transfers were from water to insects and from water to fish. Uptake of U from water by fish and insects was less than uptake of $^{226}$Ra and $^{210}$Pb. Fish feeding near the sediments took up more radionuclides than did pelagic fish. The critical pathway was considered to be sediment to insects, to forage fish, to whitefish, to man. The large fish had an internal radiation dose estimated at 1 to 2 rad•yr$^{-1}$. One serving of fish per week for a year would provide a dose of two percent of the annual limit set by the International Commission of Radiological Protection (Swanson 1985).

# CHAPTER EIGHT

# IRON ORE

## Introduction

Global production of iron ore is almost 1,000 Mt per year, of which Canada typically accounts for about four percent (Franz et al. 1986; Boyd 1992). This places Canada seventh among producers, after the former Soviet Union, Brazil, Australia, China, India, and the United States. Despite a downward trend in production over the past decade, iron ore is still Canada's sixth most valuable non-petroleum mineral, as Table I.1 indicates.

At the end of 1992, there were only four major iron ore mining operations in Canada, compared with 19 in 1974 (Ripley et al. 1982; Boyd 1993). The three operations accounting for most of the production were all open pit mines located in the Labrador Trough of northern Québec and Labrador. The ores they mined were primarily specular hematite and magnetite, as well as a small amount of goethite and limonite (see section on Orebodies). Except for a very small fraction that was direct-shipped as ore, approximately two-thirds of the ores were marketed in the form of pellets, and the rest was marketed as concentrate. The remaining underground mine, located near Wawa in northwestern Ontario, produced, for the owner's steel operations in nearby Sault Ste. Marie, a small amount of sintered iron ore from siderite (EMR 1992). British Columbia also produces iron concentrate as a by-product of the mining of other metals, and a plant near Havre St. Pierre, Québec, makes pig-iron from ilmenite mined primarily for titanium (Boyd 1993).

## History

Canada's first iron ore mine was established in 1738 near Trois-Rivières, Québec, to exploit bog iron deposits. In 1820, when technology had emerged to mine magnetite, the deposits near Marmora, Ontario, were developed. By 1886, there were 13 mines in operation, distributed across Nova Scotia, Québec,

Ontario, and British Columbia (Boyd and Campeau 1988). Growth of the industry was quite slow until 1954, when development of mines in the Labrador-Québec border region began. Subsequent expansion was rapid: by 1982, there were ten mines in operation, and production that year was 34.5 Mt (EMR 1984).

In the early days of iron-mining in Canada, ores were of sufficiently high grade that they did not require concentration before smelting. These "direct-shipping" ores contained at least 65% iron; pure hematite and magnetite, by contrast, contain 70%–72% iron. Most known Canadian high-grade ores have now been exhausted. Ores currently mined in Canada have contents of iron between 23% and 40%, with a mean value of about 32% (Sirois and MacDonald 1983; Boyd and Johnson 1991).

## Orebodies

Comprising more than five percent of the Earth's crust, by weight, iron occurs in many forms (although only rarely in its native state), of which eight are suitable for mining:

- the *oxide ores*: hematite, magnetite, goethite and ilmenite

- the *carbonate ore*: siderite

- the *silicate ore*: chamosite, and

- the *sulphide ores*: pyrite and pyrrhotite.

In Canada, more than 90% of metallic-iron production derives from hematite, magnetite, and siderite ores. The balance comes from ilmenite as a by-product of titanium mining and from polymetallic sulphide ores as by-products of copper and nickel mining (Boyd 1992; EMR 1992). The deposits being mined are essentially specular hematite with mixtures of granular quartz and magnetite. Conventional open pit methods are used to provide the feed for nearby beneficiation plants. After concentration, most of the ore is pelletized before shipping.

Since the 1990 closure of two mines in the Kirkland Lake region of Ontario, virtually all of the country's iron ore production has been derived from the Labrador geosyncline. Other potential sources of iron ore are the Precambrian Shield just north of the Great Lakes, and the Pacific coast. A very small amount of iron has also been produced as a by-product of sulphide base metal operations in Ontario (Sirois and MacDonald 1983). This operation closed in the mid-1980s; current shipments are from stockpiles.

## Extraction and Processing of Ore

The quantity of iron ore extracted is only slightly less than that of sulphide ores. After extraction, Canadian ore is crushed, ground, and concentrated in the ratio, on average, of 2.5:1; in other words, the weight of the mill feed is 2.5 times that of the concentrate. Somewhat more than half of the ore is waste rock, and about the same quantity of tailings is produced by the concentration stage. Iron ore processing produces about 20% of the total quantity of tailings from Canadian mineral processing. Some of the processed ore is shipped as concentrate. It is necessary to regrind some of the concentrate to permit separation of the minerals. This very fine concentrate is then agglomerated, either by pelletizing or sintering in preparation for further processing—in most cases, into pig iron and steel (Boyd 1992, EMR 1993b).

The conversion of iron ore to iron is achieved in one of two ways: either by the traditional method of reduction to pig iron in blast furnaces or by the "newer" method of direct reduction—which bypasses the blast-furnace stage—into solid forms (Miller 1976). Direct reduction was used in prehistoric times and has recently been reintroduced as an environmentally preferable method for use by smaller operations.

In 1993, all of Canada's iron ore mining operations carried out some sort of beneficiation at or near the minesite, ranging from crushing alone to comminution, flotation, and agglomeration (Scollan 1994). The secondary stages of processing will be discussed only peripherally in this work. They are usually carried out away from the mine and mill sites, and involve the addition of considerable quantities of scrap iron, coke, ferroalloys, and fluxes.

Metallic-ore beneficiation, as described in Chapter 1, produces no chemical changes in the ore. It introduces no new substances except those process reagents used in flotation and heavy-media separation. The beneficiation of iron ore consists of several phases. After crushing, the ore is ground and classified with or without the addition of water and/or grinding media. This stage produces a small amount of dust, which is discharged to the atmosphere. If the comminuted ore is of sufficiently high grade, it may be shipped directly to a smelter for further processing. Normally, however, concentration is required, which may be accomplished through the use of spirals, cyclones, flotation, or magnetic methods (Table 1.3). Currently, the method used most often in Canada is gravity separation, using spirals and plain water. A considerable amount of the iron contained in the ore is lost during the concentration process (Sirois and MacDonald 1983).

Almost all of the Canadian iron ore extracted in 1992 came from open pit mines. Solid waste from the extractive phase alone is nearly half the tonnage of ore (Boyd and Perron 1993). Tailings from the beneficiation stage are somewhat greater. Subsequent processing of the ore, in a blast furnace and converter, results in slag wastes approximately one-fifth the weight of the iron produced (EMR 1993b).

The impurities commonly found in iron ores have been described by Miller (1970). These are the substances other than iron that must be discharged as tailings when the iron is extracted from the ore. While most of them are released during steel-making and are therefore not included in this book, they will be named here, and the pertinent emissions, where known, will be indicated. The major constituent impurities are the common "rock" substances: silica, alumina, lime, and magnesia. These scavenge other impurities during pyrometallurgical processing. The natural content of these materials must almost always, however, be adjusted by controlled amounts of added substances to achieve the proper proportions. Silica and aluminum form an "acid" slag, while lime and magnesia form a "basic" slag during smelting. Fluxes added during smelting are chosen to neutralize, at least partially, the gangue material in the ore. All of these substances are widespread in their occurrence, and pose little threat to the environment, except when finely ground and released to air or water. The minor constituents often found in iron ore include phosphorus, sulphur, titanium, vanadium, zinc, copper and, more rarely, chromium, nickel, arsenic, lead, tin, and cadmium. Amounts of these substances vary considerably from one ore to another, and their environmental effects will be described later in this chapter.

The iron itself is present in the ore in a variety of combinations with such substances as oxygen, carbonate, and water. The higher the grade of ore entering the mill, the smaller the amount of waste that will be produced per unit of concentrate. The higher the grade of concentrate produced by the mill, the lower the amount of blast furnace slag that will be produced at the smelting stage. Higher grades of ore and concentrates also result in a lower consumption of smelter coke and in lesser amounts of residual emissions. There is considerable scope for recovery of the residual iron in the mill tailings whenever market conditions make it a viable proposition (Collings 1980a).

## Environmental Effects

The land disturbance associated with iron ore mining operations is mainly attributable to the open pit method of mining and to the disposal of waste rock and mill tailings. Some data for a specific iron ore mine in the Labrador Trough were presented by Marshall (1982). The mine had estimated reserves of 900 Mt, a production capacity of 140 kt·d$^{-1}$, and a life span of 20 years. It was ultimately expected to produce an open pit with an area of 464 ha, a waste rock dump covering 480 ha, and a tailings area of 218 ha.

An estimate of the total land disturbed by iron ore mining in Canada up to 1970 gave an overall figure of 18,000 ha (Rabbitts et al. 1971). This included 4,000 ha resulting from the draining of Steep Rock Lake in northwestern Ontario. Another estimate (Marshall 1982) indicated that, by 1975, the disturbed area had not increased. An inventory of mine wastes taken in the mid-1970s, based on satellite imagery (Murray 1977), found roughly 6,000 ha of disturbed land, of which only 6% had been revegetated. This survey did not attempt to be comprehensive, nor did it include waste dumps smaller than 10 ha in area. Since revegetation is undertaken only once sites are closed or impoundments reach capacity, adverse conclusions should not necessarily be drawn from the small proportion of areas reclaimed in this manner. Relatively little of the waste material is used, although it has potential for road construction, as concrete aggregate, and in brick manufacture (Collings 1977).

The large quantity of ore mined and the almost exclusive use of open-pit methods mean that surface disturbance is the most obvious effect of iron ore mining, although the discharge of particulate residuals to the atmosphere and hydrosphere has the potential to affect a greater area.

Although almost all current iron ore operations are confined to the relatively remote Labrador geosyncline region, many older operations were located in more southern ecosystems. A magnetite deposit near Marmora in southeastern Ontario, for example, was surface-mined between 1955 and 1978 (Carter 1984). The deposit was overlain by 5 m of overburden and 30 m of limestone. Access to the orebody was achieved by removing about 22 Mt of overburden and limestone, and approximately 28 Mt of ore were removed for pellet production. Operations left a pit 40 ha in area and 220 m deep, which has not been reclaimed: it is partially filled with water, and its sides are too steep to allow vegetation to develop. Waste rock from the mine is currently used for roadmaking.

After the ore itself, water is the second most significant input into an iron mill operation. The gross water use of iron ore mills has been estimated at between 5 and 25 tonnes of water per tonne of ore processed (Scott and Bragg 1975). This amount includes water for concentration as well as agglomeration. The potential for water reclamation depends upon the concentration method used; typically, it is about 75% for hematite ores and 95% for magnetite. These figures may be compared with the survey conducted in 1986 by Environment Canada (Table 1.4) that attributed a gross water use of 916 Mt to iron ore mining and milling. This amount is more than a third of the total water use of the entire Canadian mining industry. The gross consumption—24 tonnes of water per tonne of ore produced—is at the upper end of the range indicated by Scott and Bragg (1975). Environment Canada's survey shows that 97% of the water was used for processing and that 75% of it had been recycled. Water use in the concentration stage is equivalent to 1.25 tonnes of water per tonne of ore and is considered appropriate for the amount of reclaim used in the industry (Sirois and MacDonald 1983).

Virtually all of the water not reclaimed is discharged to the hydrosphere, carrying with it any dissolved and suspended substances. Examples of chemical analyses of iron ore mine and mill wastewater are presented in Table 8.1. Although many are from British Columbia, which is currently only a minor iron ore producer (and none are very recent), they do give some idea of the range of hydrospheric residuals attributable to iron ore operations, particularly without strict environmental controls. An obvious attribute of most of the analyses is the near-neutral pH levels. The only exceptions are a basic discharge to the sea from a mill in the Queen Charlotte Islands and an acidic discharge from an old tailings pond in southern British Columbia.

**Table 8.1** Chemical Analyses of Water from Iron Ore Mines and Milling Operations in Canada (parts per million except for pH)

| | Mine water[a] | Mine water[b] | Direct-discharge tailings effluent[b] | | Marine-discharge tailings effluent[c] | | Tailings pond effluent[b] | Tailings recycle water[a] |
|---|---|---|---|---|---|---|---|---|
| pH | 6.4 | 7.8 | 7.5 | 7 – 9 | 7.8 | 10.4 | 3 – 4.5 | 7.3 – 8.5 |
| dissolved solids | | | | | 31,900 | 32,000 | | |
| suspended solids | | | 57,000 | | | | 40 | 12 |
| total solids | | | | 260,000 | 71,000 | 77,900 | 4,890 | |
| hardness | | | | | 2,700 | 11 | | 250 |
| alkalinity | | | | | 13,000 | 11 | | |
| arsenic | | < 0.005 | | | | | 0.02 | |
| copper | 0.1 | 0.01 | 0.24 | 0.4 | 44 | 41 | 0.32 | 45 |
| iron | 1.3 | 0.05 | 0.48 | 1.0 | 6,700 | 230,000 | 455 | 2.8 |
| lead | 0.1 | 0.01 | 0.15 | < 1.0 | < 3.7 | < 12 | 2.5 | |
| magnesium | | | | | 2,100 | 870 | | 0.25 |
| manganese | | | | | 180 | 230 | | 8 |
| nickel | 0.1 | | | | 5.1 | 6 | | |
| sodium | | | | | | | | 60 |
| zinc | 0.1 | 0.01 | 0.09 | 0.15 | 7.8 | 12 | 13 | |
| cyanide | | | | | | | 3.5 | |
| sulphate | 320 | | | | 1,750 | 35 | 1,000 | |
| xanthates | | | | | | | 1.7 | |

Blank entries indicate lack of data rather than absence of substance. Separate columns under headings are for different mines.

Source: [a]Scott and Bragg (1975); [b]Mining Association of British Columbia (1972); [c]Hoos (1975).

Direct discharge of tailings, a practice rarely used today, results in a large quantity of dissolved and suspended solids. Rabbitts et al. (1971) described the suspended solids problems of iron ore mining as the result, for the most part, of colloidal iron from limonitic ores, finely divided silica from milling, and fine iron oxide dust from cooling water in sintering plants. Many of the samples in Table 8.1 show values for dissolved substances that would have been greatly in excess of current guidelines for aquatic life and wildlife. Particularly noticeable is the very wide range in the contents of iron and copper, which spans several orders of magnitude.

Residuals from iron ore mining are rich in iron; however, the low toxicity of the metal and the small amount produced in comparison with natural emissions suggest that the effects are not likely to be significant on a large-scale geographical basis. A 1975 report by the Study Panel on Assessing Potential Ocean Pollutants (SPAPOP) found typical dissolved-iron concentrations to be about 30 $\mu g \cdot l^{-1}$ in freshwater and 3.4 $\mu g \cdot l^{-1}$ in marine environments. A well-established inverse relationship between metal release and pH (Tipping et al. 1986) means that most iron and other metals will not only dissolve, but may be scavenged from solution at pH values typical of iron ore mining effluents.

Both atmospheric and hydrospheric effects may be significant on local or regional scales. Although the areas in which iron ore is mined in Canada are not heavily populated, they do contain salmon-spawning waters, as well as lakes and rivers that are licensed for commercial fishing. All but the northernmost mines are located in important forest-resource areas.

Iron is an essential nutrient for both plants and animals. In some circumstances—marine systems, for example—iron may be in sufficiently short supply to limit plant production. At the other extreme, very high concentrations produced in rivers and lakes downstream from iron ore operations have been known to kill salmon roe and to have harmful effects on plankton and benthic organisms (Sirois and MacDonald 1983). Because of the apparently minor toxic effects of iron ore residuals, they have been the subject of considerably less research than have the residuals from other mining sectors, such as sulphides.

The minor toxic effects of waterborne iron ore residuals may explain why a more recent study than that of Horn (1966)—on the environmental effects and impacts of iron ore mining in Canada—could not be found. At that time, Horn surveyed three of the major iron ore operations along the Labrador geosyncline, all of which are still in production today. The major sources of wastes were judged, at that time, to be drainage water from the open pits, townsite sewage, and mill and pellet-plant wastes. Since all three of the mines had been in operation only a few years at the time of Horn's report, no problems had yet arisen from mine drainage. However, difficulties were foreseen as the sizes of the mines, and their potential drainage, grew with time. Sewage disposal was considered to be well managed by treatment plants constructed at each townsite.

The major environmental threats were thought to be solid and liquid wastes from the milling and pelletizing operations. At one of the operations, process water was removed from Lake Wabush, near the Québec/Labrador border. In spite of some recycling, there was a release of 60 Gl•yr$^{-1}$ from the concentrator (20% solids) and 13 Gl•yr$^{-1}$ from the pellet plant. The water and tailings—containing quartz, iron oxides, and calcium and magnesium carbonates—were discharged directly into Lake Wabush. Although most of the material settled out rapidly, enough remained in suspension to discolor the lake water. Lake Wabush drains into Lake Shabogamo and, through a number of other lakes and rivers, about 1,200 km to the Atlantic Ocean. A survey conducted in 1965 showed clear evidence of waste contamination near the discharge source, but not beyond 15 km downstream. The carbonate content of the discharge produced an increase in water hardness throughout Lake Wabush. Turbidity increases extended to less than 10 km downstream. The pellet-plant discharge was found to be the main source of turbidity, due to a high concentration of finely ground material. Turbidity had been reduced through the use of higher-capacity thickeners; nonetheless, subsequent installation of a magnetic separation plant to process the tailings was expected to result in an increase in colloidal-size particles because of the need for further grinding.

All stages in the processing of iron ore produce atmospheric dust emissions (Rabbitts et al. 1971; SPAPOP 1975; Joyner 1985; EMR 1993b). Since the ore undergoes little change along its path from mine to product, the atmospheric emissions associated with each phase have similar compositions, both to each other and to the ore itself.

Gross Canadian atmospheric emissions in 1985, for the iron ore industry as a whole, were estimated by Kosteltz and Deslauriers (1990); they suggested 48 kt of carbon monoxide, 103 kt of total particulate matter, and 117 kt of sulphur dioxide. The Study Panel report (1975) attributed 42% of worldwide atmospheric iron emissions to natural crustal weathering processes, 28% to human-induced weathering, 13% to the iron and steel industry and 10% to coal combustion.

Typical emissions from the extraction phase were described in Chapter 1. Ore-handling and coarse crushing and screening are expected to produce approximately 0.05 kg of particulates per tonne of ore processed. Estimates of uncontrolled crushing and grinding emissions range from 0.23 (Rabbitts et al. 1971) to 0.93 kg·t⁻¹ (SPAPOP 1975). The EMR study (1993b) observes that, now that emission controls are in place at all Canadian operations, emissions are at the lower end of the 0.23-0.93 kg·t⁻¹ range.

While sintering was once the most common method of agglomeration, less than 20% of iron ore is currently sintered in Canada (EMR 1993b). The process produces a high proportion of the industry's total particulate emissions, however, and as a result of the high sulphur content of pyrite-siderite ore, sintering also produces a considerable fraction of the industry's sulphur dioxide emissions. Carbon monoxide emissions from sintering are estimated at approximately 22 kg·t⁻¹ of output (Joyner 1985).

The only Canadian sinter plant currently operating—in northwestern Ontario near Wawa—was responsible for environmental damage in the area as a result of using high-sulphur coal as fuel for the process. It is located in the mixed-forest zone, near the southern boundary of the boreal forest (Figure 2.3). According to Rabbitts et al. (1971), more than 100,000 ha of forest were already affected by these emissions twenty years ago. An ecological study of the region (Gordon and Gorham 1963) traced out areas of noticeable vegetation damage to a distance of 30 km downwind from the prevailing wind direction, with severe damage as far as 20 km from the source. Within 8 km downwind from the source, lake waters had pH values as low as 3.2 to 3.8 and soils were high in sulphate and noticeably eroded. The strongly unidirectional nature of local winds had confined most damage to a single quadrant. A later study of the same area used color-enhanced satellite imagery to delineate vegetation damage zones (Murtha 1974), which matched quite closely those determined thirteen years earlier.

Most Canadian ore was pelletized rather than sintered. The best available estimate of uncontrolled pellet plant particulate emissions is approximately 4 kg·t⁻¹ (Joyner 1985). An assumption of 90% control reduces this figure to 0.4 kg·t⁻¹. Pellets are balls formed by mixing water, binder, and coke breeze (fine coke particles recovered from coke oven emissions) that are indurated in grate furnaces to make them hard enough to withstand shipping. Induration is an energy-intensive process, consuming approximately 1,000 MJ of energy (primarily in the form of heavy fuel oil) per tonne of pellets (Sirois and MacDonald 1983). As with the sintering process, atmospheric emissions of sulphur dioxide and carbon dioxide are produced by the heating of the ore, as well as by the combustion of the fuel in the pelletizing process. The potential exists for recovering most of the heat lost from the process and for using alternative energy sources to reduce atmospheric emissions that arise from fossil-fuel combustion.

## Reclamation

The scarcity of recent information extends to the reclamation of areas disturbed by iron ore mining. As indicated earlier, relatively few of these disturbed areas have been reclaimed, presumably because of their remoteness from cities and because of the lack of any serious pollution problems.

The difficulty of revegetating material with low fertility in a region of harsh climate has been addressed by Jasper et al. (1988). They found that revegetation success could be enhanced substantially through the addition of specific nutrients and the inoculation of soil with mycorrhizal fungi and rhizobia.

## Summary

Most Canadian-produced iron now comes from low-grade ores mined by open-cast procedures, with a small proportion from sulphidic and ilmenitic ores. There is a legacy of wastes and damage from sulphur emissions produced by mines that have closed; it is often difficult, however, to distinguish between the impacts of former mines and mills.

Large quantities of solid wastes are produced, and much of their storage area has yet to be reclaimed. Water use is high, but much is recycled. Released wastewaters, despite settling, cause turbidity and, in some cases, have been toxic to aquatic organisms. Dust emission occurs both when iron is mined and during processing. Despite controls with up to 90% efficiency, the quantity of atmospheric emissions is large.

# CHAPTER NINE

# CARBON PRODUCTS: COAL, PEAT, GRAPHITE, AND DIAMOND

## Introduction

When the organic remains of plants and animals decompose under anaerobic conditions such as those found in deep wetlands and are subjected to compaction over a long period of time, they undergo a gradual lithification. The effects of microbial action, temperature, and pressure produce metamorphic changes with time, as indicated in the following sequence: plant material→peat→lignite→sub-bituminous coal→bituminous coal→semi-anthracite→anthracite→graphite→diamond.

Most of the components of this sequence, from the more decomposed peat through to anthracite, are used primarily for generating electricity and the production of steel.

The original organic material was composed primarily of the elements carbon, hydrogen, and oxygen. It also contained minor amounts of sulphur and nitrogen, and traces of silicon, aluminum, calcium, magnesium, iron, sodium, potassium, phosphorus, and titanium (Hills and Jones 1981). While some of these trace elements are present in coal itself, the majority are found in waste material that can be separated from coal by physical methods. Trace radionuclides are also present in coal. As the organic material ages, its physical properties and chemical composition change: it becomes darker and harder, its water content decreases, and its carbon content and caloric value increase (Landheer et al. 1982; Romaniuk and Naidu 1987). This process is called coalification or carbonification. Sulphur combines with metals present in the organic material to form metallic sulphides—most commonly iron pyrites (Kelly et al. 1988)—and volatile components are driven off (Whiteway 1990).

## Coal

The coal group of energy minerals consists of five types: lignite, sub-bituminous, bituminous, semi-anthracite, and anthracite. Although all ranks of coal occur in Canada, as illustrated in Table 9.1, only the first three varieties have been produced during the past decade, with lignite coming only from Saskatchewan and sub-bituminous only from Alberta (Aylsworth and Shapiro 1993). As Table I.1 indicates, coal is one of Canada's most important mineral resources, and Canada ranked eleventh among coal-producing nations in 1992. Of the total world production of 4.5 Gt, Canada's share was less than 1.5% (Hoddinott 1993). The primary markets for Canadian coal are thermal power generation and metallurgical coking.

Lignite (brown) coal is a brownish-black coal with a caloric value of less than 19.3 MJ•kg$^{-1}$ and accounts for approximately one-sixth of Canada's total coal production. Sub-bituminous is a black coal characterized by a higher carbon and lower moisture content than that of lignite. It represents about a third of Canada's saleable coal output (Aylsworth, Shapiro and Lomas 1994). The best use for lower-priced lignite and sub-bituminous coals is as fuel for power-generating plants.

Bituminous (soft) coal is dark brown or black in color. Bituminous coals, at 26.7 MJ•kg$^{-1}$, have twice the caloric value of lignites; however, they also have the highest content of volatile compounds (14%), including sulphur. The volatile components are transformed, upon combustion, into atmospheric residuals. Bituminous coal, which is mined in Nova Scotia, New Brunswick, Alberta, and British Columbia, comprises the largest sector of coal mining in Canada and normally accounts for close to half of the country's coal production (Aylsworth, Shapiro and Lomas 1994). Bituminous coals are mainly used for smelting.

Anthracite, or hard coal, is the most desirable in terms of limiting atmospheric residuals. Anthracite is also easier to treat and to handle because it produces less waste and dust. However, it is scarcer and more costly than lower grades. Since the 1979 closure of the Canmore Mines at Banff National Park, Alberta, very few anthracite coals have been produced in Canada.

## History

The earliest discovery of coal in Canada was reported in 1672 by Nicholas Denys in his description of the North American coast. His writings refer to what is now called the Sydney Coalfield, located on Cape Breton Island, Nova Scotia. In 1720, the first coal was produced using conventional mining methods there at Cow Bay. Only four years later, coal was first exported from Cape Breton to Boston. By 1830, the first shaft was sunk in the Sydney Coalfield, beginning Canada's long tradition of underground coal mining. Coal was not discovered in western Canada until 1857, when Sir James Hunter located coal in the Souris River

**Table 9.1** Canadian Coal Production by Type and Origin, 1985–1992 (000 tonnes)

| | 1985 | 1986 | 1987 | 1988 | 1989 | 1990 | 1991 | 1992 |
|---|---|---|---|---|---|---|---|---|
| **Bituminous** | | | | | | | | |
| Nova Scotia | 2,800 | 2,695 | 2,925 | 3,540 | 3,512 | 3,415 | 4,134 | 4,486 |
| New Brunswick | 560 | 490 | 533 | 542 | 520 | 548 | 498 | 399 |
| Alberta | 7,841 | 6,994 | 7,202 | 9,561 | 9,907 | 9,153 | 10,312 | 10,508 |
| British Columbia | 22,994 | 20,359 | 21,990 | 24,911 | 24,840 | 24,581 | 24,962 | 16,922 |
| Total | 34,195 | 30,538 | 32,650 | 38,554 | 38,779 | 37,697 | 39,906 | 32,315 |
| **Sub-bituminous** | | | | | | | | |
| Alberta | 16,871 | 18,225 | 18,537 | 19,910 | 20,918 | 21,252 | 22,242 | 23,020 |
| **Lignite** | | | | | | | | |
| Saskatchewan | 9,672 | 8,281 | 10,020 | 12,148 | 10,816 | 9,407 | 8,981 | 10,027 |
| **Total Production** | 60,738 | 57,044 | 61,207 | 70,612 | 70,513 | 68,356 | 71,129 | 65,362 |

Source: Godin (various years); Statistics Canada.

Basin in southwestern Manitoba. In 1879, the coalfield at Crowsnest Pass, British Columbia, opened up coal mining in western Canada. Commercial mining in southern Saskatchewan began in 1880, to supply coal for home-heating and steam locomotives. Coal was mined underground on the prairies until 1927, when this method was replaced by horse-drawn stripping and hauling methods.

### Orebodies

Currently, there are seven coal-producing regions in Canada: Nova Scotia, New Brunswick, Saskatchewan, Alberta Plains, Alberta Foothills, Alberta Mountains, and British Columbia. Figure 9.1 indicates the locations of coal mines operating in 1993.

The Sydney Coalfield of Nova Scotia is located in the Atlantic Maritime terrestrial ecozone and the Boreal ecoclimatic region. This coalfield contains eleven major seams, ranging in thickness from 1.0 to 4.5 meters. Much of this coalfield is submarine and is suited to underground mining methods. Typically, the coal of this region is high-volatile bituminous: low in ash, high in sulphur.

The Minto Coalfield, New Brunswick's only major coalfield, is a large (32 km by 13 km) single-seamed coal deposit. The coalfield is contained in the Atlantic Maritime terrestrial ecozone and the Cool Temperate ecoclimatic region. The occurrence contains high-volatile bituminous coal.

The Saskatchewan coal region contains two coalfields: the Estevan and the Willow Bunch. These lignite coal occurrences lie in the Prairie terrestrial ecozone and the Grassland ecoclimatic region. The four sub-bituminous coal deposits of the Alberta Plains—Battle River, Sheerness, Wabamun, and Wetaskiwin—lie in the Prairie terrestrial ecozone and the Grassland ecoclimatic region. These coalfields, as well as those in New Brunswick and Saskatchewan, are flat-lying formations that lend themselves well to strip-mining methods.

The Alberta Foothills region contains two major coalfields: the Coalspur and the Obed Mountain Coalfields. These two coalfields are characterized by

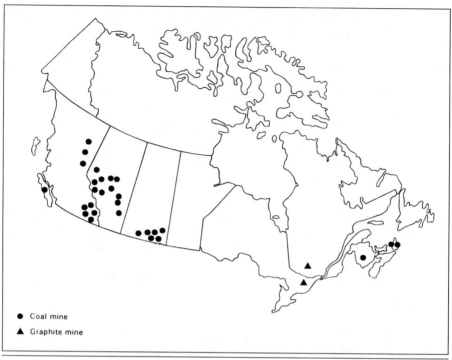

**Figure 9.1** Canadian coal and graphite mines, 1993. Source: Coal Association of Canada (1993); Aylsworth, Shapiro, and Lomas (1994).

high-volatile bituminous deposits. The rolling, semi-mountainous topography and associated geology of these coalfields facilitate the use of open-pit mining methods.

Low- to high-volatile bituminous deposits are typical of the Alberta Mountains and British Columbia coal regions. With the exception of the Smoky River Mine, which incorporates both open-pit and underground methods, this coal is mined almost exclusively by open-pit methods. Both coal regions lie in the Montane Cordillera terrestrial ecozone and the Cordilleran ecoclimatic region.

## Extraction and Beneficiation

More than 90% of coal in Canada is mined by open-pit or strip-mining methods. Two of the underground mines are in Nova Scotia and are actually undersea mines exploiting part of the Sydney Coalfield (Konda and Kochhar 1986). There is also one underground/open-pit operation in Alberta.

Canada's underground mines are highly mechanized, using sophisticated machinery and longwalling methods (Singhal and Sahay 1986). The relatively low structural strength of the ore, its highly inflammatory nature, and the gases that frequently accompany it, such as methane, make underground mining riskier than surface methods.

Among surface-mining methods, strip-mining using large shovels and draglines is most suitable for shallow deposits in flat terrain. This method permits relatively easy replacement of the soil cover and the return of the land either to its previous use or to one comparably productive. Faulted deposits in hilly terrain, by contrast, are usually mined by open-pit methods and are not reclaimed to the same extent as those in flat terrain.

The beneficiation process for coal consists of five steps: breaking and screening of the raw coal, and removing slate and rocks by hand; removing undesirable constituents, using methods such as jigs, tables, heavy-medium vessels, cyclone, and froth flotation; dewatering and drying of the cleaned coal; treating and recycling or discharging the process water; and thickening and disposing of the solid wastes.

The main purpose of coal beneficiation, or cleaning, is to remove from the coal non-carbonaceous mineral material, as well as such ash-producing substances as rock, clay, and pyrite (Paine and Blakeman 1987). The preparation coal receives before it is sent to market depends on the intended end use of the coal, the quality of the run-of-mine product, and the economics of transporting waste materials (Romaniuk and Naidu 1987). A great deal of mined coal, for example, is moved directly to adjacent power-generating stations without any processing; the remainder is crushed and screened before being sent. Coal destined for metallurgical use and export is generally cleaned to remove the undesirable non-carbonaceous constituents that would result in slag.

For ores containing pyrite (that is, the eastern ores), conventional cleaning processes can remove 30%–60% of the sulphur and reduce the ash content to between 3%–4%. For clean (western) ores, these techniques can reduce the ash content to 7.5%–13%. This intensive cleaning is necessary to minimize sulphur dioxide emissions in the processing and combustion of coal and to reduce emissions of carbon dioxide. Promising research is being conducted to produce a clean coal with 2%–8% ash and less than 1% sulphur (Mikhail, Salama and Humeniuk 1992).

## Environmental Effects of Coal Production

Underground mines are known to produce relatively small amounts of waste. The main sources of residuals in mine drainage water are the coal itself, the associated gangue material, and blasting agents.

Surface subsidence is a common result of the underground mining of many shallow coal deposits. In Canada, because of the almost exclusive use of opencast methods, subsidence is a problem more of abandoned mines than of existing or proposed operations. Such is the case not only among former underground mines, but for undersea operations as well. An example of the former type is room and pillar mining, which was carried out under the present city of Lethbridge, Alberta, between 1872 and 1965 (Sladen and Joshi 1988). This has necessitated geotechnical investigations to evaluate potential subsidence hazards and to permit the development of guidelines for precautions in construction. Since very little is known about undersea subsidence, a program has been established to monitor the seafloor of the Sydney coalfield in order to evaluate the extent of subsidence following mining activities there (Aston 1989).

A satellite-imagery survey carried out in the mid-1970s (Murray 1977), while not comprehensive, found 12 kha of land disturbed by coal mining at a total of 26 sites; 40% of this had vegetative cover. Most of the disturbance was associated with the removal of overburden at strip-mining sites. An analysis reported by Marshall (1982) indicated that the size of the area directly disturbed by individual Canadian coal mines ranged up to a high of 1,430 ha; the average was 335 ha. Although this amount is greater than that of any other type of mining, most of the disturbance is due to the laterally extensive—but more easily rehabilitated—strip-mining of relatively shallow deposits in the Prairies.

The cleaning of coal presents a number of environmental problems (Sirois and MacDonald 1983). The first is to manage the large quantity of solid wastes produced by the extraction and beneficiation stages. About 19% of all raw mined coal in Canada becomes waste material lost in the beneficiation process and must be discharged as tailings. By rank, waste ranges from zero for lignite (all of which is used for power generation) and 1% for sub-bituminous coal to 29% for bituminous coal.

As production increases, the amount of wastes produced will rise. These wastes may be subject to spontaneous combustion, erosion from wind and water, and, in the case of eastern Canadian mines, acid drainage. Because many contain large amounts of fine particles that both inhibit drainage and foster instability, alternatives to surface dumping as methods of disposal are therefore highly desirable. For example, coarser refuse can be used for underground-mine backfill or for tailings-dam construction as well as for other industrial purposes, provided that the waste is non-reactive (Wood 1983; Roberts and Masullo 1986; Dunbavan 1990; Hart 1990). Finer materials are not as useful unless they are stabilized by mixing with, for example, cement, lime, pulverized fly ash or polymers (Atkins et al. 1986; Stewart et al. 1990). Some of the fine coal particles in the waste may be recovered by mixture with a light oil and subsequent formation into spherical agglomerates using a balling disc (Capes et al. 1970).

The choice of site and the design of the dump area (with respect to local topography and water-drainage patterns) are very important in cases where surface dumping is unavoidable. It is easier to design waste dumps with maximum stability and minimum impact on downstream water quality in mines where the rock and coarse wastes from extraction and the finer wastes from beneficiation are integrated (Das et al. 1990). An argument in favor of surface dumping is that old waste dumps can be, and often are, re-mined for recovery of their contained minerals, such as zirconium and cobalt (Mikhail et al. 1981; Romaniuk and Naidu 1987).

Major environmental challenges facing the coal mining industry are to reduce the amount of water used and to improve the quality of water discharged to the environment. Beneficiation uses the greater part of the water required for coal production (Barrie and Carr 1988). The water is used primarily to clean metallurgical coal and, to a lesser extent, to suppress the dust of thermal coal. Most separation processes are carried out in aqueous media, many with the addition of process reagents. The choice of method is dictated both by economic considerations and by the quality of the coal, especially particle-size distribution.

An estimate of water use in coal mining for western Canada alone (Barrie and Carr 1988) was 31.2 Gl for 1990, most of it in British Columbia. Projections indicate the possibility of water shortages in the Plains regions of Saskatchewan and Alberta by 2000 if the current rate of usage continues. A 1986 survey by Environment Canada assessed the gross water use of coal mining at 43.1 Gl, of which 72% was recycled (Table 1.4); 98% of this water was used for bituminous coal mining and cleaning (Tate and Scharf 1989b). Only 8 Mt of water was actually discharged; it contained impurities derived from substances in the coal itself, as well as reagents associated with the beneficiation process. These include both flotation reagents (as discussed in Chapter 1) and such heavy-medium separating substances as magnetite, which are used to process, respectively, 12% and 54% of beneficiated coal. The remainder of the coal is cleaned in water (Sirois and MacDonald 1983).

While water recycling is practiced at all coal-processing plants in Alberta and British Columbia, discharges still occur as a result of seepage from water impoundments, periodic overflows, and the release of accumulated process reagents (Paine and Blakeman 1987). A number of alternative cleaning methods, some of which use less water and produce fewer residuals, are under development.

In common with other types of mines, both open-cast and underground coal operations have the potential to disturb the flow and quality of surface and of sub-surface water. Coal mines differ from other mines, however, in that they may affect much larger horizontal extents of land (Stoner 1983). In addition to the extraction of coal, other operations—such as removal of overburden and mine dewatering—may have substantial influences on the quantity and distribution of groundwater (Barrie and Carr 1988).

The trace-metal composition of coals depends to a large extent on the grade and the source region. As Table 9.2 indicates, Nova Scotia coal, for example, is noticeably higher than average in aluminum, arsenic, chromium, cobalt, iron, lead and manganese, and lower in barium, strontium and uranium. Saskatchewan lignite is lower in titanium and vanadium and higher in zirconium.

The contamination of coal-mine wastewaters is related directly to the composition of the coal and to the vegetation from which it was derived (Tables 9.2

**Table 9.2**   Trace-Element Content of Canadian Coals (parts per million)

| Component | Saskatchewan Lignite | Alberta Sub-bituminous | Alberta Bituminous | British Columbia Bituminous | Nova Scotia Bituminous |
|---|---|---|---|---|---|
| Aluminum | 14,034 | 14,513 | 13,414 | 16,476 | 19,004 |
| Antimony | 1.02 | 0.63 | 0.74 | 0.75 | 0.79 |
| Arsenic | 1.57 | 1.16 | 1.67 | 1.73 | 28.99 |
| Barium | 811 | 4,81.5 | 497.6 | 414.8 | 136.54 |
| Beryllium | 1.65 | 0.5 | - | - | 0.96 |
| Boron | 1,134.2 | 1,029.6 | 640.4 | 818.5 | 1010.6 |
| Cadmium | - | 7.5 | - | 8.4 | - |
| Chromium | 9.23 | 3.98 | 9.26 | 13.05 | 34.36 |
| Cobalt | 1.77 | 1.07 | 1.70 | 1.69 | 21.60 |
| Copper | 11.76 | 6.03 | 12.08 | 13.77 | 15.50 |
| Iron | 3,537 | 3,921.7 | 1,772.6 | 2766 | 18,682 |
| Lead | 9.21 | 6.76 | 8.06 | 5.21 | 25.88 |
| Manganese | 47.87 | 85.5 | 15.7 | 19.85 | 111.54 |
| Mercury | 0.074 | 0.12 | 0.10 | 0.06 | - |
| Molybdenum | - | - | - | 76.3 | 157.6 |
| Nickel | 10.1 | 7.95 | 5.63 | 75.42 | 18.54 |
| Selenium | 3.5 | 2.86 | 3.6 | 3.35 | 3.89 |
| Silver | 0.83 | 1.8 | - | 3.3 | 1.47 |
| Strontium | 401.8 | 331.6 | 331.8 | 405.1 | 162.8 |
| Thorium | 3.74 | 3.73 | 2.61 | 1.57 | 3.85 |
| Titanium | 849.5 | 376.7 | 1,226.6 | 1,144 | 1,092.8 |
| Uranium | 1.88 | 1.17 | 1.68 | 1.15 | 0.86 |
| Vanadium | 11.98 | 6.07 | 32.34 | 35.6 | 30.4 |
| Zinc | 51.04 | 25.67 | 35.58 | 27.7 | 32.0 |
| Zirconium | 124.4 | 75.2 | 44.42 | 36.8 | 46.54 |

Source: Paine and Blakeman (1987).

and 9.3). However, it must be emphasized that, although high concentration of trace elements in coal may be mirrored in the wastewater, it does not occur in every case; it depends on the rank of coal and its degree of oxidation, as well as on solute composition and the pH of contacting water, among other things. For example, organic sulphur (that is, sulphur contained in organic molecules) has been shown by Casagrande et al. (1989) not to contribute to acid generation in mine drainage.

The National Coal Wastewater Survey, carried out by Environment Canada during the early 1980s (Paine and Blakeman 1987), assessed the quality of effluents from a large number of coal mines across Canada (Table 9.4). Data from the different mines showed a wide variety in the composition. Most striking in these results were the generally high concentrations of iron, and often those of aluminum, barium, boron, manganese, and strontium as well. Total suspended solids cited in the Survey varied from less than 1 to more than 22,000 mg•l⁻¹; pH normally falls in the near-neutral range, except for one or two sites in Nova Scotia. Concentrations of sulphate ions ranged from less than 10 to more than 2,500 mg•l⁻¹. The use of ammonium nitrate blasting agents (Barrie and Carr 1988) may account for the elevated concentrations of ammonia and nitrate seen in many of the samples. The potentially harmful effects on the natural environment of much of this wastewater requires that it be treated before discharge.

**Table 9.3**  Composition of Canadian Coals (percentage)

| Component | Saskatchewan Lignite | Alberta Sub-bituminous | Alberta Bituminous | British Columbia Bituminous | Nova Scotia Bituminous |
|---|---|---|---|---|---|
| Carbon | 50.06 | 54.20 | 80.28 | 77.50 | 58.50 |
| Oxygen | 14.33 | 14.75 | 3.87 | 3.63 | 4.91 |
| Hydrogen | 3.13 | 3.17 | 4.42 | 4.23 | 3.95 |
| Nitrogen | 0.95 | 0.85 | 1.10 | 1.11 | 1.24 |
| Sulphur | 0.45 | 0.33 | 0.36 | 0.37 | 2.39 |
| Ash | 12.27 | 11.31 | 9.21 | 11.59 | 16.95 |
| Calorific Value ($MJ \cdot kg^{-1}$) | 19.00 | 20.84 | 32.30 | 31.18 | 22.57 |

Source: Paine and Blakeman (1987).

Etter (1971) examined the quality of stream water in a coal-mining area located in the upper foothills of western Alberta and concluded that sediment, including coal particles, had moved from the mined land into the drainage system both during spring runoff and at intervals later in the year. Effects from the mining activity on water and sediment quality included turbidity, suspended and dissolved solids, iron, apparent color, and the occurrence of coal in bottom sediments. Oil contamination and odor were also noted; impacts on aquatic organisms, however, were not studied.

Water quality has also been affected in springs and streams on the eastern slopes of the Rocky Mountains near Grand Cache, Alberta (Hackbarth 1981). Water drained readily through spoil piles in this area, emerging from springs and seeps and draining into the streams. Total dissolved solids increased to four times the background values, but relative amounts of cations remained constant. This increase in total dissolved solids, then, is due solely to an increase in anion concentrations; in water from mined sites, this rise is accompanied by a shift from bicarbonate to nitrate and sulphate. Enhanced oxidation of sulphide minerals in the soil increases sulphate (by comparison with background levels), but sulphide oxidation does not reduce stream pH, mainly because the spoils contain approximately 14% carbonate and only 0.07% total sulphur. Iron and manganese commonly exceed standards set for drinking water; levels of copper and lead, on occasion, also exceed acceptable levels. However, similar levels are observed in some streams unaffected by mining. Despite changes in the quality of drinking water, the water chemistry in Sheep Creek, the major receiving stream, changes only slightly between sites upstream and downstream from the area of coal mine spoils.

The release of fine sediments into stream ecosystems can adversely affect aquatic organisms, particularly the reproductive success of salmonid fishes. Literature reviewed by MacDonald and McDonald (1987) indicated that the reduction of egg and fry survival in response to high levels of fine sediment have been demonstrated in field and laboratory studies. Coal mining activities along the Fording River in southeastern British Columbia increased the amount of fine sediments downstream, as compared with water upstream from the mine sites (MacDonald and McDonald 1987). The measured levels of fine sediments would reduce the production of salmonids if downstream areas were used as spawning habitats.

The greatest risk of sediment effects on rivers occurs when sediments are released during low flow periods, or are released in large amounts over extended time periods (Bietz 1989). The total suspended sediments in water can contain some large, rapidly settling particles, which harm fish. Lethal sediment levels rarely occur, but sublethal effects are relatively common in many rivers; such effects include reduced growth rates, decreased disease resistance and downstream displacement. For example, suspended sediment levels greater than 100 mg•l$^{-1}$ produce sublethal effects in arctic grayling. Coal sediment loads of 82 mg•l$^{-1}$ damage invertebrate populations. Solids settling to the bottom can suffocate fish eggs and newly hatched embryos, especially among salmonid fish dependent on gravelly river bottoms for egg laying.

Guidelines of the United Nations Food and Agriculture Organization indicate the probable effects of suspended sediment levels on fisheries:

| Suspended Sediment Level | Effect on Fisheries |
|---|---|
| less than 25 mg•l$^{-1}$ | no harmful effects |
| 25 to 80 mg•l$^{-1}$ | small decreases in numbers and growth rates |
| 80 to 400 mg•l$^{-1}$ | good fishery cannot be supported |
| greater than 400 mg•l$^{-1}$ | few or no fish |

Data summarized in Bietz (1989) show that coal mining and washing can produce suspended solids at the high end of this scale. Bietz is sceptical about the capacity of current settling pond technology—even with the use of flocculants—to lower total suspended solids in Alberta mining effluents to a level commensurate with set guidelines of 50 mg•l$^{-1}$.

During mining operations, the water table must often be lowered below the coal-bearing rock formation so that excavation can take place more readily. In some cases, the regional water supplying aquifer is excavated on a large scale or even removed completely, necessitating a new water supply for the area. Such is the case at the Highvale Mine area in Alberta, a rural residential area with minor agricultural activity. Pre-mining domestic water requirements for the region were met by a sandstone aquifer (the Paskapoo Formation) in the bedrock overburden above the coal-bearing unit. This aquifer will be disrupted by mining; what remains of the aquifer will likely yield water of a quality so poor and a transmissivity so highly variable that it will no longer be suitable for exploitation. Hydrogeological studies have not been successful in finding another source of groundwater sufficient to meet the needs of the region; an alternative to groundwater must be found (Trudell 1986).

Discharges and seepage from open pits, stockpiles, and waste rock dumps can contain large amounts of nitrogen, which derives primarily from the blasting agents ammonium nitrate and fuel oil. When released into surface waters, the nitrate can affect water quality and, indirectly, aquatic habitats. The high cost of removing nitrate from the large discharges of water from surface coal mines motivates the search for alternatives to conventional treatment.

For example, wetlands have been examined for possible treatment of mine wastewater at a coal mine west of Campbell River, British Columbia (Whitehead et al. 1989). A natural wetland—which is adjacent to the main settling pond and vegetated by hardhack, sweetgale, and sedge—receives drainage from all disturbed areas of the mine. An area of the natural wetland was converted to an experimental wetland by means of an artificial off-stream impoundment. In its first year of operation, the wetland system removed an average of 87% of the total nitrogen from wastewater. Low temperatures in late autumn reduced nitrogen removal, but the inhibition was less than expected.

Coal is mined from the Tertiary Ravenscrag Formation at Coronach, Saskatchewan, providing fuel for a thermal power station. Relatively large quantities of excess water must be pumped from within and above the lignite-bearing strata in order for mining to proceed. This dewatering, combined with sump drainage and water diversion at the mine site, significantly affects both groundwater and surface water flow in the Poplar River Basin on the border between Saskatchewan and Montana (IJC 1981). A number of contaminants derived from the coal seams could be carried along during this dewatering (Collerson et al. 1991). Uranium, for example, is commonly associated with lignite; uraniferous lignites of economic grade are found in the Williston basin of North Dakota. Uranium content of coal in Saskatchewan ranges from 2.6 to 5.0 mg·kg$^{-1}$ at Coronach to 17 mg·kg$^{-1}$ at Estevan, and up to 100 mg·kg$^{-1}$ in the coal deposits of Cypress Hills. Very few data are available concerning concentrations of trace elements and radionuclides in the dewatering water and the lignite. Of particular concern are those trace elements known to be concentrated in lignite, such as barium, molybdenum, sulphur, arsenic, selenium, uranium, and thorium, as well as decay products of $^{238}$U and $^{232}$Th. Collerson et al. (1991) pointed out the need for additional information on several subjects: for example, the potential bioaccumulation of coal-derived radionuclides in terrestrial and aquatic food chains in the ecosystems of the Poplar River drainage basin is currently unknown. In general, lignite coal mining has the potential to release radionuclides and metals into ecosystems through mine dewatering or the leaching of coal stockpiles.

Values of pH in coal effluents vary regionally (Table 9.4). Radford and Graveland (1978) report that pH levels of effluent from abandoned and operational coal mines in the Crowsnest Pass area of Alberta generally exceed 7. The effluent also shows relatively high levels of iron (0.4 to 31.2 mg·l$^{-1}$), which were reduced substantially by settling ponds, whether natural or artificial, between the

**Table 9.4** Canadian Coal-Mine Wastewater Analyses.

| Location | Source | Coal type | Total Suspended Solids | Total Dissolved Solids | pH | $NH_3$ | $NO_3^-$ | $SO_4^{2-}$ |
|---|---|---|---|---|---|---|---|---|
| Sydney, Nova Scotia | mill | bituminous | 320-3,900 | 8-1,700 | 3.0-7.8 | 0.34-1.0 | 0.05-0.3 | 460 |
| Lingan, Nova Scotia | mine | bituminous | 370-22,900 | 35-23,100 | 5.5-6.0 | 5.3-5.9 | 0.04-1.6 | 750 |
| Thunder Bay, Ontario | storage | bituminous/lignite | 4-0 | 1510-1,890 | 7.5-8.3 | 0-0.18 | 0.02-0.05 | 300-347 |
| Bienfait, Saskatchewan | mine | lignite | 1-812 | 2,100-9,460 | 8.0-8.8 | 0-10.2 | 0.02-0.86 | 794-2,706 |
| Coronach, Saskatchewan | mine | lignite | 4-6 | 850-1,100 | 6.9-8.1 | 0.05-0.97 | 0.02-0.41 | 209-270 |
| Luscar, Alberta | mine/mill | bituminous | 10-69 | 343-890 | 8.2-8.4 | 0.08-0.84 | 0.31-3.8 | 37-320 |
| Wabamun, Alberta | mine | sub-bituminous | 25-1,700 | 360-1,100 | 8.2-8.5 | 0.01-0.18 | 0.08-0.57 | 30-95 |
| Grande Cache, Alberta | mine/mill | bituminous | 240-770 | 1-70 | 8.0-8.4 | 0.01-0.06 | 0-8.8 | 50-370 |
| Elkford, British Columbia | mine/mill | bituminous | 5-5,850 | 5-573 | 7.5-8.4 | 0-7.25 | 0-26.3 | 23-206 |
| Vancouver, British Columbia | storage | bituminous | 49-374 | 48-107 | ND | 0.08-0.10 | 0.10-0.11 | 62-82 |
| Delta, British Columbia | storage | bituminous | 380-740 | 159-525 | 6.4-6.9 | 0-0.12 | 0.15-0.24 | 6-8 |
| Water-quality guidelines [1] | | | | 500 | | 0.01 | | 10 |

All values are expressed in mg·l⁻¹ except pH. ND indicates value not determined.
[1] Upper limit for human consumption.

Sources: Paine and Blakeman (1987); Task Force on Water Quality Guidelines (1987).

effluent source and the receiving streams. The effluents reduced the standing crops of benthic insects; the zone of influence, however, was small, mainly confined to small tributary streams.

Sulphur is an undesirable residual of many coals. Its oxidation leads to the production of acid drainage from mines and waste heaps. Burning the coal produces sulphur dioxide, which contributes to acid deposition. These topics are discussed more fully in Chapters 4 and 5. The average sulphur content of Canadian coals ranges from 0.4% for most western coals to more than 2% for those from Nova Scotia (Tables 9.2 and 9.3) and even higher for New Brunswick (Landheer et al. 1982). On average, about half of the sulphur is associated with the coal itself and half with the associated inorganic materials (mainly in the form of pyrite). The latter may be removed relatively easily by beneficiation, depending upon its grain size and on its distribution in the coal. Such techniques as chemical comminution (Howard and Datta 1977) may be used to assist in the liberation of inorganic sulphur without causing excessive size reduction of the coal. No satisfactory commercial method has yet been developed, however, for the removal of chemically bound organic sulphur (Dibbs and Marier 1973). Although such removal is possible using chemical or biological methods, these techniques are expensive to use and have not yet been fully developed (Sirois and MacDonald 1983; Paine and Blakeman 1987; Barrie and Carr 1988).

Because of their low sulphur content, the mining of Rocky Mountain coals produces few problems with respect to acid drainage. By contrast, coal-mining activities have increased the alkalinity of some streams in the Alberta foothills; further increases are likely in the future (Slancy 1971). The main environmental problem seems to be erosion; this process leads to turbidity and sedimentation in those streams that drain from mined areas. The U.S. Geological Survey has determined that the sediment yield from strip-mined portions of a watershed can be 10–1,500 times the amount from undisturbed land. The problem is especially critical in areas with high rainfall and fine-textured, easily erodible overburden. A solution was proposed by McCarthy (1971) for the state of Washington. When gravitational settling proved inadequate due to the colloidal nature of the eroded material, the addition of a flocculant and a two-pond settling system prevented siltation of the receiving stream. Stabilization of mine spoils by revegetation eventually slows down erosion and sedimentation. However, the problems described above commonly occur both during the period before vegetation becomes established and afterward, for example, along roadways.

The extraction process, particularly by open-pit methods, results in a significant amount of dust emissions to the atmosphere (Joyner 1985). These are produced by blasting, bulldozing, dragline operation, loading, and hauling, as well as by wind erosion of active storage piles. This type of emission depends on the characteristics of the coal itself (including rank and degree of oxidation), on the intensity of human activity (such as weight and speed of a transporting vehicle), and on weather factors such as humidity and wind speed. The main losses are likely to result from wind erosion of fine material from exposed storage piles.

Coal preparation leads to fugitive dust emissions from roadways, stock piles, refuse areas, conveyor belts, and preparation facilities. Many of these are now minimized by the widespread use of enclosed storage silos and covered conveyors. The thermal drying of coals separated in aqueous media produces major emissions, as do air separators (Joyner 1985). Typical emission factors for drying are 10 kg of particulates per tonne of coal processed without controls; this figure drops about half for cases in which air cyclones are used to recover coal fines. In addition to particulate matter, which consists mainly of large particles with short atmospheric residence times, there are also emissions of sulphur dioxide and nitrogen oxides, both of which generate acid rain when in the atmosphere, and such volatile organic compounds as methane, a toxic greenhouse gas. The additional use of Venturi scrubbers and mist eliminators can reduce emissions to less than 1%. The only coals currently dried in Canada are those bituminous coals produced in Alberta and British Columbia that are destined for export or for metallurgical use (Romaniuk and Naidu 1987). Another source of atmospheric pollution is the spontaneous combustion of in-mine coal or heaps of stored waste. Although this occurs infrequently, it can be very difficult to manage (Couper 1990; Nieman and Meshako 1990).

Since regulations with respect to both acid drainage and sulphur dioxide emissions from coal combustion are becoming increasingly stringent, new methods of reducing sulphur content are being sought. The problem of sulphur pollution has been partly responsible for the substitution of western coal for eastern coal, which is higher in sulphur (Environment Canada 1986). The benefit of the lower sulphur content of western coal is partly offset by its lower caloric value and by the effects of large-scale surface mining on the plains and in mountainous areas; therefore, it is still necessary to make eastern coal less environmentally harmful.

Since the review of the methodology over two decades ago (Dibbs and Marier 1973), there have been developments in the desulphurization of coal by use of a number of advanced physical, chemical, and biological methods (Howard and Datta 1977; David et al. 1991). Of particular note are the removal of pyritic sulphur by means of a high-gradient magnetic separator (Mathieu 1981) and the use of microorganisms, such as *Thiobacillus thiooxidans* and *Sulpholobus acidocaldarius*, to oxidize the sulphides, facilitating their removal. It is important to remember, however, that transferring the sulphur from the marketable coal to the solid wastes merely transforms a potential atmospheric emission problem into one of acid drainage if the waste is in the sulphide form. If the sulphur has been oxidized previously in the desulphurization process, it will not generate acid and, thus, will not pose a threat to the environment. More detail on acid generation can be found in Chapters 4 and 5.

## Reclamation

Grimshaw (1986) suggested that proper reclamation could not only return an area to its pre-mining condition, but might also, in some cases, "improve" upon the original landscape. This might involve altering surface topography in order to render drainage more effective, diverting rivers and streams, introducing new features into the landscape, or making the area more suitable for agriculture. It should also be noted that small amounts of coal are recovered through the reclamation of old mine waste dumps (Paine and Blakeman 1987). While these views may well apply to more-recent strip-mining operations that are designed and carried out to minimize environmental effects, they may be inappropriate for open-pit operations in mountainous terrain, or even for strip-mining in level terrain mined without concern for reclamation (Trost 1972). Rehabilitation is also more difficult when the overburden is saline or acidic, when the soil is thin or poor in nutrients and micro-organisms, or when climate is arid (Committee on Mineral Resources and the Environment 1975). Procedures for dealing with such situations are discussed more fully in Chapter 4.

In Saskatchewan, lignite coal has been mined by surface methods in the Estevan coalfield since 1930, resulting in the disturbance of several thousand hectares of land. The strip-mines are located within the broad Souris River Valley and on adjacent glacial till plains. The most common materials in the spoils are glacial till of loam to clayey loam texture and silty to clayey bedrock materials (Anderson et al. 1975). The till has low to moderate amounts of soluble salts and sodium, in contrast to the high sodium levels of the bedrock materials. Nitrogen and phosphorus contents are extremely low and probably limit growth on most sites. Montmorillonite is the dominant clay mineral which, when coupled with the high sodium levels of the bedrock materials, results in strong surface crusts that limit the infiltration of moisture and restrict rooting of vegetation.

Many native prairie species have invaded the area, particularly on older, non-saline spoils. Weed species such as kochia, Russian thistle, and perennial sowthistle are most common on recent spoils. Alkali grass and wild barley are dominant on saline spoils and on those high in sodium; sweet clover and bromegrass are also often present. Soils with extremely high soluble salts and sodium generally have little or no vegetation.

Guidelines established for the Estevan mining area (Saskatchewan Environment 1984) outline the objectives of reclamation. Operators are expected to return to productive use those lands disturbed by surface mining. This is accomplished by ensuring the stabilization of soils and the establishment of self-sustaining vegetative cover as soon as possible, preferably concurrently with ongoing mining operations. Land classes 1 through 4, under the Canada Land Inventory Soil Capability for Agriculture, are to be returned to agricultural uses by ensuring adequate grading of slopes, proper drainage, topsoil replacement, and soil amendments to reestablish adequate soil quality. On land unsuitable for agriculture,

**Table 9.5** Plant Species Recommended for Coal Mine Spoils in the Estevan Area

| Nonsodic | Moderately Sodic | Highly Sodic |
|---|---|---|
| Alfalfa | Streambank wheatgrass | Streambank wheatgrass |
| Altai wild rye | Altai wild rye | Altai wild rye |
| Tall wheatgrass | Yellow sweet clover | Russian wild rye |
| Intermediate wheatgrass | | |

Source: Godwin and Abouguendia (1986).

revegetation is to be undertaken to provide shelter and food for wildlife. In order to build a sound basis for future reclamation efforts, records are to be maintained of all pertinent activities.

The quality of forage produced on revegetated coal spoils near Estevan is similar to that from vegetation in the same ecological zone (Godwin and Abouguendia 1986). Grass-legume mixtures give good forage and enhance species diversity of herbivores. Amendments with hay and straw resulted in better establishment of grasses by improving infiltration and permeability of soils. Although plant responses to chemical amendment of spoils were poor, there was evidence that some amendments reduced the sodium adsorption ratio (SAR). The recommended optimum treatment for highly sodic spoils (of an SAR greater than 9) included organic amendments (hay or straw) combined with chemical amendments ($CaCl_2$) and supplemental irrigation. Late fall seeding at high rates was recommended. Species selection differed depending on spoil characteristics (Table 9.5).

The bituminous deposits in the mountainous areas of Alberta and British Columbia, whether mined by open-pit or underground methods, experience serious environmental problems. Because of the complex topographies of both the land surface and the coal deposits themselves, major difficulties include erosion, slope instability, and poor drainage water quality. Mountainous areas are often characterized by steep slopes, short growing seasons, severe climates, and thin and easily disturbed soils. Hence, there is a high potential for erosion, stream sedimentation, and landslides, all of which can disrupt stream channels. Ongoing mine reclamation is very challenging because damage occurs easily and is slow to heal. Both wilderness and recreational areas are affected in forest or alpine tundra zones.

Bighorn sheep are an important animal species in the mountainous regions, and the amount and quality of winter range are of great importance in maintaining healthy populations. Surface mining can destroy winter range where open pits, overburden dumps, haul roads, and processing plants are located. Etter (1973) emphasized that attempts to reclaim winter range must take into account the complexity of the habitat bighorn sheep require; few of the native plant species they use are commercially available for seeding, and introduced forages are used as replacements, with species selected according to site characteristics. Animals use south-facing slopes in winter; soil should first be replaced there, graded 10 to 20 degrees, and rapidly stabilized with grasses and legumes. Hydroseeding appears well-suited for revegetation on terrain that is rugged and characterized by variable surface conditions.

Coal mining in the Alberta foothills near Hinton has significantly altered the landscape; what was once a closed forest of spruce, fir, and lodgepole pines is now open terrain with both active and abandoned pits and reclaimed areas interspersed with parts of the original landscape. Two hundred bighorn sheep occupy a mine lease in the district, concentrating themselves in areas providing rocky escape terrain—the high walls of exhausted pits, for example, which they also use for bedding sites, travel routes, sources for mineral licks, and occasional lambing—in proximity to areas with quality forage (MacCallum 1989). Despite active mining in the area, the sheep remain. Maintenance of these high walls as part of bighorn sheep habitats must be recognized as an important aspect of reclamation.

Reclamation of a coal-mined area in the Hinton area includes conversion of an exhausted pit into a mountain lake to be used for sport fishing. This was made possible by the pit's location in a natural creek valley, where a water supply was guaranteed. Water diverted from a nearby lake provided extra water flow, as well as a source of benthic and planktonic organisms to colonize the new lake. Resloping of part of the steep pit walls provided shallow water required for development of a littoral zone; this is important for supplying diverse and abundant populations of organisms to form the base of an aquatic food chain. Pondweed and northern water milfoil were transplanted into the new lake from a natural lake nearby, and fish habitats were improved. Initial water quality in 1986 was poor, with high turbidity, low oxygen, and eutrophication; by 1988, however, water quality was much improved. High nitrogen levels, from the use of ammonium nitrate/fuel oil as blasting agents during mining, were not considered a major problem. More biological improvements were planned for 1989, including further introductions of macrophytes and benthic invertebrates, as well as trout species (Acott 1989).

Moose and deer frequent a coal test-minesite in west-central Alberta revegetated with alfalfa, clover, lodgepole pine, white spruce, and willow (Roe and Kennedy 1989); moose and deer were found at the site within one year of its revegetation. Willows were consistently browsed for the five or six years during which the study was conducted, suggesting that planting this genus would attract moose. Seed mixes with a high proportion of legumes tended to attract deer. Moose diet was made up of willows (88%) and spruce (8%); deer diet was primarily *Cornus* (31%), legumes (23%), *Equisetum* (15%) and *Populus* (6%).

Macyk (1974) carried out revegetation trials in a coal mining area near Grand Cache, Alberta. After coal was extracted, the area was backfilled and levelled; salvaged topsoil was spread on the surface. Problems were encountered in revegetating steep slopes, due to instability (creep) of the surface, and the lack of favorable microsites for seed germination. Containerized lodgepole pine and white spruce seedlings had a maximum survival rate of 54%. The main factor reducing survival was frost-heaving.

Experiments at the Byron Creek Collieries, southeast of Sparwood, British Columbia, have demonstrated that unconsolidated runoff sediment can be used as a soil amendment to increase the success of revegetation (Kennedy and Kovach 1987). Runoff material includes fine particles of mudstones, shales, coal, and other sedimentary rocks carried from mining areas by water and collected in interceptor ditches around the mine site. This inert, coarse-textured material, which has a low nutrient status, was spread on the surface of a waste dump slope of 21 to 26 degrees, then seeded with a mixture of grasses and legumes and fertilized. Red fescue and orchard grass dominated the plant cover after establishment. The soil amendments improved the production of seeded forages, as well as the survival of lodgepole pine seedlings.

## Peat

Peatlands are areas with extensive accumulations of organic material and are often waterlogged. Peat develops when the accumulation rate of dead plant material exceeds its decomposition rate. This accumulation of partially decayed plant material is the result of certain climatic and physical conditions, generally cold and wet (Moore and Bellamy 1974; Keys 1992). Canadian peatlands are typically 5,000 to 10,000 years old, having been formed since deglaciation.

Peatlands contain peat of varying age. Younger peat, located at the surface of the wetlands, tends to be relatively undecomposed and light yellowish-brown in color, with a high fiber content. It is used primarily for horticultural purposes and to purify industrial wastewater (Coupal 1985; Bergeron 1992). It is also used in the production of paper towels, metallurgical coke, and activated charcoal. Because of the cellular structure of peat and its high capacity for ionic exchange, it is ideally suited for use as a natural filter for acidic and other drainage waters and as an absorbent for oil spills (Coupal 1985).

More mature peat, which is almost black in color, is the variety that is used for fuel. Fuel peat is widely used as an energy source in many European countries and has been considered as an alternative low-sulphur fuel for power generation in Canada. Peat can be burned directly or converted to coke, synthetic natural gas, or methanol. While its calorific value, on a dry-weight basis, is about the same as that of coal, peat has the environmental advantage of a considerably lower sulphur content than that of most Canadian coals. For domestic use, it can be sold in briquettes and produces very little dust.

It is estimated that 50 Mt of peat accumulate in the natural environment in Canada every year. That current application for peat harvesting utilizes only 700 to 800 kt·yr$^{-1}$ demonstrates the difficulty of harvesting peat economically (Keys 1992).

## History

Prior to the Second World War, peat production in Canada was limited to a few small plants in Eastern Canada, one in Alberta and two bogs in British Columbia. Production was insignificant, as a result of a low domestic demand for the product for agricultural use. With the advent of war, however, demand increased; by 1944, 32 Canadian bogs were producing peat for export to the United States.

## Orebodies

Worldwide, most peatland lies in the circumboreal zone of coniferous forest, or taiga, of northern Europe, Alaska, and Canada. Peat is found in bogs, swamps, and marshes. Peatlands cover approximately 12% of the Canadian land surface and are found in every province, as shown in Table 9.6. Most of them, however, are not useable for fuel (Bergeron 1993). Some occur in areas with permafrost.

Peat harvesting requires thick deposits of high-quality peat. The orebody must be at least two meters deep, with an area of at least 50 ha, a good potential for drainage, and suitably dry weather during the harvesting season.

The bulk of the peat produced in Canada comes from Québec (39%), New Brunswick (36%), and Alberta (12%); most is of the horticultural variety (Bergeron 1993). Although Ontario is estimated to have 23 Mha of peatland (Keys 1992), much of it is underlain by permafrost, which makes recovery very difficult (Guillet 1985).

## Extraction and Processing of Peat

The usual method of mining (or harvesting) peat is to drain the area, remove the vegetation, level and loosen the surface soil to allow it to dry, and then remove the peat by vacuum or mechanical means (Boffey 1975). The operation is limited to non-permafrost areas and is practicable only under dry weather conditions during the summer months. The surface must be kept bare as long as the operation continues, to permit continued drying.

The peat-harvesting area would increase considerably if peat were used as fuel. It is estimated that as much as 40 kha would be required to supply a single gasification plant using the harvesting method described above (Boffey 1975). In addition, the operation of equipment and provision of access roads could increase the disturbed area by 10%–20%; a much larger increase would result from the flooding and waterlogging of adjacent lands due to peat-bog drainage. Other environmental effects include the lowering of the land surface as a result of the peat removal, and an exacerbation of flooding and water-quality problems in the mined region.

Winkler and DeWitt (1985), in their review of the environmental effects of such activities in the United States, attributed the obscurity of information

**Table 9.6**  Distribution of Canadian Peatlands

| Province | Peatland Area (ha x $10^3$) | Percentage of Area of Province | Percentage of Canadian Total |
|---|---|---|---|
| Alberta | 12,673 | 20 | 11.4 |
| British Columbia | 1,289 | 1 | 1.2 |
| Manitoba | 20,664 | 38 | 18.6 |
| New Brunswick | 120 | 2 | 0.1 |
| Newfoundland | 6,429 | 17 | 5.8 |
| Northwest Territories | 25,111 | 8 | 22.6 |
| Nova Scotia | 158 | 3 | 0.1 |
| Ontario | 22,555 | 25 | 20.3 |
| Prince Edward Island | 8 | 1 | 0.01 |
| Québec | 11,713 | 9 | 10.5 |
| Saskatchewan | 9,309 | 16 | 8.3 |
| Yukon Territory | 1,298 | 3 | 1.2 |
| Canada | 111,327 | 12 | 100 |

Source: adapted from Keys (1992).

regarding peat harvesting to the small scale of traditional operations. By contrast, a more modern process of large-scale peat utilization is currently under consideration as a result of shortages in fossil-fuel energy sources. Winkler and DeWitt advise against this utilization of peat, on the basis that it is non-renewable and would have many damaging effects on the environment, including the release of toxic metals and organic pollutants from the peat, eutrophication of surface waters, increased runoff and changes in groundwater supply, and increased air pollution and fires. Peat harvesting is fundamentally different from other types of mining, in that it disturbs very large expanses of land.

An assessment of the hydrological effects of a peat-mining operation in Newfoundland was conducted by Panu (1989). These effects included a two- to five-fold increase in peak water flows from the peat-harvested area, significant changes in runoff volume and quality, and a considerable increase in the suspended peat load of the runoff water.

Peat has been mined for centuries in Ireland, intensively for decades (Aldwell 1990). Earlier considered wastelands, the bogs are now a focus of conservationism, especially as over 90% of the "raised bogs" have been destroyed. Apart from loss of terrestrial habitats, peat harvesting has resulted in siltation of waterways. While non-toxic, the silt smothers spawning grounds and threatens drinking water supplies. Reclamation of the mined-out peatlands is in the form of afforestation, conversion to cropland, and some flooding to create lakes for recreational purposes.

Most literature about human effects on peatlands deals with those that are drained for agricultural use, forestry, or development. According to Sjörs (1980), the greatest ecological impact of peatland destruction is the reduction in the numbers of specialized plants and animals that occur there. Conservation is required to save representative types for the future, especially in temperate zones.

The necessity for peatland conservation was emphasized by Maltby (1992) in a volume devoted to sustainable use of wetland resources.

## Graphite

Graphite, also known as plumbago and black lead (although it contains no lead at all), is a natural form of carbon and has many desirable physical and chemical qualities. It is an excellent conductor of heat and electricity, extremely resistant to acids and chemically inert; it has a wide variety of uses, ranging from automotive parts to dry-cell batteries and pencils (Boucher 1992). Graphite's highly refractory nature is the primary reason for its use in foundry facings and in the metallurgical and refractory industries.

In 1993, Canada's two graphite operations (Figure 9.1) produced about 22 kt, which represented slightly less than two percent of world production (Boucher 1994a). Most of this production was exported to the United States.

### History

Major graphite production in Ontario began in 1870 at the Port Elmsley deposit. The Black Donald mine at Calabogie in eastern Ontario started operation in 1884 and continued until reserves were exhausted in 1954 (Prud'homme 1986).

### Orebodies

Graphite is found in one of two forms. The first of these is crystalline flake graphite, which is disseminated in such calcareous or siliceous metasediments as gneisses, schists, and marbles. Deposits of this type have been found in Saskatchewan, Ontario, Québec, Nova Scotia, and New Brunswick. The second variety is vein graphite, which occurs in the form of circular or massive vein accumulation along the contacts of intrusive rocks with limestones (Prud'homme 1986). Canadian graphite deposits occur principally in eastern Ontario and Western Québec. They are associated with gneisses and marbles.

### Extraction and Processing of Graphite

The beneficiation process involves comminution with an impact crusher, followed by two stages of flotation, then filtration. Ball-mill grinding is required for only a small fraction of the ore.

In the absence of any specific information concerning residuals of the extraction and beneficiation processes, it can be assumed that they are likely to resemble those of coal mining and milling.

## Diamonds

Diamond is a mineral composed entirely of crystallized carbon. It is formed in magnesium-rich rock-melts at depths of 150 km or more below the Earth's surface. Extremely high pressures, provided by carbon dioxide gas, are required, as are temperatures in excess of 1,400°C. The host rock may be forced to the surface in the form of kimberlite pipes; under erosive action, this material may release diamonds into glacial tills or alluvial gravels. However, the major sources of mined diamonds remain these vertically oriented kimberlite pipes with dimensions approximating one kilometer in diameter.

Although there are as yet no mines in Canada, a large part of the central and northern regions of the country is underlain by the North American Craton, which is very old and forms the nucleus of the North American continent. Cratons contain most of the global distribution of kimberlite ores (Boucher 1993a). Prospectors have located diamondiferous kimberlite pipes in the Northwest Territories, Saskatchewan, British Columbia, Alberta, Ontario, and Québec.

A site at Lac de Gras, 260 km northeast of Yellowknife in the Northwest Territories, with proven diamond ores has been purchased and may be mined in the near future (Ulman 1994). The discovery of diamondiferous kimberlite pipes in the area in 1991 prompted widespread claim-staking and exploration for diamonds across Canada (*Mining Review* 1993). Diamond exploration expenditures in 1992 were nearly $19 million, a dramatic increase from $7 million in 1991 and $5 million in 1990, about 70% of it in the Northwest Territories (Cranstone and Bouchard 1994). Diamondiferous kimberlite pipes have also been located near Fort à la Corne, Saskatchewan. Both this site and Lac de Gras have produced gem-quality macro-diamonds (*Mining Review* 1993). According to Boucher (1994a), the geology indicates a strong possibility that diamonds may be discovered in Canada in commercial quantities.

Diamond is the hardest substance known. It also has the highest thermal conductivity and has a very low electrical conductivity. Diamonds can be divided into two classes: industrial diamonds, which are mainly used in abrasives and cutting applications, and gemstones.

Because it is not reactive, diamond requires only physical separation from its ore and produces no direct residuals. The main source of any residual material is the kimberlite host rock that is a magnesium-iron silicate and is probably environmentally benign. The most significant effects would probably be surface disturbance from both exploration and production, as well as the waste material associated with the beneficiation of the ore. Exploration is likely to disturb wilderness areas. It is likely that affected areas could be reclaimed with relative ease, assuming that such plans are incorporated fully into the original mining plan.

## Summary

The environmental effects of the mining of carbon products are quite variable and depend in large part upon the location of the mines as well as on the product. For both coal and peat, the greatest problem is the large extent of surface mining and the subsequent drainage problems. This is also a great concern in open-pit coal mining, especially in mountainous regions, where drainage plays a major role in dictating slope stability and erosion. Revegetation of these areas is especially important, but is generally problematic, due to the thin, easily disturbed soils.

Reclamation of strip-mining practices has been more successful in Saskatchewan than in other regions, due to the shallow nature of the deposit and the flatness of the topography. However, high salinity of soils in the vicinity of the mines has led to some problems with revegetation.

The greatest challenge facing coal mines in eastern Canada is the high sulphur content of the product. Until desulphurization techniques are improved, it is likely that deposits in the west will continue to be chosen over those from the east, in spite of the facts that the underground mining operations in the east pose a lesser environmental threat and that the coal is of higher caloric value.

# CHAPTER TEN

# POTASH AND OTHER SALTS

## Introduction

This chapter examines the mining of the chlorides and sulphates of sodium and potassium. Several other salts, including gypsum, anhydrite, limestone, and magnesite, are mined either as structural materials or as sources of metallic elements. The more important of these are treated separately in Chapter 11. As Tables I.1 and I.2 indicate, potash and other salts together make a significant contribution, in terms of both value and quantity, to Canadian mineral production.

Changes in sea level, current direction, and embayment size are all examples of changes that occur over geological time and that allow for the deposition of evaporite deposits such as potash and other salts. When these changes occur, sedimentary deposition basins, such as those within embayments, may be cut off from the main section of a sea so that they no longer receive water. When this happens, the water evaporates over time, leaving behind as a deposit the salts it contained. Less-soluble salts, such as calcium sulphate, precipitate first in these situations, followed by those that are more soluble. Potassium and magnesium chlorides are the last to be deposited (Haryett 1983). Due to these precipitation mechanisms, evaporite deposits are usually layered, with strata of different salt types.

## Potash

The term "potash" refers to the oxide and carbonate compounds of potassium that were, at one time, produced from wood ashes. The main commercial potassium minerals are sylvite, carnallite, kainite, and langbeinite, with sylvite accounting for the majority of the ore. Potash production is still often expressed in terms of the potassium oxide ($K_2O$) equivalent, although potassium chloride

259

(KCl) is the actual chemical form of the mineral sylvite. This book uses only weights of KCl, which can be converted to $K_2O$ equivalents by multiplying by 0.63.

There were ten operating potash mines in Saskatchewan in 1993 and two in New Brunswick. (Two Saskatchewan mines used solution-mining, a process described in Chapter 1.) Eighty-five percent of production came from the Saskatchewan potash deposit, the Prairie Evaporite, which lies at a depth in excess of 1,000 meters and represents 40% of known world reserves. In Saskatchewan alone, known deposits are adequate to supply world demand for at least 2,500 years.

Although potash is used mainly as an agricultural fertilizer, roughly 5% is used in the manufacture of matches, dyes, television tubes, pharmaceuticals, and detergents (Haryett 1983).

## History

The first Canadian potash mines began construction in Saskatchewan in the late 1950s. Continuous production of the mineral began in 1962. During the next decade, the industry expanded rapidly and by 1970 production capacity was close to 12 million tons of KCl (Boyd 1976). By 1978 there were ten mines in operation in Saskatchewan (EMR 1982). In New Brunswick, activity in the industry first began in 1971, with the discovery of the Penobsquis deposit in the eastern part of the province. Extensive exploration was done in the area in 1975 to further delineate this deposit and another nearby at Cloverhill; mining at the Penobsquis site began in 1983 and at the Cloverhill site in 1985 (Howie 1988).

## Extraction and Processing of Potash

All Canadian potash mines are underground operations, most using borers and continuous conveyors. Three mining methods are in use: room and pillar, chevron pattern (in which the mined areas are allowed to collapse and the mine openings to close), and cut and fill.

Of Saskatchewan's two potash solution mines, the one at Belle Plaine has used solution methodology since it first began production thirty years ago. The other, near Saskatoon, was converted to solution methods in 1989 after years of flooding problems. A heated brine is circulated through the underground workings, dissolving the potash ore and forming a brine solution, which is then pumped to the surface to cool in ponds covering 30 ha. There, the solutes precipitate and are dredged and processed (Scott 1991).

## Rock Salt

Rock salt is also referred to by its mineral name, halite (NaCl). The main global markets for salt are as a chemical raw material (60%), table salt (20%), a road de-icer (10%), and in animal feed and water treatment (10%). Canadian consumption, however, is dominated by its uses as a de-icer and in the chloralkali industry (Morel-à-l'Huissier 1994a). There were 19 salt operations in Canada in 1993, with about 60% of product coming from Ontario.

## History

In 1866, while drilling for oil at Goderich, Ontario, Sam Platt discovered a deposit of very pure NaCl. The following year, Canada's first commercial salt operation started, with the salt being extracted by brining. In 1917 large deposits of rock salt were discovered at Malagash, Nova Scotia. This became the site of the first underground salt mine in the country. The mine operated from 1919 until 1959, when the present mine at nearby Pugwash was opened (EMR 1986).

## Extraction and Processing of Rock Salt

Seven of the mines operating in 1993 were conventional, producing solid rock salt, either as primary product or as a by-product of potash. The balance were brining operations, using solution-mining or extracting salt from potash tailings. Twenty-five percent of brine production was evaporated for table salt, fisheries, and water-conditioning uses. The remainder was used by the chloralkali industry and for de-icing. Two other operations produced salt in conjunction with potash mining. Typically, 70% of production is in the form of rock salt, 25% as brines, and the remainder as evaporated salt. The chloralkali industry processes salt chiefly into caustic soda, chlorine, sodium carbonate, and calcium chloride (Morel-à-l'Huissier 1994a).

## Sodium Sulphate

Sodium sulphate ($Na_2SO_4$) is used primarily as a component of detergents (40%), and in the pulp and paper (35%), glass (5%), and textile-dyeing industries (Morel-à-l'Huissier 1992). While it is produced around the world in a variety of ways, including as a by-product of various chemical operations, the largest single source of sodium sulphate is naturally occurring brines and evaporites. At the end of 1992, there were six Canadian plants in operation, all in Saskatchewan (Morel-à-l'Huissier 1993).

The repeated yearly cycle of water evaporation, salt saturation, and subsequent precipitation in the shallow lakes of Alberta and Saskatchewan has resulted in fairly thick deposits of sodium sulphate in the lake sediments. These lakes have restricted drainage and contain a number of salts in solution. During the dry, hot summers, the salts become more concentrated as the water evaporates and eventually precipitate out of solution as the water becomes saturated. Thick deposits of the salt form in the lake beds as this yearly cycle continues.

## History

The sodium sulphate deposits in the alkali lakes of Saskatchewan were discovered in 1918, in the course of extensive exploration for a domestic source of potash to replace suspended supply from Germany. An erroneous report on the presence of potash led to a rush of claim staking. The leaseholders were disappointed to find that there was no potash on their claims, and many simply allowed them to lapse. A few, however, realized that their claims did contain large quantities of Epsom and Glauber's salts, the hydrated forms of magnesium and sodium sulphate. From 1921 to 1924, the Dominion Department of Mines made a complete survey of the sodium sulphate occurrences in western Canada. The conclusion that the deposits were of great commercial interest led to their subsequent development (Tomkins 1954).

## Extraction and Processing of Sodium Sulphate

Sodium sulphate is extracted either directly or by reaction with potash to produce potassium sulphate, resulting in a sodium chloride residual ($Na_2So_4 + 2KCl \rightarrow K_2SO_4 + 2NaCl$) that is returned to the lake. The energy required is mainly in the form of natural gas, and the releases to the atmosphere and lithosphere are ordinarily expected small.

Sodium sulphate is also produced from natural brines (MacWilliams and Reynolds 1973). This product comes mainly from alkaline lakes in Saskatchewan and Alberta (Morel-à-l'Huissier 1992). Lake brines are concentrated by pumping them into reservoirs during the summer, where continued evaporation leads to near-saturated conditions. In the autumn, when the solution cools, crystallization begins; the least soluble sodium sulphate leaves solution first. The remaining brine, which contains other salts such as sodium and potassium chloride, is pumped back into the lake, and the crystals are harvested and stockpiled.

The mining of sodium sulphate produces somewhat different residuals than do any of the operations previously described in this chapter (MacWilliams and Reynolds 1973). Since the material is contained either in the lake sediment or in the water itself, the mining process essentially consists of water evaporation, leaving the mineral behind. Requirements are heat for extraction and beneficiation processes and fresh water to compensate for evaporation. Any surplus water is usually returned to the parent lake. Residuals are related to the changing chemistry of the individual lake.

The potassium sulphate facility at Big Quill Lake, Saskatchewan, came into production in 1992 after operating as a pilot plant for several years. It was planned to have a 50-year lifetime. Of the 170 ha of total land it expects to require, approximately 10 ha would be severely disturbed. By contrast with a sodium sulphate operation, the production of potassium sulphate will release sodium chloride into the lake, gradually changing its chemistry over a period of years.

## Environmental Effects

The environmental effects of the mining and milling of potash and other salts are mainly local. The most important are surface disturbances caused by waste disposal and land subsidence, and changes in air, soil, and water quality resulting from the dispersal of salt particles.

Since most of these mining operations are located in agricultural regions of Canada, there is the potential for conflicts between the two industries. Undoubtedly, the most prominent environmental challenge posed by the mining of these substances is the effective containment of large quantities of waste sodium chloride and brines, to prevent their dispersal to adjacent land and water bodies.

The extraction stage itself produces very little waste, whereas beneficiation results in the rejection of almost two-thirds of the ore as tailings (Godin 1992). This material is composed primarily of sodium chloride, with small amounts (5%–10%) of iron oxide, insoluble clays, carbonates, and sulphide minerals. In Saskatchewan, the tailings are usually stockpiled on the surface—a result of the lack of a major market for the salt, the difficulty of returning it underground, and the much lower cost of surface containment. The beneficiation of potash consumes a great deal of energy—principally in the form of natural gas fuel—about half of which is used for product drying. Almost 25% of the natural gas consumed in Saskatchewan in 1982 was used by the potash industry (Saskatchewan Energy and Mines 1984); this amounted to a total energy use of approximately 30 PJ—almost 0.9 TJ per tonne of extracted ore.

While the underground mining of rock salt produces little direct waste, it does create somewhat more waste than potash does. However, salt mill tailings are considerably less, amounting to just over one sixth of the mined ore (Godin 1992). Hydrospheric and atmospheric releases are similar to those of potash mining and milling. The main environmental effects of brining operations are the potential for groundwater contamination and subsidence, and atmospheric emissions from the drying process.

## Surface Disturbance

Land subsidence is a common effect of underground mining of sedimentary rocks and is particularly prevalent in potash and salt extraction. This subject has already been described generally in the section on extraction in Chapter 1, but also has some attributes which relate more specifically to the mining of salts. Most bedded deposits are weaker structurally than igneous and metamorphic rocks. When they are overlain by weak material (as is the case with potash) or are located at shallow depth (as is the case with salt), there is a high potential for surface subsidence. The problem has occurred in many parts of the world, including Germany, England, the United States, and Canada (Martinez 1971). Subsidence may be induced not only by the direct extraction of material, but also by the dissolution of deposits by groundwater, which gets access to the material through mining activities. Dissolution by groundwater occurs only if the intruding groundwater is not already saturated; groundwater of the deep aquifers in the potash region of Saskatchewan is often quite close to, if not at, saturation. The use of room-and-pillar extraction methodology—whereby approximately 60% of the ore is left in place to provide support (Strathdee 1994) —minimizes subsidence as long as extraction continues. After decommissioning, however, any unsaturated groundwater intrusion may lead to the dissolution of the pillars and thus to subsidence. At the deep potash mines currently in operation in Saskatchewan, subsidence, if a problem at all, is fairly minor, with sinking occurring quite slowly and over a broad distance (Saskatchewan Potash Producers Association: personal communication 1993).

In one of the potash mines of New Brunswick, it is common practice to backfill the mined rooms once excavation in that particular area is complete. At the other New Brunswick mine, backfill is placed in the active stopes and is used as part of the mining cycle. Backfilling at these mines is more prevalent than in Saskatchewan because regulations prohibit surface storage of tailings. Therefore, some salt is backfilled, and some is dissolved and discharged into the sea. In 1986, studies were begun to look at hydraulic backfilling as a more efficient alternative to the dry method that had been practised. It has been shown that, to reduce subsidence, the backfilled tailings must have a certain density and stopes must be filled to a certain degree; this degree was not being met using dry backfilling methods. It was found that hydraulic backfilling could attain the required density placement and was much more efficient at completely filling stopes and rooms. Plans were made to replace completely the dry backfilling methods with the hydraulic method by 1989 (MRAZ Project Consultants Ltd. 1989).

## Solid Wastes

The waste produced by potash mining is mainly common salt. Tailings typically amount to approximately 65% of the weight of ore produced (Godin, various years). Somewhat less than half is contained in ponds and the remainder in tailings piles (Saskatchewan Potash Producers Association: personal communication 1993). In the thirty years during which potash has been mined in western Canada, more than 300 Mt of waste salt have accumulated around the ten operating mines. The total land area occupied by waste was estimated to be 3,400 ha in 1989 (Saskatchewan Environment 1991).

The relationship between waste volume and area covered depends on the handling of slimes, the process technology used, and the stacking height of the piles—usually between 20 and 60 meters—which is limited by the strength of the underlying soils (Hart 1989; Strathdee 1994). It has been estimated that by the time the potash deposits have been exhausted, the total wastes accumulated will have reached 20 Gt, covering 160,000 ha (Hart 1986). However, this estimate is based on the assumption that absolutely all the evaporite deposits in the Prairies will be mined out, that waste management techniques will remain unchanged, and that the waste fraction will remain constant. Thus, it represents the worst-case scenario. The trend toward the use of higher stacking (or coning) reduces the required waste area.

## Liquid Wastes

According to Strathdee (1994), each Mt of potash product results in 1.98 Mt of tailings and 0.85 Mt of brine. A pond is required to cool and clarify the brine before the water is reused in the mill. For a typical potash mine, the brine-pond size is about 100 ha in area and one to two meters deep (Reid: personal communication 1993). An analysis of the chemical composition of the brine from a mine in southeastern Saskatchewan (Vonhof 1983) showed a pH of 7.6, with 200 parts per thousand of chloride, 89 of sodium, 40 of potassium, 8 of magnesium, 2 each of calcium and sulphate, and other substances in lesser amounts. Saskatchewan mines are not permitted to discharge any effluent to surface water.

In Saskatchewan, 10%–50% of dissolved salt waste and excess brine, usually about 0.32 Mt each year for a typical potash operation (Strathdee 1994), is injected into deep aquifers below the orebody, where the water is already highly saline. The remainder is stored in large tailings ponds. The amount depends upon local precipitation, the surface area of the tailings pile and brine pond, and the volume of freshwater used to aid dissolution (Reid: personal communication 1993). Accidental spills of brine during pressurized injections have salinized farmland around potash refineries (Struthers 1991).

The brine ponds at all the potash mines have brine levels above the surrounding ground surface. Thus there is a tendency for the brine to move downward and outward to the surrounding areas. Consequently, active measures to contain the brine are essential to prevent contamination of shallow groundwater, soil, and surface water around the waste disposal sites. Brine containment practices have been focused on retarding horizontal movement in surficial aquifers. They include using compacted clayey-till liners for brine ponds; lowering brine pond elevations to reduce hydraulic head and, consequently, the potential for flow; using slurry trenches or sheet pile cut-off walls; constructing extensive systems of dikes, backed up by passive or active containment using receptor drains and ditches; and employing containment wells (Figure 10.1).

In addition to shallow horizontal movement at all the mines, there is a potential for downward movement of brines to deeper groundwater at some of the minesites, particularly those that are located in groundwater recharge areas. A comprehensive study of this vertical brine migration (Maathuis and Van der Kamp 1994) concluded that vertical migration is very slow at most of the minesites because of the low permeability of the underlying tills and clays,

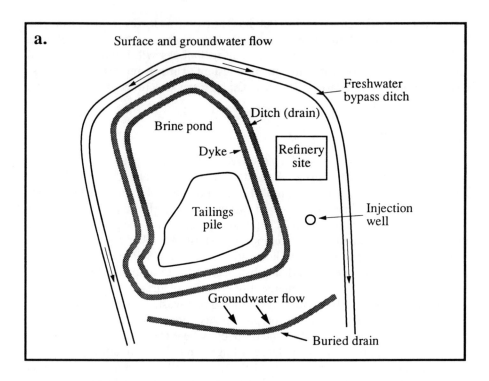

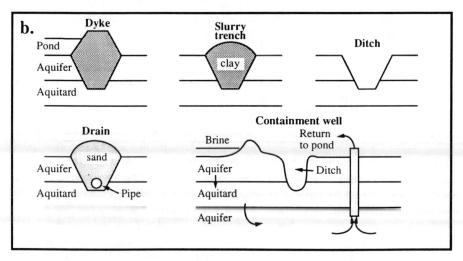

**Figure 10.1** Example of a waste management site (a) and potash brine containment structures (b) for a potash refinery (mill). Adapted from Hart (1989).

combined with small downward gradients at some of the mines. Diffusion is the dominant mechanism for deeper migration in unfractured tills and clays. Fractures in the shallow tills have permitted deeper migration at some of the minesites, and the density of the brines (about 1.2) has also led to enhanced downward movement. Brines will remain within the clay and till formations beneath the disposal areas for a very long time, spreading slowly by diffusion. Deep confined aquifers beneath some of the minesites may eventually become contaminated after hundreds of thousands of years. Control measures such as containment wells may need to be implemented in the future.

Monitoring of existing contamination is accomplished through recording of water levels and sampling of monitoring wells set up around the mine perimeters. Geophysical surveys, both surface and downhole, are being used increasingly to map out and monitor the migration of brine in the subsurface. Geophysics is very effective because the high electrical conductivity of the brine allows sensitive detection.

## Emissions to Air

The mining and milling of potash ore involve the release of a moderate amount of particulates to the atmosphere. This is relatively easy to control. The main sources of airborne particulates are stack emissions associated with the drying of potash products (MacWilliams and Reynolds 1973; Pretty et al. 1977). They contain sodium, potassium and chloride (Schutzman 1983), all of which are common in nature, are readily mobile in the environment, and have sources other than potash mining. Potassium, for example, is a major nutrient required by all plants. Even at relatively high concentrations, it is non-toxic in plant tissue. Sodium, on the other hand, is generally a non-essential and toxic element in plants. Chloride is readily absorbed and translocated by plants. It is required by plants in very small quantities, and is more toxic than sodium.

Particulate size distributions range from 0.1 to 50 mm in diameter. Approximately two-thirds of the particulates are in the suspended particulate range—less than 10 mm (Schutzman 1983). The high sub-micron content (about one quarter) requires a large energy input in the scrubbing process, which affects the design of emission control equipment.

Saskatchewan's Clean Air Act limits the concentration of salt particulates in air emissions to 0.57 g·m$^{-3}$ (Saskatchewan Environment 1991). Although environmental damage around potash refineries is not expected at this emission level, tree damage is reported. Injury to trees has been well documented but was confined to an area within a few kilometers of the refineries (Townley-Smith and Redmann 1980). Sodium and chloride tend to concentrate in the terminal buds of woody plants. The resulting effects on shoot growth include a distinctive "clubbed" pattern of branch growth.

The effect of salt dust is strongly influenced by environmental conditions, which explains why injury tends to be episodic (Redmann 1983a). Dew and high humidity increase injury from salts deposited on plants. Leaf surface character-istics—such as cuticle structure, stomatal distribution, and leaf wettability—all influence the uptake of salts. Salts can cause injury when applied to woody shoots in winter, as well as during the growing season.

In a research program sponsored by the Potash Corporation of Saskatchewan, injury by salt ions to aspen leaves, twigs, and buds was determined at different times of the year using a membrane leakage test (Redmann and Haraldson 1986; Redmann and Ryan 1988). At equivalent concentrations, more injury was caused by chloride salts than by sulphate salts. Sodium salts were slightly more damag-ing than potassium salts. Injury to leaves was greatest before mid-summer and least in fall. Twigs were generally undamaged by salt treatment in fall and winter, but were injured by chloride salt treatments in spring and summer. Buds were easily injured by salt treatments during periods of bud break in the spring and of new bud formation in late summer. This was observed after laboratory applica-tions of salts to field-collected buds (using the membrane leakage test), and confirmed in separate experiments involving the application of salts to intact twigs in the field, followed by observation of actual bud survival and develop-ment.

The lack of standardized techniques and variation in environmental condi-tions in different studies make it difficult to compare the relative resistances of plant species to salinity (Redmann 1983a). Herbaceous plants accumulate more sodium and chloride than woody species do, but are less likely to be injured. Among tree species, green ash and aspen are relatively sensitive, while Russian olive and caragana are more tolerant (Redmann 1983b; Redmann et al. 1986). Many perennial forage crop species are highly resistant to soil salinity and airborne salts. Injury by potash dust to annual field crops has not been reported (Hart 1983). To date, soils in the vicinity of potash mines have not generally shown significant increases in salinization, structural deterioration, or nutrient imbalances (Rennie 1983).

Salination of sloughs and lakes would result in loss of diversity of plants and animals (Maltby et al. 1992), as freshwater organisms have specific tolerances of $Na^+$ and $Cl^-$ ions. Among the most sensitive are amphibians, such as leopard frogs.

## Reclamation

There is no appreciable market for the huge amount of waste salt produced by potash mining, although efforts are being made to develop uses (Morel-à-l'Huissier 1993). A small fraction is used for road de-icing and other purposes, and some of the brine is returned underground. The bulk of the material is stored on the surface, where it must be contained indefinitely in an economically and environmentally acceptable manner (Kent and Clifton 1986). Its high susceptibility to dissolution may pose major threats to adjacent groundwater and soils if the waste is not adequately contained. The risk can be obviated somewhat by using waste salt as backfill when mining is completed.

Possible methods of managing the solid and liquid residuals produced by conventional underground potash mining include storage underground or on the surface, or discharge into the ocean. Underground disposal of solids is being used successfully at some mines in New Brunswick and Germany exploiting folded ore deposits. Unfortunately, the flat nature of the Prairie evaporite deposits makes this an uneconomic option in Saskatchewan, at least at present. If this limitation could be overcome, a large fraction of the waste could be accommodated, although it is not an option for abandoned mines that have already collapsed. Approximately 15% of Saskatchewan brine is stored underground (Hart 1985). This method could be expanded to include the use of slightly saline groundwater to dissolve surface salt and return it underground. The effects on receiving aquifers would have to be assessed.

Marine disposal of both solids and brine would be, perhaps, the least damaging to the environment if the material could be transported to a suitable site with high dispersal characteristics. While this may be feasible for the New Brunswick mines (and, in fact, is used by one of them), transportation costs for Saskatchewan mines would be prohibitive. This method may have lethal effects on some marine and estuarine organisms (Hutcheson 1983) because they are adapted to a narrow range of salinity.

Where solid wastes have been stored on the surface, reclamation will ultimately be required for all existing and future disposal sites in order to prevent continued dissolution of the exposed salt. This requires capping the material with a barrier to salt migration that will remain effective for hundreds of years. Although no potash mine in Saskatchewan is expected to be decommissioned in the near future, reclamation proposals are being developed (Hart 1989; Saskatchewan Environment 1991). Field-scale reclamation of potash tailings and associated areas contaminated by brine has not been attempted in Canada. In an effort to revegetate highly saline soil near a potash brine pond, Thorpe (1989) found that upward migration of salt rapidly contaminated capping layers of soil or sewage sludge 30 cm-thick added to the surface. Modelling of salt migration showed that even 150 cm of soil amendment would be rapidly contaminated by capillary movement of salt from underlying saline layers into the overlying soil. A capillary barrier is therefore essential to prevent the upward migration of salts. In Germany, plastic liners and layers of construction rubble have been used successfully for this purpose (Reid: personal communication 1993).

In order to be effective, any cap over tailings must restrict not only upward migration of salts, but also downward movement of precipitation. Water reaching the underlying tailings would dissolve the salts and cause surface instability. A soil cap, stabilized with permanent vegetation, would have to remain effective for centuries after abandonment. Capping might be most appropriate for reclamation

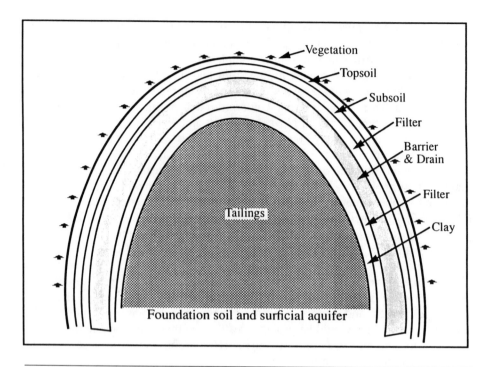

**Figure 10.2** Example of a multilayered soils cap over a potash heap, showing cap and pile drain interaction with foundation soils. Adapted from Hart (1989).

of brine ponds or salt-contaminated soils where some penetration of precipitation to the underlying saline layer would be acceptable. By contrast with tailings piles, dissolution and instability would not occur. Hart (1989) concluded that, considering the nature of potash tailings, unless a relatively complex system of multiple layers is used, it is doubtful that capping would be suitable for permanent reclamation (Figure 10.2). Recent studies done by the Saskatchewan potash industry, however, have had promising results using a single-layer cover of sand mixed with polymerized bentonite (Saskatchewan Potash Producers Association 1993); their results indicated that this type of cover appears to perform better than the multi-layered one.

Some attempts have been made to establish vegetation on potash tailings in Europe. To date, however, no reclamation has been carried out at North American sites (Hart 1985). Successes in Germany are the result of the development of a leached layer—up to 1 meter thick and free of sodium chloride—on the surface of the tailings pile, which supports vegetation. Because the insoluble component in Saskatchewan tailings is small, a leached layer will be slow to develop, and capping will be required before revegetation can be achieved. Vegetation would stabilize the surface by reducing erosion and downward movement of precipitation through the soil cap. Hart (1985) recommended the quick establishment of a grass cover, with species selection based on tailings location, slope, and aspect. The effect of plant roots on the stability of multiple layer caps over potash tailing is a subject area that deserves more research (Hart 1985).

## Summary

Potash and other salts make a substantial contribution to the Canadian mining industry, particularly in Saskatchewan. The transfer of soluble salts transferred from subterranean sites to mine wastes is a serious environmental concern because the salts can damage crops, well-water, and both terrestrial and aquatic habitats. The proportion of waste to saleable materials generated creates large accumulations of solid waste. This problem can be mitigated by such methods as backfilling or selling of waste salt, but surface storage is still necessary for a considerable volume. Because the salts are highly soluble in water, surface tailings impoundments must be properly designed and constructed in order to prevent contamination of surface and ground waters.

Deep-well injection of brine is an attractive method of dealing with excess volumes and may be developed further in the future. In New Brunswick, another option for brine disposal may be discharge into the ocean. Studies will be necessary to ensure that the organisms living in discharge areas can survive changes in salinity caused by large volumes of brine.

# CHAPTER ELEVEN

# INDUSTRIAL MINERALS

## Introduction

This chapter describes the mining of most industrial minerals. An industrial mineral may be defined as any rock, mineral, or other naturally occurring substance of economic value, exclusive of metallic ores, mineral fuels, and gemstones. The mining of these minerals does not have environmental impacts at a national or even at a regional scale. It does, however, create local surface disturbance, disrupted drainage systems, and particulate emissions at the many and widely dispersed surface mining operations.

The group of industrial minerals referred to as "structural minerals" includes asbestos, cement, clays, dimensional stone, lime, and aggregates, mainly sand and gravel. Other industrial minerals discussed in this chapter are barite, fluorspar, gypsum and anhydrite, mica, nepheline syenite, silica and silicon, and talc, steatite, and pyrophyllite. For some of these—asbestos, cement, clay products, sand and gravel, stone, and gypsum—their relative contribution to the value and/or quantity of Canadian mineral production may be found in Tables I.1 and I.2. The rest, however, are produced in such minor quantities that they do not make a significant contribution.

## Structural Materials

Structural materials have low unit values and are produced in large quantities, as shown in Tables I.1 and I.2. They are the largest component of mining in Canada, with the greatest number of operations and the highest production output (Godin 1992). With the exception of lightweight aggregates and specialty clays, the widespread occurrence of these structural materials is one of their main features. As a result, factors such as transportation cost, worker availability, and potential environmental effects tend to influence the locations of mines. Deposits are usually surface-mined or quarried and are processed as near as possible to their markets. Since the main markets are in urban areas, quarries and open pits tend to be clustered on the outskirts of cities. Remote mines may have associated quarries to supply road materials and other aggregates.

The use of structural materials in Canada probably dates back to the first arrival of explorers to the continent. It is thought that the first lime and hydraulic cement were made at Hull, Québec, as early as 1830 and that hydraulic cement was also manufactured in Ontario at Kingston—probably for construction of the Rideau Canal System—as well as in Thorold and Queenston, and in Québec's Gaspé County around the same time. The first official records of the cement industry in Canada refers to production from a Québec limestone in 1856. The first production of Portland cement was by C.B. Wright and Sons of Hull, Québec, and Napanee Cement Works of Napanee, Ontario, both in 1889. The first plant in the western provinces opened in Vancouver in 1898. In the early part of the twentieth century, the industry expanded quite rapidly. At the end of World War I, there were 21 cement plants in Canada (Stonehouse 1973).

By the beginning of the twentieth century, the clay and ceramics industry was also active. In 1905, investigations were begun into the extent of clay and shale deposits that could be of use in the manufacturing of clay products. Exploration began in Manitoba, then moved to the Maritime provinces and Alberta in subsequent years (McLush 1909).

## Environmental Effects of Production

The raw ingredients for almost all structural materials are mined using open-pit methods. The manufacturing of cement, lime, and clay products requires the use of kilns for calcining. This section gives a general overview of the environmental impacts of producing structural minerals. Information specific to a particular commodity is provided later in this chapter in the context of each individual commodity or commodity group.

The main environmental problem facing the structural materials-mining industry is the disposal of waste materials. Millions of tonnes of fine-grained limestone dust are stockpiled in limestone quarries (Sirois and MacDonald 1983), awaiting an economic use or a disposal solution. Other major industrial-mineral waste stockpiles include gypsum wastes in Nova Scotia (Collings 1980a) and asbestos wastes in Québec, British Columbia, and Newfoundland (Collings 1977; 1978; 1980a). Many wastes can be re-mined for values left behind by earlier milling practices. Others have potential uses in construction and road building and as mineral fillers (Collings 1980b).

Wastewater is a major source of hydrospheric emissions from quarrying operations. Water is used primarily for beneficiation—for transportation of the material through the processing equipment, as well as in the actual treatment. Water is also used extensively to control dust emissions. Total water use can range from 1 tonne to more than 20 tonnes (in small dredging operations) per tonne of product (Lund 1965). A typical gross water use in a sand and gravel operation would be approximately 2 tonnes of water per tonne of sand and gravel produced. About half is recycled.

Quantities of suspended particles of sand, silt, and clay that remain in the water after processing operations vary by weight from approximately 1%–20%. For dredging operations in existing stream beds, the level of water-borne waste can be as high as two-thirds the amount of sand and gravel recovered (Lund 1965). Suspended solids and colloidal fines have often been found to have harmful effects on benthic communities, plankton, and the reproductive capability and species composition of fish populations in receiving streams. Particulate loadings in wastewater from stone-crushing facilities are probably significantly less than those from sand and gravel operations.

Many pits and quarries use settling ponds and similar systems to remove some of the solids before reuse or discharge to the environment. The level of reuse depends upon the abundance and cost of the water as well as upon regulations controlling emissions or land use. Normally, there is no recycling in dredging operations, and water is discharged back to the lake or stream from which the materials were taken.

Ontario's Municipal/Industrial Strategy for Abatement (MISA) Program established effluent quality standards in 1994 for the industrial minerals sector. The limits were set on the basis of a year-long monitoring routine required of all industries and municipalities in the province. Pits and quarries were only required to sample once during the year. For 35 of the 46 sites tested, all tests gave non-lethal results. A cement plant in Picton indicated toxic results, probably due to pH level, which reaches as high as 12 in the effluent at this particular plant. Suspended solids levels in the effluent at this plant and at one in Bowmanville were also fairly high. High pH was also a problem at two different lime plants; levels were as high as 10 and 11 in these effluents (Westlake et al. 1992).

Emissions to the atmosphere from the mining and processing of structural materials are seldom toxic. Consequently, their environmental effects have received little attention, although a large number of operating and abandoned pits and quarries are in very close proximity to major cities. Most of the dust produced by industrial mineral operations is derived from widely dispersed sources (fugitive dust) and is difficult both to quantify and to control. Sand and gravel operations were assessed as emitting more than 120 kt of particulates to the atmosphere in 1972 (Environment Canada 1976). At the time, this amount represented 11% of all industrial particulate emissions. These emissions have the same characteristics as the products themselves. A survey conducted in 1984 (Kosteltz and Deslauriers 1990) attributed slightly more than 200 kt of particulate emissions to mining and quarrying as a group—12% of the industrial total.

These data indicate that 500 g of particulates are emitted per tonne of product. Approximately 90% of these emissions originated from material-handling operations and from stockpile losses (Environment Canada 1973). The use of wetting agents and dust covers may be quite beneficial in some cases. Although high levels of silica dusts can produce silicosis after prolonged and severe exposure (primarily among workers in the industry and therefore beyond the scope of this book), non-toxic particulates are mainly a nuisance in the atmosphere (Thrush 1968).

Brandt and Rhoades (1972) attempted to assess the long-term effects of airborne particulates (limestone dust) on structure and composition of forest communities. In comparisons with a control site, significant changes were found in the seedling, shrub, sapling, and mature tree strata of the experimental site. Dust-tolerant species were expected to assume ultimate dominance in the experimental site if dust accumulation continued.

Dust and gases from calcination and fuel combustion can be some of the more serious environmental effects of industrial-mineral mining. Most atmospheric emissions from lime production originate in the lime kilns. If uncontrolled, vertical kilns produce an estimated 4 kilograms of particulates per tonne of lime produced, and rotary kilns 180 kilograms (Joyner 1985). Many operations use cyclones and electrostatic precipitators, which can typically reduce kiln dust emissions to 2–3 $kg \cdot t^{-1}$. Gaseous emissions—including carbon dioxide (700 $kg \cdot t^{-1}$), carbon monoxide (1 $kg \cdot t^{-1}$) and nitrogen oxides (0.1-1.4 $kg \cdot t^{-1}$)—are rarely controlled. Particulate emission factors for clay mining in cement production have been suggested as 17 $kg \cdot t^{-1}$ for storage operations, 35 $kg \cdot t^{-1}$ for drying and 38 $kg \cdot t^{-1}$ for grinding (Joyner 1985).

Associated with the particulate emissions is the consumption of a large quantity of fuel for calcination—5 GJ per tonne of product. The comminution of industrial minerals is also energy-intensive (Cienski and Doyle 1992). The processing of lime and limestone required the least energy per unit of production.

## Reclamation

Considerable surface disturbance is unavoidable when using open-pit mining methods. Although reclamation is now the rule rather than the exception (Robertson 1986), there are many thousands of abandoned industrial-mineral pits and quarries across the country that have not been reclaimed to any degree. Unless reclaimed, many sand and gravel pits remain sources of sediment and dust; others are hazardous to animals, including humans, and attract the dumping of waste and scrap.

Boivin (1981) surveyed mines and quarries in an area of southern Québec. Of the 290 he identified, 34% were active quarries and 52% abandoned quarries. Kryviak (1987) estimated that, in Alberta, 18,000 ha of land were disturbed by more than 2,700 gravel pits, only 100 of which had been reclaimed. (Many of these pits, although almost depleted, contain enough material to dissuade their owners from closing them down.) However, the reclaimed area is increasing as a result of the efforts exerted under Alberta's *Land Surface Conservation and Reclamation Act* (1973) and subsequent regulations.

If reclamation takes place at the completion of mining activities, pits and quarries are relatively easy to rehabilitate. At present, reclaiming even a moderate percentage of Canada's abandoned pits would be a colossal undertaking, though not an impossible one (Klco 1990).

The 800 ha Heritage Ranch, belonging to the City of Red Deer, and Lethbridge's 100 ha Pavan Park are two examples of gravel pit reclamation successes in Alberta. An excellent example of a reclaimed limestone quarry is the Butchart Gardens in Victoria, British Columbia (Dufour 1982). Many of these projects, rather than attempting to fill existing pits, have used them to create artificial water bodies for recreational use. The primary reclamation objectives were to restore surface stability, to improve the aesthetic quality of the area, and to restore the area as a functioning (albeit artificial) ecosystem. Recent regulations and guidelines have been created with a view to minimizing future disturbance on this scale. A comprehensive set of guidelines for the construction of pits and quarries in Canada's Yukon and Northwest Territories was prepared by MacLaren Plansearch (1982). The Redland Quarries case study at the end of Chapter 4 describes the proposed conversion of a mined-out limestone quarry in southern Ontario into a landfill site.

European methods of reclamation directed to several goals, including recreational lakes and wildfowl habitats, were described by Bradshaw (1990).

## Asbestos

The term *asbestos* refers to approximately 30 fibrous hydrated silicates that are chemically inert and mechanically separable into filaments (Hoskin 1993). Chrysotile (white asbestos), a member of the serpentine mineral group, crocidolite (blue asbestos), amosite (brown asbestos), and anthophyllite, a member of the amphibolite group, are the varieties that are commercially important. Chrysotile ($Mg_3Si_2O_5(OH)_4$), the most commonly used, is the only type of asbestos mined in Canada. It has long, thin fibers that are highly resistant to heat. It is also the least dangerous to human health because its sensitivity to acid allows it to dissolve in the lungs to some extent, rather than remain as an irritant. The fibrous nature of asbestos makes it an extremely valuable material. Depending on the fiber length, it can be used in textiles, packings, brake linings, insulation, and in asbestos-cement products such as pipes (Morel-à-l'Huissier and Hoskin 1992).

Canadian asbestos mining commenced in the Thetford Mines area of southern Québec in 1878, with an annual production of approximately 50 tonnes (Vagt 1976). The industry expanded fairly rapidly, with new mining operations opening in British Columbia, the Yukon, Newfoundland, and Ontario, as well as

Québec. A great deal of this expansion occurred in the 1950s, with the addition of four new mines (Scollan 1994). The bulk of Canadian production comes from the Thetford Mines region of Québec; the Appalachian belt of ultra-mafic rocks in the Eastern Townships is one of the world's major sources of chrysotile asbestos, where the mineral occurs in large veins in serpentinized peridotite bodies (Vagt 1976).

The number of asbestos mines in Canada had declined from fourteen in 1974 (Ripley et al. 1982) to four mines in Québec and one tailings processing mill in Newfoundland at the end of 1993 (Morel-à-l'Huissier 1994). Of these four mines, three used open-pit methods, producing 78% of the total asbestos ore and most of the waste rock and overburden.

At present, Canada supplies about 16% of the world market of asbestos, with a production of 510kt in 1993. While both production and market shares have declined steadily during the past two decades, there is some evidence that a production plateau has been reached (Morel-à-l'Huissier 1994b).

Although inhalation of asbestos fibers was recognized as a possible cause of lung cancer as far back as 1935, definition of the health risks associated with the mineral was more recent. The U.S. Environmental Protection Agency (EPA) attempted to prohibit the use of asbestos, announcing its intention to do so in 1979. This resulted in a sharp decline in asbestos production between 1979 and 1984, with Québec shipments dropping by 45% (DRIE 1985), and forced three of the province's seven mines to close. Since then, operations in the Yukon, British Columbia, and Ontario have also ceased. Although the EPA's 1986 Rule under the *Toxic Substances Control Act*, banning 94% of commercial uses of asbestos, was overturned by an appeals court in 1989, the stigma attached to the product's image by negative media attention and public concerns about safety—largely unwarranted under proper applications—has kept markets from reviving.

Since the early 1970s, world health authorities have developed a better understanding of the risks of using asbestos. It is now recognized that the risk depends on such factors as fiber type and length and the density of the material in which it is used. The current consensus is that for certain applications, asbestos is likely to be safer than possible alternatives. In others, however, such as spray applications, it is a hazard because exposure cannot be controlled.

Canadian ore grades about five percent asbestos on average. After extraction, the ore is crushed and dried, then separated by a cyclic crushing, screening, and aspiration process (Sirois and MacDonald 1983). The final product is graded by size, bagged, and shipped.

## Environmental Effects

The main environmental effects of the mining process are surface disturbance, solid wastes, and dust emissions. There are also minor hydrospheric residuals associated with the discharge of mine drainage.

A typical open-pit asbestos mine (Marshall 1982) with a surface area of approximately 100 ha and a depth of 200 m discards as waste rock approximately 55% of material extracted from the mine. Data presented by Godin (various years) indicate that processing asbestos produced tailings equal to approximately half the tonnage of ore extracted. The total accumulation of tailings in the Thetford Mines region was estimated at approximately 1,000 Mt in 1980. Underground asbestos mines typically produce more waste rock than ore.

Asbestos fibers in soil, water, and air derive from the natural weathering of serpentinite, a rock consisting almost entirely of the mineral serpentine; they are also produced by mining activities and by the use of products that contain asbestos (Shugar 1979). Serpentine is an important factor in the etiology of asbestos-related human disease in some agricultural communities. Aside from its natural occurrence, serpentine has, in some cases, been added to soil for the fertilizing effects of its magnesium content.

The only impacts of asbestos on plants, reported by Shugar (1979), were toxic effects on bacteria in close contact with the fibers. Soils containing asbestos dust were found to have low populations of microflora and fungi, although the same effect could be simulated using additions of $MgSO_4$. The difficulty of revegetating asbestos waste dumps is similar to the difficulty encountered by natural vegetation in serpentine barrens: the barrens and tailings have low levels of nutrients, a high proportion of magnesium to calcium, and high concentrations of nickel and chromium (Brooks 1987).

Emissions to the atmosphere are mainly particulates, which are produced by most of the operations involved in the mining, crushing, drying, milling, and shipping of asbestos ore and end-product. Some emissions also result from the handling of waste rock and tailings. Dust from tailings dumps is considered the major environmental problem facing the asbestos industry (Rabbitts et al. 1971). Airborne particulates pose a serious health concern to mammals. The Canadian emissions inventory of 1985 (Kosteltz and Deslauriers 1990) attributed slightly more than 4 kt of particulate emissions to asbestos mining and milling. This represented approximately 0.5% of the total industrial emissions of particulates. An earlier inventory (Environment Canada 1974) estimated that 40% of particulate emissions derived from asbestos mining and most of the balance from milling. Aside from their content of asbestos fibers, these emissions also contain high concentrations of chromium. Dust emitted from asbestos mining and milling operations has approximately the same composition as the material being handled. Therefore, dust produced by mining operations will contain at least 5% asbestos. Dust from the latter stages of milling is predominantly asbestos.

An effective program of dust measurement and control has been in place in the asbestos industry for at least 25 years (Horn 1966; Hutcheson 1971). Controls have included waterspraying of roads during dry periods; the use of dust-containment devices around drills, crushers and grinders; dust removal from aspiration air used in the drying process; and the reduction of fugitive dust, both that generated within the mill and that from waste and tailings dumps. Cyclones, electrostatic precipitators, and wet bag collectors have all been used to contain dust, which is best agglomerated before disposal to prevent subsequent dispersal.

Since the majority of asbestos processing in Canada has used a dry-milling process, water is generally required only for cooling and for supporting ancillary operations (Table 1.4). This excludes mine water discharge. Asbestos mine water is typically neutral or slightly alkaline and contains some silt and asbestos particles in suspension (Rabbitts et al. 1971). The various pathways of asbestos through the hydrologic cycle (Figure 2.2) have been studied by Schreier and Taylor (1981). They noted that a small proportion of the Canadian population was exposed to asbestos concentrations in drinking water in excess of $10^8$ fibers per liter (FPL). In two cases, the concentrations were twenty times that value. Runoff from asbestos-bearing rock and soil carries the fibers into streams and lakes and, eventually, to the oceans. This is believed to be the greatest input of fibers to the hydrosphere: concentrations of up to $10^{11}$ FPL have been measured in natural river systems (Schreier and Taylor 1980). The dumping of taconite tailings, which contained asbestos minerals, into Lake Superior has resulted in concentrations as high as $87 \times 10^6$ FPL in some areas of the Lake. By comparison, background levels fell below the detection limit of $10^5$ FPL (Durham and Pang 1976).

Aquatic organisms, such as mussels and fish fry, are affected through both habitat disruption and incorporation of fibers into tissue. A major concern is the extent to which these organisms will pass the fibers up the food chain. Studies on a large number of animals have shown health effects similar to those in human beings (Shugar 1979).

By its nature, dry milling is a dusty operation. The logical alternative, a wet-milling process, was tested as early as the 1920s. A number of wet mills have operated at various times, both in Canada and abroad (Sirois and MacDonald 1983). Nevertheless, in 1990 all Canadian asbestos mills used dry-concentration methods. The smallest, in Newfoundland, switched to a wet-milling process in 1991. Although wet milling would eliminate most atmospheric emissions from the mill, it would replace them with hydrospheric residuals, unless these were adequately controlled.

### Reclamation

The total land disturbed by asbestos mining in Canada to 1975 was estimated by Rabbitts et al. (1971) at 2,800 ha. A satellite-imagery survey conducted in the mid-1970s corroborated this estimate, locating 29 sites with a total disturbed area of 2,271 ha. Of this amount, ten percent had been revegetated (Murray 1977).

After 60 years of asbestos mining in southeastern Québec, tailings covered an area of 550 ha; waste rock and overburden covered an additional 900 ha (Moore and Zimmermann 1972). Establishment of vegetation on the tailings is important for reducing potential health hazards associated with release of asbestos fibers by wind and water erosion; methods to recover less-desirable components of asbestos fibers from waste rock and tailings are also being developed. The tailings have a low macronutrient content, low calcium, low water-holding capacity, high pH, high magnesium, and large amounts of both nickel and chromium. These properties are similar to those of other serpentine-derived soils that occur in Québec (Brooks 1987).

Serpentine barrens are characterized by dwarfed plants and low species richness and sometimes by metal-tolerant endemics (Shaw 1989). Because serpentine severely limits plant growth, untreated asbestos tailings are largely devoid of vegetation. The main problem for plant growth on asbestos tailings is inadequate mineral nutrition. Metal toxicity is low, probably because the high pH keeps the metals out of solution. Seeded vegetation persisted for three years after treatment of tailings with 10 t·ha$^{-1}$ of NPK fertilizer and 40 t·ha$^{-1}$ of manure or sawdust (Moore and Zimmermann 1972). Sewage sludge applied at 100 t·ha$^{-1}$ has also been used successfully as an organic amendment (Moore and Zimmermann 1979). Introduced grasses such as Altai and Russian wild ryes and volunteer species from the surrounding area have become established on the tailings, but planted tree seedlings have not been as successful. Even after treatment, however, seed germination and plant growth were reduced by low soil water potentials during dry periods. The cost of successful revegetation was estimated to be as high as $3,460 per hectare (1977 dollars).

### Health Concerns

As early as 1935, a cause-effect relationship was established between asbestos-dust inhalation and lung cancer. Since then, a number of other human ailments have also been found to relate to the ingestion of asbestos fibers; these include asbestosis, pleural calcification and plaques, and peritoneal mesotheliomas (Shugar 1979; Jaworski 1980). The effects of fiber size on animals are not clear. When inhaled, asbestos fibers become trapped in the upper or lower respiratory tract, where they act as an irritant. Fibers that reach the lungs may pass through the respiratory membrane and enter the circulatory or lymphatic systems (Jaworski 1980). Although smaller fibers can enter into more intimate contact with cell membranes and can more effectively induce mesothelioma, they are more easily engulfed by macrophages and thus eliminated from the body (Shugar 1979).

Fibers may also be ingested by drinking contaminated water or eating contaminated food; significant concentrations of asbestos fibers have been found in rainwater, snow, lakes, rivers, and groundwater in Canada, but very little is known about the long-term health hazards of water-borne fibers (Schreier and Taylor 1980). Once the fibers have been deposited in the lungs or other soft tissue, they cause the formation of "asbestos bodies," which are fibers that have been engulfed by protein and iron. These bodies are commonly found within cancer tumors (Jaworski 1980).

Asbestos-related diseases have been most prevalent among workers exposed to very high concentrations of fibers, and working conditions have improved considerably in recent years. In view of the ubiquity of the fibers in air, soil, and water, some controversy exists as to the magnitude of the threat asbestos poses to workers, as well as to the general public, when proper air and water controls are in place. There is a more moderate view, however, which points to the examples of other hazardous substances; some synthetic mineral fibers used to replace asbestos are more durable than the natural mineral and, thus, represent a worse irritant to the lungs because they persist longer. Regulations can place exposure limits at sufficiently low levels; if these laws are properly monitored and enforced, asbestos can be used relatively safely (Hoskin 1993; Morel-à-l'Huissier 1994b).

## Cement

The term *cement* refers to Portland cement, which is used almost entirely for construction purposes. Residential, non-residential, and public-related construction sectors consume the material in almost equal shares. In 1993, Canada had 26 widely distributed cement plants in operation, which produced over 8.6 Mt of cement and clinker products (Vagt 1994b).

In the cement industry, electrical energy represented 30% of total operating costs, and comminution accounted for 70% of electrical energy use, as opposed to 50% for the other industrial minerals. Heating the kilns required 75% of the energy (Vagt 1992a). The rate of energy use has been reduced since 1974, and the industry is exploring several new ways of increasing energy efficiency. The most promising are through the use of waste fuels, recovery of waste heat, innovative modification of process and equipment, and a conscious effort at each plant to ensure that all systems are working properly.

Settleable dusts emitted from the kilns of cement plants are a significant component of total particulate emissions in Canada. Total Canadian particulate-matter emissions from cement operations were estimated in 1972 at 159,000 tonnes (Environment Canada 1976). This figure was based on an uncontrolled emission factor of 135 $kg \cdot t^{-1}$, and on the assumption of 88% emission control. This was equivalent to a controlled emission factor of 16 $kg \cdot t^{-1}$ of product. At the time, this amount represented over 11% of the total industrial emissions of particulates.

Cement dust landing on plants causes effects that vary from beneficial to lethal (Darley 1966; Lerman and Darley 1975). The differences depend on the size and composition of particles and atmospheric conditions during deposition. Dry dust may form a crust when it becomes moist, and dust deposited during moist conditions may create a toxic alkaline solution. Dust that does not form a crust can be harmlessly blown away. Lerman and Darley (1975) concluded that leaf injury or death of plants is due to blocking of light necessary for photosynthesis, plugged stomata inhibiting gaseous exchange, and necrosis of leaves. Cement dust applied directly to leaves caused small reductions in growth and metabolic activity (Brooks 1980).

Plants reported as being affected by cement dust include deciduous and coniferous tree species (Darley 1966), understorey plants, and nonflowering plants (Malhotra and Blauel 1980). Diversity can be lowered in the ground layer of vegetation and in epiphytic lichens.

Cement dust is alkaline, and can increase soil pH, limiting the growth of species of plants adapted to more acidic conditions (Malhotra and Blauel 1980). There is no apparent effect on the pH or on vegetation composition in calcareous soils. Soils downwind from a cement plant in Wyoming had unusually high calcium carbonate concentrations (Stiller and Reider 1979).

## Clays

Clay minerals are all secondary minerals—that is, they are formed by the weathering of other minerals. They are fine-grained minerals, composed of hydrous aluminum phyllosilicates with small amounts of iron, alkalis, and alkaline earths. Each of the different clay minerals have separate geological occurrences, properties, and, consequently, uses. Clays may be classified into two categories: specialty clays—including bentonite and montmorillonite—and kaolinitic clays—including ball clay, refractory clay, stoneware clay, and kaolin. The latter class is used primarily for the manufacture of brick and tile, ceramics, cement, and refractory mortar. Bentonite, a swelling clay, is used primarily as a sealant in well drilling. Fuller's earth, a clay consisting largely of montmorillonite, is important for its high adsorptive capacity; it is used in the pharmaceutical industry and as a purifier (Andrews 1993). In 1992, Canada's 32 operations, located in most provinces, produced a wide variety of clays. The major producers were Ontario, Québec, British Columbia, and Alberta. Until 1992, bentonite was mined in Manitoba, near Winnipeg, and in Alberta, at Rosalind.

Common clay, which is a mixture of rock flour and a number of clay minerals—primarily of the illite type—is used in the manufacturing of brick, tiles, and cement. Common clays and shales are found in all parts of Canada, with shales being dominant in the Maritime provinces, and unconsolidated clays of glacial origin more common in the prairies and in Ontario. Kaolin occurs in almost all provinces, although the deposits are seldom of commercial size. In Ontario, there is a large deposit of kaolinized sand along the Missinaibi and Mattagami rivers southwest of James Bay; the deposit covers an area of 10,000 km$^2$. The most significant deposit in Manitoba is located in the Kergwenan area south of Ste. Rose du Lac. In Saskatchewan, the most important deposits are the Wood Mountain area in the south-central part of the province and the Eastend-Shaunavon area in the southwest. Both kaolin and fire clay are found at Wabamun, Alberta, although only the kaolin is of economic importance. In British Columbia, the most important deposit is found at Lang Bay, in the southwestern part of the province. The only significant occurrences of ball clay are in Saskatchewan, in the Whitemud and Ravenscrag formations. The clay is mined at Estevan, Rockglen, Flintoft, and Readlyn. The Whitemud formation is also a source of stoneware clay and fire clay. Good-quality fire clay also occurs in a number of other provinces: in Nova Scotia in the Musquodoboit Valley and at Shubenacadie, in Ontario in the James Bay lowlands, and in British Columbia, where it is quarried on Sumas Mountain.

## Dimensional Stone

Dimensional stone is quarried and used for building, for ornamental purposes, and for monuments. The most widely used types of stone are limestone, granite, sandstone, slate, and marble. The dimensional stone classification of "granite" actually includes a number of other crystalline rock types; consequently, "granites" are found in a number of geological environments. *Marble* is metamorphosed limestone and is not as common as its sedimentary predecessor; substantial marble occurrences of economic importance are found in western Newfoundland and throughout Ontario. *Slate* is metamorphosed shale or mudstone; economic deposits are not common—only one slate deposit is mined in Canada, in Newfoundland.

In 1992, 50 Canadian quarries produced 89 Mt of dimensional stone: of this amount, 21% was granite, 73% limestone, 3% sandstone, 2% slate, and less than 1% marble (Vagt 1994b). Most of the stone-producing centers were in the Boreal Shield and Mixed Wood Plains ecozones (see Figure 2.3), primarily in southern Québec, with a few in each of the other provinces except Alberta and Saskatchewan.

## Lime

Limestone is rock containing calcite ($CaCO_3$), with or without small quantities of dolomite ($CaCO_3 \cdot MgCO_3$). Limestone, dolostone, and sandstone are all common types of sedimentary rocks. Dolostone consists mostly of dolomite, mined as a source of calcium, strontium, and magnesium. Limestone and dolostone may be calcined to produce either high-calcium lime ($CaO$)—often called quicklime—and/or dolomitic lime ($CaO \cdot MgO$)—frequently called dolime. Lime may be marketed directly, or after being reacted with water, as slaked or hydrated lime. There were 20 lime-producing operations in Canada in 1993, of which ten were located in Ontario, most of them at the western end of Lake Ontario (Scollan 1994). Fifteen of the operations produced quicklime, three of them dolime, and three a combination of both. Demand for lime is increasing because of its application in environmental control practices. Chapters 3 and 4 describe its use in the treatment of acid drainage, for instance.

The quarrying of dolostone in Canada likely dates back to colonial times. Specific mining for the production of metallic calcium or magnesium, however, is a much more recent occurrence. Because the chief use of these metals is in metallurgical works, their production is most likely restricted to the 20th century. Indeed, the only calcium processing plant is the one at Haley in northern Ontario, which opened in 1942 (Scollan 1994); however, studies of magnesite ($MgCO_3$) beneficiation date as far back as 1920. Calcium, strontium, and magnesium frequently occur in association with each other. The only Canadian source of calcium and strontium, for example, is a mine in Ontario that also produces magnesium.

Calcium, in spite of its great lithospheric abundance, has few uses in its elemental form. Because of its high reactivity, however, it may be used to reduce several metallic oxides, and to de-oxidize, de-sulphurize and de-gas steels and cast irons (Ignatow et al. 1991). Canada's calcium production is less than 900 $t \cdot yr^{-1}$, which supplies approximately a third of global requirements. Canada is the world's leading producer of strontium. In its carbonate form, strontium is used primarily in television picture-tube construction; in its nitrate form, it is used in fireworks. Metallic strontium is most widely used as a modifier in aluminum/silicon casting alloys, for increased ductility and strength. Magnesium is used for casting and wrought products as well as in the production of aluminum alloys. Demand and production, therefore, are very closely linked to that of aluminum products.

Strontium occurs commonly as the mineral celestite, and less commonly as the mineral strontianite. Both of these minerals form as chemical sediments, and thus are associated with sedimentary rocks such as limestone, dolostone, and evaporites. At the dolostone quarry near Haley, the celestite occurs in a different form: as slab-like masses of radiated, columnar, or fibrous crystals surrounded by dolomite (Dawson 1985). Because of the association with dolostone and limestone, strontium is mined along with calcium and magnesium derived from the dolomite and calcite minerals.

Magnesium is also derived from the minerals magnesite and brucite, which generally occur in association with limestone and dolostone. Another potential source that has been considered is the asbestos tailings at Thetford Mines, Québec, which are extremely rich in magnesium.

Globally, most magnesium is produced by the electrolysis of seawater and brines (Sirois and MacDonald 1983). However, all four Canadian producers use the silicon-thermic reduction of dolomite or magnesite ores. The open-pit operation in Haley produces both calcium and magnesium metals, using the same process for the two metals: dolomite or limestone is calcined in a rotary kiln to convert it to dolime or quicklime, which is then reduced in an electric vacuum retort (Pidgeon process) to the pure magnesium or calcium metal. Calcium reduction requires the addition of aluminum metal to the vacuum retort; magnesium reduction requires ferrosilicon.

Environmental concerns regarding magnesium production have been summarized by Sirois and MacDonald (1983). These include the high energy consumption of the Pidgeon process, which challenges innovation to produce a more energy-efficient method, and the development of a satisfactory way to extract magnesium from asbestos tailings, which would also reduce the accumulation of this undesirable residual. Effluent samples collected from the Haley magnesium processing plant for purposes of Ontario's MISA program were found to be nontoxic.

## Mineral Aggregates

The term *mineral aggregates* includes sand and gravel, crushed and pulverized stone, and lightweight aggregates. Sand and gravel deposits are quite common; those that occur near the surface are generally a result of the widespread glaciation that occurred in North America during the Pleistocene Era. Because of the large number of small operations and the overlaps with dimensional stones and clays, an accurate count of sources of these materials is difficult to obtain. A recent estimate (Vagt: personal communication 1994) put the number at 400 separate mineral aggregate operations.

Most sand and gravel is used for road construction and concrete aggregate. Crushed stone is used for the same purposes; it is also used for construction of railway beds, earth dams, and breakwaters. According to Vagt (1994c), demand has increased steadily since World War II. Environmental impacts have been growing concomitantly, and all jurisdictions, particularly provincial and municipal, are paying more attention to aggregate operations, particularly with regard to reclamation.

Lightweight aggregates—including vermiculite, perlite, and expanded clays and shales—are used primarily in horticulture and in the manufacture of concrete blocks, ceiling tiles, gypsum plaster and loose insulation. Currently, the only raw materials mined in Canada for use in lightweight aggregates are normal clays and shales. Canada does not mine raw perlite, pumice, or vermiculite.

## Other Industrial Minerals

All of the following group of minerals are extracted from oxidized ores, including oxides, carbonates, silicates, and sulphates. Most of these minerals are mined on a small scale, usually using open-pit methods. In 1993, minerals of this group represented roughly 47 mines scattered across the country (Scollan 1994).

There have been very few studies of the environmental effects of the mining operations described in this section, mainly because of the small size of the operations and their wide dispersal, and because of the low toxicity of the ores. There were few national mining surveys that studied the environmental effects of these small operations: one was conducted in the early 1970s (Rabbits et al. 1971) and another ten years later (Sirois and MacDonald 1983). The earlier survey estimated that the total land disturbance from these types of mines was approximately 40 ha, which was expected to increase to 46 ha by 1975. It was expected that approximately one-third of this area would be revegetated. A satellite survey (Murray 1977) found 133 ha disturbed, of which only 3 ha had been revegetated. All three surveys included only a few of the types of mines covered in this section.

## Barite

Barite, or barium sulphate ($BaSO_4$), is used primarily in Canada as a weighting agent in drilling muds. To a lesser extent, it is also used as a filler in paints, paper, and textiles.

In Canada, barite was first extracted in 1866 from a mine at Bass River, Nova Scotia. Barite is found in almost every province; however, economically significant deposits are restricted to Newfoundland, Nova Scotia, Ontario, and British Columbia. Between 1940 and 1978, almost all of the barite mined came from one of the largest operations in the country: the Magnet Cove Barium Corporation, near Walton, Nova Scotia. Mining initially took place at the surface, but later was moved underground. Although the mine and quarry are now flooded, the mill at Walton still operates.

Most Canadian barite deposits occur in the form of veins. The mineral is often found in association with lead-zinc sulphide deposits. In 1993, barite ore was mined both at the surface and underground (Scollan 1994). Some of the barite produced was recovered from the tailings of lead-zinc mining activities. Beneficiation involves washing to remove adhering clays, crushing, and concentrating. Concentration may be achieved by a number of methods, depending on the size of the ore particles. Gravity methods of concentration are quite successful due to the high specific gravity (4.5) of barite. For larger particles, such techniques as heavy-media drums and jigging are used, while for fine particles, flotation is used. Once concentrated, the barite is then ground to the required size.

## Fluorspar

Fluorspar is another name for fluorite or calcium fluoride ($CaF_2$), and it is the most important source of the element fluorine. The mineral is used as a metallurgical flux, and in the ceramics and chemical industries (Prud'homme 1990).

Fluorspar is usually associated with other minerals such as quartz, calcite, dolomite, or barite and is common to a wide range of geological environments. Vein deposits have been the most important sources of fluorspar ore in Canada. They occur mainly in Newfoundland, Nova Scotia, Ontario, and British Columbia. The chief deposits of economic importance lie in Newfoundland, in the Burin Peninsula, where the ore has a composition of 35% $CaF_2$ (Prud'homme 1990). The large fluorspar deposit there was first discovered in 1843, but mining did not commence until 1933, when the St. Lawrence Corporation of Newfoundland Inc. started operating a surface mine.

The facility at St. Lawrence operated three open pits and an underground mine in 1990. In the late 1980s, the mine encountered environmental problems related to the adequacy of its tailings system and to dust releases during the shipment of the product. At the end of 1990, as a result of poor market conditions, the mine was forced to close. Should market conditions improve in the future, it may reopen (Kilburn 1992). This prediction is based on the expected replacement of ozone-depleting chlorofluorocarbons by hydrofluorocarbons and hydrochlorofluorocarbons. The latter two are less destructive environmentally and require larger amounts of fluorite for their manufacture.

The fluorite deposit in Nova Scotia occurs principally within a barite orebody with $CaF_2$ content varying from 17%–19%. This deposit was in production only prior to 1931, and then only for a short period of time. In Ontario, mining of a large fluorspar deposit first started in the Madoc area in 1905, ending in 1961. In British Columbia, four deposits of economic interest exist: the Rock Candy deposit in the southeast, the Quesnel Lake and the Birch Island deposits in the north-central area, and the Liard River deposit in the north. Mining in the province dates back to 1918, when the Rock Candy mine started operation.

Following mining, the ore may be upgraded by hand-sorting before beneficiation, if the particle size is relatively coarse. Beneficiation uses gravity concentration methods to separate the mineral from its common gangue minerals of calcite and silicates. However, when the gangue minerals include barite, celestite or sulphide minerals, gravity separation is not effective, since these minerals have a heavier specific gravity than fluorite. In this case, flotation is used. Although fluorite has a fairly low solubility, it is not completely insoluble, and particulate emissions may therefore be a source of fluoride contamination. Studies have shown that high exposures to the element can cause a number of health problems, including chromosome damage to plant and animal cells (Rose and Marier 1977). Fluoride is readily bioaccumulated—for example, by aquatic plants. It has been estimated that the food chain concentration factors for the element are 10:1 or more. High fluoride exposure in animals causes bone damage; however, this effect is usually nullified if high levels of calcium are present, as is the case with dissolved fluorite. Fluoride addition to drinking water for the protection of teeth is a widespread practice, precisely because of the potential for bioaccumulation.

## Gypsum and Anhydrite

Gypsum is a hydrous calcium sulphate ($CaSO_4 \cdot 2H_2O$). When calcined, it releases three-quarters of its water to form plaster of paris. This can be moulded as a hard plaster and is used in paint, paper, wallboard, plastics, and joint compounds. Anhydrite is the naturally occurring anhydrous form of gypsum; it is used in cement manufacture and as a crop fertilizer.

Gypsum and anhydrite usually occur in the form of thick, relatively pure beds within sedimentary rocks. The minerals form as a result of the evaporation of seawater. Anhydrite reacts with near surface waters to form gypsum, so it is more likely to be found in subsurface rocks, below the gypsum, rather than outcropping. Anhydrite is found in association with all the major gypsum deposits.

Gypsum was quarried in Nova Scotia as long ago as 1770 by farmers who sold it to the United States for use as fertilizer. Gypsum mining began in 1822 in southern Ontario, and a calcining plant opened in 1896 in the same area. Since that time, there have been a number of producers in Ontario; currently, there are only three. Gypsum mining began in New Brunswick at approximately the same time as in Ontario, with farmers excavating deposits in the southern part of the province for use as fertilizer. In 1912, a wallboard plant was opened, and it served the Maritimes until 1982.

Most Canadian gypsum is now produced in Nova Scotia and exported. Ontario produces about 12%, most of which is used on-site, and the balance is produced in British Columbia, Manitoba, and Newfoundland (Vagt 1994d). All of the anhydrite is produced at two facilities in Nova Scotia. The Ontario mines use underground methods, producing 17% of the ore, but only 4% of the waste rock. The open-pit operations discard approximately 40% of the extracted ore as waste rock and overburden and an additional 4% of the ore as tailings (Godin 1992).

Of the fourteen gypsum and anhydrite mines in Canada, only two—both in Ontario—were underground. Crude gypsum must be crushed prior to being processed. In the initial stages of processing, rock impurities such as limestone and dolostone, and fines such as mud and silt must be removed by sorting, washing, and screening. Air separation may also be required to separate the limestone and dolostone. Separation, preliminary crushing, and grinding are followed by calcining, or heating using kettles or rotary kilns, to remove the associated water molecules.

### Environmental Effects

The main environmental effect of gypsum mining is thought to be the surface disturbance that open pit mining entails (Thirgood 1969). At the time of Thirgood's study, there were hundreds of hectares of water-filled gypsum quarries in Nova Scotia, with depths of 8–10 m. Although the gypsum industry operated a voluntary reclamation program, many sites had little or no overburden available for use as fill.

Particulate emissions are produced by the drying and grinding of the ore and calcining it in kilns; these emissions have significant local effects. Uncontrolled emissions have been estimated at 1.3 kg·t$^{-1}$ for roller mills, 20 kg·t$^{-1}$ each for raw material driers (if used) and calciners, and 50 kg·t$^{-1}$ for impact mills (Joyner

1985). The use of fabric filters reduces these values to 60, 20, 20 and 10 $g \cdot t^{-1}$, respectively. Other sources of emissions are primary and secondary crushers, screens, stockpiles, and roads. Emissions are also produced by drilling and blasting if quarrying is required.

In addition to particulates, gaseous emissions are produced from fuel combustion, including nitrogen and sulphur oxides, and carbon monoxide.

## Mica

Micas form a group of hydrous aluminum silicate minerals that split easily into tough, flexible sheets. They are very common minerals that occur as constituents of igneous and metamorphic rocks and are valued for their special thermal, mechanical, electrical, and optical properties. Two varieties of mica are of commercial interest: muscovite (potassium mica), which is white or colorless, and phlogopite (magnesium mica), which is dark brown or amber in color. Phlogopite occurs primarily in metamorphosed limestones, peridotites, and dunites, or in magnesium-rich igneous rocks such as pyroxenites. It does not usually occur in large sheets, as is possible with muscovite, but in the form of flakes. Phlogopite is currently the only variety being mined in Canada; the mine is at Parent, Québec, 300 km north of Montréal, and the ore is milled at a plant near Montréal. Micas in Canada are primarily used as fillers in paints and caulking products.

In the nineteenth century, mica was mined from a number of small pits in eastern Ontario. In the early 1900s, the Lacy mine there was the largest mica mine, and the large Blackburn mine was located in western Québec. Both of these mines produced phlogopite, chiefly for the windows of stoves and furnaces.

Processing methods depend on whether the mineral is in the form of sheets or flakes. Preparation of sheet mica involves removing any adhering dirt or rock and trimming the edges, at which point it may be sold as is or ground to a desired size. Flake mica must be recovered by crushing and screening of the host rock, followed by concentration through flotation or gravity methods. Once concentrated, the mineral is sieved and ground to the required size specification.

## Nepheline Syenite

Nepheline syenite deposits are quite common in Canada, but the occurrences of those with economic potential for mining are more rare. Commercial deposits of the rock contain at least 20% nepheline and 60% feldspar. The deposits occur in Ontario, Québec, and British Columbia.

Nepheline syenite was first produced in Canada in 1935 from the Blue Mountain deposit northeast of Toronto, Ontario, which is the only deposit currently mined. There are two quarries in operation at the site.

Nepheline syenite is used in the manufacture of glass, in ceramic glazes, in tiles, and as a filler in paints and plastics. Production of nepheline syenite rose from 200 kt in 1958 to 600 kt in 1974. Output remained constant at this level until 1980, at which time it dropped significantly. This drop is attributed to the recycling of waste glass.

Nepheline syenite is mined from an open pit. The ore is crushed and ground to a particle size of about 840 mm, at which point the iron-bearing minerals can be removed by magnetic separators. The resulting product is then sized to standard specifications. It may also be ground more finely if it is to be used in ceramic applications.

## Silica and Silicon

Silica ($SiO_2$) has, by far, the greatest production of all the minerals in this subgroup of industrial minerals, 35% of it mined in Québec and 39% in Ontario (Boucher 1993b). Silica occurs naturally as quartz in sands, sandstones, and quartzite throughout Canada, but few deposits are pure enough for commercial use. This abundant mineral has a wide range of uses, mainly in foundries, but also in glass-making and non-ferrous smelting. Silica is also used in roofing, cleansers, fertilizers, and pulp and paper products, as well as in sandblasting, in the chemical industries, and as an abrasive.

Information regarding early silica mining in Canada is not well documented. It seems reasonable, however, to assume that the earliest mining was likely for use in glass production. The technology of glass making was brought to the continent by the colonists; while English glassmakers had progressed to lead crystal in the 18th century, France was still making common glass, which is composed essentially of sand mixed with vegetable ashes. Thus, this common glass was the first type to be introduced to Canada, in the 18th century. It would appear that glassmaking began early in Canadian history, and, presumably, that silica mining began at approximately the same time.

In 1993, there were 16 open-pit mines in operation, in every province except Prince Edward Island; three of these were associated with plants that reduced some of the silica to ferrosilicon or to elemental silicon.

Silica milling facilities typically carry out crushing, screening, washing, and drying operations. As the most abundant component of the Earth's crust, silica itself can hardly be considered an undesirable pollutant, although silica dust is a hazard to human health. The large number of open-pit silica mines in Canada produce considerable surface disturbance, overburden waste, sediment-laden runoff, and atmospheric dust emissions. Most operations use some form of control to reduce such emissions. The silica-processing industry is also a major consumer of energy (Table 11.1).

Silicon is processed from mined silica, usually by the reduction of quartzite using an electric-arc furnace. In Canada, almost all ferrosilicon is used by the iron and steel industries. Pure silicon is exported and used in the production of alloy castings and silicon metals. Little information is available regarding residuals

**Table 11.1** Energy Used in the Comminution of Industrial Minerals in Canada,1990

| Mineral | Production (Mt) | Energy use (TJ) | Energy/tonne (MJ/t) |
|---|---|---|---|
| asbestos | 0.69 | 670 | 980 |
| cement | 12 | 2,050 | 170 |
| talc | 0.14 | 20 | 150 |
| nepheline syenite | 0.6 | 70 | 120 |
| silica | 2.1 | 140 | 70 |
| gypsum | 8.2 | 160 | 20 |
| potash | 7 | 140 | 20 |
| lime | 2.3 | 20 | 8 |
| limestone | 87 | 350 | 4 |
| total | 120 | 3,620 | 30 |

Source: Cienski and Doyle (1992)

from the furnace process, trace-element constitution of the input materials to the furnace, the amount and composition of solid residues, or water use. However, the high-purity quartzite used for silicon production would have oxygen as its main residual; the combustion of such substances as coal and coke would produce carbon oxides and, undoubtedly, some sulphur oxides and toxic metals.

## Talc, Steatite and Pyrophyllite

Talc, and its massive form steatite, are hydrous magnesium silicates, $Mg_3Si_4O_{10}(OH)_2$. They are extremely soft, generally white minerals that are chemically inert, hydrophobic, and organophilic. They have high melting points and low thermal and electrical conductivities. Talc's main use is in the paint, pulp and paper, cosmetics, plastic, and chemical industries. Steatite, also called soapstone, is used for sculptures. Pyrophyllite, $AlSi_2O_5(OH)$, is a hydrous aluminum silicate with properties similar to talc; it is used in ceramics, refractories, and insecticides.

Talc is derived from the hydration of non-aluminous magnesium silicate rocks under intensive metamorphism; the most common host rocks are dolostone and ultramafic rocks, where the mineral occurs as small veins, tabular bodies, or irregular lenses. When talc occurs massively, it often contains other mineral impurities. In this form, it is referred to as steatite. Pyrophyllite results from hydrothermal alteration associated with low- to medium-grade metamorphism of acid igneous rocks; the most common host rocks are rich in aluminum.

Commercial production of talc began in 1896 in the Madoc area of southeastern Ontario and continues today from the original two sites. In British Columbia, sporadic talc mining occurred in deposits scattered along the central and eastern portions of the province in the early 1900s. Mining of pyrophyllite also began at around the same time in the Fox Trap area of Newfoundland.

In 1992, Canadian production of these minerals came from five operations in Ontario, Québec, and Newfoundland (Bergeron and Andrews 1993). Talc is mined by underground methods when the orebodies are gently dipping or flat, and by open-pit methods for steeply dipping bodies. Production proceeds at a slow pace with both of these methods. The slippery nature of the talc mineral also poses some unusual problem: special slip-reducing tires and chains must be used on all machinery, and haulage slopes must not be steep.

Beneficiation of talc involves pre-concentration and milling. Milling involves crushing and rolling to produce a fine-grained product that is required for use in fillers or extenders. The mills use a variety of grinding, classification, and flotation operations. Mills may also have oil or gas-combustion chambers for drying of the ore. Fluid energy mills are usually used when very fine grinding is required of the talc. Dust control is necessary for milling.

For paper-grade or cosmetic-grade talc, flotation is required for production. The ore is first ground, then floated, usually at a pH of 9.0 using an alcohol frother. Sodium silicate may be used as a slime dispersant or quartz depressant, while calcium hypochlorite is used for destroying bacteria when the talc is for cosmetic uses. Following flotation, the talc concentrate is dried and ground further.

## Summary

The most recent assessment of ores and wastes, in 1989 (Godin 1992), attributed 2 Mt of ore to "miscellaneous non-metals." Associated with this mining are 1.9 Mt of waste rock and overburden and 0.5 Mt of tailings. It is apparent, then, that the volume of waste generated from this portion of the mining industry is not one of its major environmental problems. Particulate emissions would seem to be the most serious problem associated with the mining of industrial minerals; particulate matter derived from this type of mining causes problems both in the atmosphere and in the hydrosphere. Untreated or uncontrolled wastes from asbestos and fluorite operations and from cement plants are toxic and/or carcinogenic. Research needs include methods of lowering alkalinity in cement plant effluents and reducing the huge energy demands for processing raw materials.

# APPENDIX

# Conversions and Abbreviations

## Unit Conversions

The units used in this book are those of the International System (SI) or are derived from them. The basic units include the meter (m), kilogram (kg), second (s), and degree celsius (°C). The following conversions are provided for the convenience of readers who may not be familiar with the system.

*Length*
    1 meter (m) = 3.28 feet
    1 kilometer (km) = 0.621 mile

*Area*
    1 square meter ($m^2$) = 10.76 square feet
    1 hectare (ha) = $10^4$ $m^2$ = 2.47 acre
    1 square kilometer ($km^2$) = $10^6$ $m^2$ = 247 acre

*Volume*
    1 litre ($\ell$) = 0.220 imperial gallon = 0.264 U.S. gallon
    1 cubic meter ($m^3$) = 1.31 cubic yard = 35.3 cubic foot
    1 cubic kilometer ($km^3$) = $10^9$ $m^3$ = $10^{12}$ $\ell$

*Weight*
    1 kilogram (kg) = 2.20 pound (avoirdupois)
    1 tonne (t) = $10^3$ kg = 1.103 short ton = 0.984 long ton

*Energy*
    1 joule (J) = 1 watt-second (W · s)
    1 MJ = 948 British Thermal Unit (BTU) = 0.278 kilowatt hour (kWh)

*Energy equivalents*
    1 tonne uranium in a CANDU reactor = 690 TJ
    1 tonne uranium in a light-water reactor = 450 TJ
    1 cubic kilometer of natural gas = 38 PJ
    1 tonne of oil = 42 GJ

1 tonne hard coal = 28 GJ
1 tonne soft coal = 14 GJ
1 barrel of oil = 5.5 GJ

*Radioactivity*

Activity
1 becquerel (Bq) = 1 disintegration per second = 37 nCi
1 curie (Ci) = $3.7 \cdot 10^{10}$ disintegrations per second

Absorbed Dose
1 gray (Gy) = 1 joule per kilogram (J/kg) = 100 rad

Dose Equivalent
[absorbed dose multiplied by a quality factor, Q, reflecting the capability of each type of radiation to cause ionization; accepted values of Q are x-, gamma and beta rays –1; fast neutrons –10; alpha particles –20]
1 sievert (Sv) = 1 Gy $\cdot$ Q = 100 rem

Exposure
[used for gamma radiation only]
1 roentgen (R) = $2.58 \cdot 10^{-4}$ coulomb per kilogram (C/kg)

# Abbreviations and Acronyms

| | |
|---|---|
| AECB | Atomic Energy Control Board |
| AQI | Air Quality Index |
| AVHRR | advanced very high resolution radiometer, a satellite remote-sensing tool with a resolution of about 1 km (see also LANDSAT) |
| CANMET | Canada Centre for Mineral and Energy Technology |
| CFCs | chlorofluorocarbons (q.v.) |
| DIAND | federal Department of Indian Affairs and Northern Development |
| EIA | environmental impact assessment |
| EIS | an environment impact statement |
| EMR | Energy, Mines and Resources Canada |
| GIS | geographic information system, which is a computer-based mapping tool for integrating many types of information, such as geophysical, hydrological, meteorological, biological, and other types of data |
| LANDSAT | a series of Earth resources satellite imagers having a resolution better than 100 m, which cover each point on the surface every 18 days |

MEND          the tripartite industry-federal-provincial government Mine Environment Neutral Drainage program

MISA          the Ontario government's Municipal/Industrial Strategy for Abatement program

MITEC         the Mining Industry Technology Council

NPK           the nitrogen, phosphorous and potassium contents of a fertilizer for application to soils

NRCC          the National Research Council of Canada

NUTP          the National Uranium Tailings Program

PSI           acronym for Pollutant Standards Index, a composite air pollution indicator (similar to Pindex)


a             year

ppb           parts per billion

ppm           parts per million

ppt           parts per thousand

# GLOSSARY

**acid deposition**: Precipitation (including rain, snow, and mists) having below-normal pH, caused mainly by sulphur and nitrogen oxide atmospheric pollution.

**acid drainage**: Drainage water from mine workings, waste and tailings, made acidic in most cases by the oxidation of sulphide minerals.

**adit**: A nearly horizontal access to a mineral deposit for mining or exploration purposes.

**adsorption**: The deposition of molecules on the surface of a solid or liquid.

**aerobic**: In the presence of oxygen.

**aerosol**: A system of colloidal particles that are small enough to remain suspended in water or air for an appreciable period of time.

**agglomeration**: An ore-concentration process based on the adhesion of pulp particles to water, or the briquetting, nodulizing or sintering of a concentrate for further processing.

**aggregation**: The grouping together of soil particles into stable structures.

**air-lift agitator**: A device employed in metallurgical processes, using compressed air to stir or shake.

**air-pollution potential**: An index of the dispersal capabilities of the atmosphere in a particular region at a particular time.

**alluvial**: Deposited by a stream or running water.

**amphibolite**: A crystalloblastic rock consisting mainly of the mineral amphibolite and plagioclase, with little or no quartz.

**anaerobic**: In the absence of oxygen.

**andesite**: A dark-colored, fine-grained extrusive rock that may contain large grains of sodium-rich plagioclase and one or more of the mafic minerals such as biotite, hornblende and py-roxene. The ground mass is generally of the same composition as the large grains.

**angiosperm**: A plant characterized by having a flower as its sexual reproductive structure.

**anion**: An ion that carries a negative charge.

**anorthosite**: A intrusive igneous rock consisting almost entirely of the mineral plagioclase.

**anthropogenic**: Produced by human activity.

**aquifer**: A body of rock or unconsolidated sediment that is sufficiently permeable to conduct groundwater and to yield economically significant quantities of water to wells and springs.

**arsine**: Hydrogen arsenide, $AsH_3$, an intensely poisonous colorless gas.

**asbestosis**: An irreversible disease characterized by diffuse stiffening of the soft tissues of the lungs with accompanying functional changes such as reduced vital capacity, reduced expansion, and obstructed bronchial passages.

**auriferous**: Containing gold.

**autoclave**: Thick-walled vessel with a tightly fitting lid, in which substances may be heated over 100°C.

**autogenous grinding**: Grinding of ore without the use of auxiliary media such as steel balls or rods.

**backfilling**: The return of mining wastes underground for disposal and/or subsidence prevention.

**barren solution**: In a metallurgical process, the solution left after the value has been removed.

**bedded deposit**: A mineral deposit having layering parallel to the stratification of the sedimentary parent rocks, and of contemporaneous origin.

**beneficiation**: The preparation of an ore for metallurgical processing, usually by means of comminution and concentration.

**benthic**: Pertaining to the benthos; also said of that environment.

**benthos**: Those forms of aquatic life that are bottom-dwelling; also, the ocean bottom itself.

**bioaccumulation**: The accumulation of certain elements in plant tissues, usually against large concentration gradients between the plant and the surrounding environment. Species that accumulate metals have been used in prospecting for ore deposits.

**biodiversity**: The number of species of organisms found in a biotic community. It may also be expressed in terms of genetic variation within species or variety of communities in the landscape. Components of diversity include richness (number of species per unit area) and evenness (degree of dominance exhibited by the species in a community).

**bioindicator**: Any organism used to indirectly indicate environmental characteristics, often in order to evaluate environmental quality or change (e.g., water milfoil are used to indicate metal contamination).

**bioleaching**: The use of organisms to facilitate leaching of an ore; an example is the use of the bacterium *Thiobacillus ferrooxidans* to break down refractory ores that are otherwise difficult to treat.

**biological polishing**: The use of plants to treat mine waste waters (see also ecological engineering).

**biomagnification**: An increase in concentration of a substance at each step up in trophic levels.

**biomass**: The mass of organic material per unit area of surface.

**biominification**: A decrease in concentration of a substance at each step down in trophic levels.

**biota**: Collectively, the organisms that inhabit a particular location.

**biotic community**: A group of interacting plants, animals, and microorganisms living in a particular habitat.

**bog**: A peat-covered wetland in which the vegetation shows the effects of a high water table and general lack of nutrients: high acidity and low oxygen saturation

**boreal**: Northern. Boreal forests are dominated by trees such as spruce.

**borrow pit**: A pit used for the sole purpose of extracting material for use in mining or construction operations located nearby; it is an aggregate operation licensed only for a specific purpose.

**Brundtland Commission**: The United Nations World Commission on Environment and Development, whose 1987 report, *Our Common Future*, introduced the notion of environmentally sustainable economic development.

**bryophytes**: Mosses, peat mosses and liver worts.

**bulk-mining**: The approach to mining that uses large equipment to remove larger amounts of ore more cheaply, in contrast to more selective mining.

**calcareous**: Containing calcium carbonate. When applied to a rock name, it implies that as much as 50 percent of the rock is calcium carbonate.

**calcination**: Heating of an ore to decompose carbonates, hydrates, etc.; surrounding air is not involved in the process as it is in roasting (q.v.).

**captive emissions**: Particulates and gases produced during particular localized operations such as crushing, conveying, calcining and smelting, which are relatively easy to capture through the use of any of a large number of devices (contrast with fugitive emissions).

**carbonatite**: A carbonate rock of apparent magmatic origin, generally associated with kimberlites and alkalic rocks.

**carnallite**: A milk-white to reddish orthorhombic mineral: $KMgCl_3.6H_2O$.

**Caro's acid**: Permonosulphuric acid, $H_2SO_5$.

**catalyst**: An agent that precipitates a process without being directly involved in it.

**cation**: An ion that carries a positive charge.

**caustic soda**: Sodium hydroxide, NaOH.

**celluloid**: A thermoplastic material made from cellulose nitrate and camphor.

**chelation**: Formation of a closed ring of atoms

by attachment of compounds or radicals to a central polyvalent metallic ion. Chlorophyll and haemoglobin are naturally occurring chelated compounds.

**chironomids**: Dipteran insects with aquatic larvae, midges.

**chlorofluorocarbons**: Synthetic substances which, although inert in the troposphere, attack the ozone layer when they move up to the stratosphere.

**claim**: A claim on mineral lands.

**clinker**: A slaggy or vitreous mass of coal ash.

**coke**: A combustible material produced by heating bituminous coal and driving off its volatile matter. It consists of mineral matter and fixed carbon fused together; it is grey, hard, porous, and as a fuel is nearly smokeless and of high caloric value.

**collectors**: Froth flotation reagents that attach themselves to specific non-floating minerals, permitting them to adhere to air bubbles and float to the surface; xanthates are the most common collectors (see also frothers and modifiers).

**colloid**: 1. A particle size range of less than 0.00024 mm, i.e., smaller than clay size. 2. Any extremely fine-grained material in suspension, or that can be easily suspended, commonly having peculiar properties because of its very high surface area.

**comminution**: The reduction of ore size through the use of crushing and grinding.

**complexation**: Combining of simple ions to form complex ionic compounds.

**concentrate**: Ore that has passed through the beneficiation stage and is ready for further processing.

**concentration**: The process of physical separation of a mineral from its gangue in preparation for chemical separation or extractive metallurgy.

**coning**: A method of tailings disposal that creates a cone-shaped heap to aid in drainage and reduce rainfall interception.

**conversion**: The final stage in pyrometallurgical processing, in which air is injected into the molten matte to oxidize most of the remaining iron and sulphur impurities.

**craton**: A part of the Earth's crust that has attained stability and has been little de-

formed for a long time. The term is restricted to continents and includes both shields and platforms.

**cuticle**: Superficial, non-cellular layer covering a plant or animal, secreted by the epidermis.

**cyanidation**: The use of a solution of sodium or potassium cyanide to recover gold.

**cytochrome**: A respiratory pigment widely distributed in aerobic organisms.

**decommissioning**: The process of permanently closing a mining operation, including rehabilitation of affected lands and plans for their maintenance.

**detritus**: The mixture of dead organic matter and microorganisms responsible for its decomposition.

**depauperate**: Stunted, impoverished.

**diamond drilling**: The common method of extracting exploratory samples of an ore body using diamond-tipped drills.

**diatom**: A microscopic single-celled aquatic plant belonging to the algae. Both freshwater and marine species occur.

**diffusivity**: A measure of how quickly a substance is dispersed when introduced into a gas or liquid; it is the reciprocal of the flow resistance.

**dinoflagellate**: A small, motile alga.

**dissociation**: The breakdown of a substance into several others by heat or solution.

**doré**: The final saleable product of a gold mine.

**drilling fluid**: A mixture of water and mud circulated through a drill for cooling and cleaning purposes.

**dry deposition**: Direct fallout of atmospheric particulates and aerosols (compare with wet deposition).

**dunite**: Peridotite consisting essentially of the mineral olivine, with accessory pyroxene, plagioclase, or chromite.

**ecological engineering**: The application of ecological principles in solving practical problems, e.g., the use of a natural or constructed wetland to treat mine drainage before release to the environment.

**ecological niche**: The role played by a species in its ecological community, in terms of time, space and functional relationships.

**ecosystem**: A group of plants, animals and microorganisms that are related with each other and with their physical environment, forming a unified whole.

**Eh**: Measurement of the oxidation-reduction potential of a solution.

**electrometallurgy**: The use of electrical current to chemically separate the components of a beneficiated ore (compare with hydrometallurgy and pyrometallurgy, q.v.).

**electroplate**: To deposit a layer of metal by chemical decomposition by passing an electric current through the substance in a dissolved or molten state.

**electrostatic precipitator**: A device for controlling air-borne particulates whereby the air is subjected to a unidirectional electrostatic field, so that the particles are attracted to, and deposited on, the positive electrode.

**electrowinning**: The removal of metals from a solution by electrolysis.

**eluate**: Solution obtained by washing to remove adsorbed substances.

**emissions inventory**: An assessment of an industry in terms of all of the emissions it produces (in contrast with source inventory).

**endemic**: Native, or confined naturally to a particular area or region; indigenous.

**epidermis**: The outermost layer of cells of a plant or animal.

**epiphyte**: A plant that grows on top of another plant species, such as some lichens and orchids.

**estuary**: The mouth of a river where fresh and salt water meet.

**etiology**: Assignment of a cause; (Med) science of the causes of disease.

**eutrophication**: The aging of a water body due to the build-up of nutrients, producing an increase in biological oxygen demand and decrease in dissolved oxygen; the process may be accelerated by the addition of industrial drainage.

**evaporites**: Deposits of salts such as potash, anhydrite and rock salt, formed over geological time by the evaporation of lakes and seas.

**exothermic**: Describes a chemical reaction in which energy is liberated.

**expert system**: A computer program that summarizes the experience of a number of experts in a particular field, which may be used by non-experts for decision making.

**extractive metallurgy**: A term for the chemical separation of a desired mineral from its mined compound; includes pyrometallurgy, hydrometallurgy and electrometallurgy (q.v.).

**fauna**: All animals of a particular area.

**fecundity**: Degree of fertility.

**feldspar**: A group of abundant rock-forming minerals of the general formula, $MAl(Al,Si)_3O_8$, where M is typically K, Na, or Ca. Feldspars are the most widespread of any mineral group and constitute 60 percent of the earth's crust; they occur in many types of rock.

**felsic**: A mnemonic adjective derive from *fel*dspar + *le*nad (feldspathoid) + *si*lica + *c*, and applied to an igneous rock having abundant light-colored minerals. Also applied to those minerals (quartz, feldspars, feldspathoids, muscovite) as a group. It is the complement of *mafic*.

**fen**: A peatland characterized by a high water table, with slow internal drainage. Nutrient levels, oxygen saturation and pH are higher than in a bog.

**flocculant**: A reagent added to water to aggregate minute suspended particles so that they may precipitate out of suspension.

**flora**: The plants belonging to a specific area. Includes bacteria in expressions such as "soil flora."

**flux**: 1. A substance that reduces the melting point of a mixture, as in making glass or ceramics, or that helps metals fuse; 2. Amount of energy or matter flowing through a cross-sectional area.

**frothers**: Chemical agents that produce the froth in a floatation cell when aerated; mainly organic substances such as amyl alcohol, cresol, toluidine (see also collectors and modifiers).

**fugitive emissions**: Particulates and gases, having very diffuse sources, such as gravel roads, waste heaps, bulldozing and blasting, which are difficult to quantify and control (contrast with captive emissions).

**fumigation**: A condition of the atmosphere in

which air is well mixed near the surface but trapped by a sharp inversion above the surface layer.

**gabbro**: A group of dark-colored, mafic intrusive rocks composed primarily of the minerals labradorite or bytownite and augite, with or without olivine and orthopyroxenes. It is the approximate intrusive equivalent of basalt.

**gangue**: The valueless rock or mineral aggregates in an ore; that part of an ore that is not economically desirable but cannot be avoided in mining. It is separated from the ore minerals during concentration.

**global scale**: A spatial scale of size ranging from a large part of a continent to the whole earth.

**gneiss**: A foliated rock formed by regional metamorphism, in which banding or lenticles of granular minerals alternate with bands or lenticles of minerals with flaky or elongate prismatic habit.

**granodiorite**: A group of coarse-grained intrusive igneous rocks intermediate in composition between quartz diorite and quartz monzonite containing quartz, oligoclase or andesine, and potassium feldspar, with biotite, hornblende, or, more rarely, pyroxene, as the mafic component.

**greenstone belt**: A region containing compact, dark-green, altered or metamorphosed mafic igneous rock that owes its color to the minerals chlorite, actinolite, or epidote.

**grid lines**: Lines cut through vegetation for soil, electromagnetic, magnetic, and gravity surveys, and for drilling (also see seismic lines).

**grinding media**: Material, such as steel balls or rods, introduced into a mill to assist in grinding an ore; as the media wear they contaminate the ore (see also autogenous grinding).

**gymnosperms**: Seed-producing plants with poorly protected seeds, generally in cones. Includes trees such as pines and spruces.

**habitat**: The range of environmental conditions to which a community or a specific organism is adapted.

**half-life**: The period of time during which half the atoms of a radioactive element or isotope will undergo disintegration. It is essentially independent of the amount of the substance.

**heap leaching**: A type of hydrometallurgy carried out by adding a lixiviant to broken ore heaped on the surface (compare with vat and in-situ leaching).

**hydraulic head**: The height of the free surface of a body of water above a given subsurface point. Water flows from high hydraulic head to low, and an increase in head difference between two points will cause an increase in flow.

**hydric soil**: A soil containing much moisture.

**hydrometallurgy**: Separation of a metal in aqueous solution from the rest of the ore, followed by precipitation in metallic form.

**hydrophobic**: Lacking strong affinity for water; describes colloids whose particles are not highly hydrated and coagulate easily.

**hydrophyte**: Aquatic plant.

**hydroseeding**: The spraying of plant seeds in a solution.

**hydrosphere**: Collectively, all of the water bodies on the surface of the earth, including oceans, lakes, rivers, etc..

**hypoxia**: Deficiency of oxygen.

***in-situ* leaching**: The most common type of *in-situ* mining (q.v.); compare with vat and heap leaching.

***in-situ* mining**: The processing of an ore without removing it from the ground, most commonly carried out by the introduction of a solvent to leach out the desired mineral.

**industrial minerals**: Minerals other than those yielding metals or serving as fuels; comprises mainly structural materials, asbestos and salts.

**ion exchange**: One of the methods used to recover the solute from the pregnant solution in an electrometallurgical operation. Replacement of an ion of one element by that of another element on an adsorptive surface.

**isotopic composition**: The distribution of isotopes (atoms having the same atomic number but different atomic weights) in a sample of an element.

**kimberlite**: An alkalic peridotite containing abundant large grains of olivine and phlogopite, in a fine-grained groundmass of calcite, secondary olivine and phlogopite, with accessory ilmenite, serpentine, chlorite, magnetite, and perovskite.

**labile**: Easily changed; mobile.

**lacustrine**: Pertaining to, produced by, or inhabiting a lake or lakes.

**leaching**: Extracting a soluble material through dissolution with a suitable solvent.

**lenticular**: Resembling in shape the cross-section of a lens, especially of a double-convex lens.

**ligand**: A substance that binds with another, such as an organic molecule with a metal.

**liquefaction**: 1. The catastrophic collapse of a pile of waste material under the influence of gravity due to internal water build-up; 2. Change of state from solid to liquid.

**lithification**: 1. The conversion of a sediment into a solid rock, involving such processes as cementation, compaction, and crystallization; 2. The lateral termination of a coal bed owing to an increase in impurities.

**lithosphere**: All of the rocks and minerals of the Earth and their derived soils, collectively.

**littoral**: Pertaining to, produced by, or inhabiting a shore or coast.

**littoral dredging**: Dredging of a shoreline.

**lixiviant**: The leaching agent in a hydrometallurgical process.

**local scale**: A spatial scale of size less than that of a large city.

**lode**: A mineral deposit consisting of a zone of veins, veinlets, or disseminations; also, a mineral deposit in solid rock as opposed to a placer deposit.

**macronutrients**: The major nutrient elements required for plant growth, namely C, H, O, P, K, N, S, Ca, Fe, and Mg.

**macrophage**: An organism larger than bacteria that ingests organisms or portions of organic matter.

**macrophytes**: Large aquatic plants.

**mafic**: Describes an igneous rock composed chiefly of dark, ferromagnesian minerals; also, those minerals. The complement of *felsic*.

**magnetometer**: An instrument for measuring magnetic intensity that is commonly used for mineral prospecting.

**manganese nodules**: Concretions found on the ocean floor at moderate to great depths, which contain sizable amounts of cobalt, nickel, copper, and manganese.

**marsh**: A wetland, dominated generally by emergent plants.

**materials flow**: An accounting of the inputs and outputs of material and energy associated with an industrial process.

**meromictic**: Refers to a lake having a highly stable stratification that prevents vertical mixing, due to layers of differing density.

**mesic soil**: Soil that is unlikely to have excess or deficient moisture during the growing season.

**mesothelioma**: A malignant tumour with a latent period of up to 60 years that involves membranes lining the body cavity, connected with overexposure to asbestos fibres.

**metal tolerance**: The ability of an organism to tolerate higher than normal concentrations of metals in its environment.

**metalloid**: An element, such as boron, silicon, arsenic or tellurium, that is intermediate in properties between typical metals and non-metals.

**meta-sediment**: A sediment or sedimentary rock that shows evidence of having been subjected to metamorphism.

**microflora**: Small, often microscopic, plants, including bacteria and fungi, especially those inhabiting the soil and acting as decomposers.

**micronutrients**: Nutrient elements that are required in very small amounts for plant growth.

**mineral aggregates**: A class of industrial minerals comprising sand and gravel, crushed stone, and lightweight aggregates

**modifiers**: Chemical reagents used to modify the normal behavior of minerals in a froth flotation circuit; includes a wide variety of substances many of which are toxic (see also frothers and collectors).

**morphometry**: The measurement of the structure of plants and animals.

**motile**: Capable of spontaneous motion.

**mulch**: A covering over the soil to provide protection from wind, temperature extremes, etc.

**muskeg**: An area covered with *Sphagnum* mosses, tussocky sedges, and an open growth of stunted trees.

**mycorrhiza**: The symbiotic association between a fungus and the root of a higher plant that usually increases the efficiency of nutrient and water uptake by the plant.

**native metal**: A metal such as copper, silver, gold, or the platinum group, that is found naturally in the elemental state.

**necromass**: Dead organic matter, such as leaf litter.

**necrosis**: The pathological death of tissue in a plant or animal. Adj: necrotic.

**norite**: A coarse-grained intrusive igneous rock containing labradorite as the chief constituent and differing from gabbro by the presence of hypersthene as the dominant mafic mineral.

**organophilic**: Having high affinity for organic molecules.

**overburden**: Material which must be removed to allow access to an orebody, particularly in a surface mining operation.

**oxidase**: Enzyme which facilitates oxidation of the substrate.

**oxidation**: The process of combining with oxygen, to form a compound such as an oxide. The term is also used more generally to include any reaction in which an atom loses electrons.

**pachuca**: A cylindrical tank with a conical bottom, widely used to circulate the pulp charge in slime leaching.

**particulates**: Solid particles and liquid droplets present in the atmosphere. Particulates have various sources, both natural and anthropogenic.

**paydirt**: An unconsolidated sediment that contains a mineral deposit.

**peatland**: A wetland or heath having an accumulation of at least 40 cm of peat; one of several types of peat-covered terrain.

**pegmatite**: An exceptionally coarse-grained igneous rock, with interlocking crystals, usually found as irregular dikes, lenses or veins.

Most grains are one cm or more in diameter. The composition of pegmatites is generally that of granite; it may be simple or complex, and may include rare minerals rich in such elements as lithium, boron, fluorine, niobium, tantalum, uranium, and rare earths. Pegmatites represent the last and most hydrous portion of a magma to crystallize and hence contain high concentrations of minerals present only in trace amounts in granitic rocks. Adj: pegmatic.

**pelagic**: Belonging to the surface of the ocean (e.g. zooplankton).

**pelletization**: The agglomerization of finely divided concentrate into pellets by rolling in a drum; a common method of preparing iron ore for further processing.

**percolation**: Slow movement of fluids through small openings within porous material.

**peridotite**: A coarse-grained intrusive igneous rock composed chiefly of the mineral olivine, with or without other mafic minerals such as pyroxenes, amphiboles, or micas, and containing little or no feldspar. Peridotite is commonly altered to serpentinite. It may also be referred to as dunite.

**periglacial**: A term applied to a region adjacent to an ice sheet.

**peritoneum**: The double membrane lining the cavity of the abdomen.

**pH**: The negative logarithm of the hydrogen-ion concentration in a solution, which denotes the acidity (below 7) or basicity (above 7) of the solution; a neutral solution has a pH of 7.

**photoemissive**: A substance is described as photoemissive if it is capable of emitting electrons when it is subjected to electromagnetic radiation. The wave-length of the radiation which will provoke such emission depends upon the nature of the substance: light provokes some metals into photoemission; other materials may require ultra-violet radiation or X-rays.

**phyllosilicate**: A class or structural type of silicate characterized by the sharing of three of the four oxygens in each tetrahedron with neighbouring tetrahedra, to form flat sheets, e.g. micas.

**phytochelatin**: A substance capable of chelating metals, produced in plant cells.

**phytotoxic**: Poisonous to plants.

**placer mining**: The removal of high-density minerals (such as native gold) from alluvial deposits by washing with water.

**plankton**: Small aquatic organisms that float or swim near the surface of marine and fresh water bodies.

**plaque**: A white, glistening, raised area of fibrous, collagenous tissue, or an inorganic surface deposit.

**pleural calcification**: In mammals, hardening of the either of the two serous membranes lining the thorax and enveloping the lungs by deposition of calcium salts.

**plume of damage**: Trail of pollutants, often roughly conical in shape with the tip as source; plumes can occur both in air or in water.

**plume rise:** Height of rise of smoke from a stack that depends on the temperature and exit velocity of the plume as well as on atmospheric conditions.

**podzolic soil**: A soil typical of cool temperate humid climate, with a strongly developed leached zone. Horizons are noticeable and an iron pan may occur.

**polychaetes:** Marine segmented worms.

**precipitation scavenging**: The absorption of atmospheric pollutants by precipitation which carries them to the ground as wet deposits.

**pregnant solution**: A value-bearing solution in a hydrometallurgical operation.

**profundal**: Pertaining to deep water.

**propagule**: A seed or other part of a plant capable of developing into a new individual.

**pulp**: A mixture of ground ore and water, capable of flowing as a liquid.

**pyrite**: An iron sulphide ($FeS_2$) often found in association with other base-metal sulphide minerals.

**pyrometallurgy**: Chemical separation of a metal from its ore using heat; the oldest and most widely used of the extractive metallurgical processes.

**pyroxenite**: An ultra-mafic intrusive igneous rock chiefly composed of pyroxene, with accessory hornblende, biotite, or olivine.

**pyrrhotite:** A magnetic iron sulphide ($Fe_{n-1}S_n$), with n between 5 and 16, that occurs commonly with nickel ore.

**radiatively active gases**: Atmospheric gases that are capable of absorbing terrestrial radiation and hence contributing to the "greenhouse effect."

**radionuclides**: Unstable isotopes of elements that decay spontaneously, emitting radiation.

**reduction:** A chemical reaction involving the addition of hydrogen or the removal of oxygen *or* the extraction of a metal from its ore in a smelter.

**refining**: The purification of a crude metal product; the next processing stage after smelting.

**refractory ores:** Ores that are difficult to treat for the removal of values.

**regional scale**: Intermediate spatial scale of size between a large city and a small continent.

**residence time:** The average lifetime of a substance moving through a system, for instance, of a pollutant in the atmosphere; it is the reciprocal of the fraction of the substance that is removed per unit time.

**residual**: The "leftovers" from an industrial process which are discharged to land, water or air; these include solids, liquids, gases, and energy.

**retort**: A glass vessel consisting of a large bulb with a long neck narrowing somewhat towards the end. In industrial processes, any vessel from which distillation takes place.

**rhizobium:** A bacterial genus which is a symbiont on roots.

**rhizome**: An underground stem resembling a root but bearing scale leaves and buds.

**rhizosphere**: The uppermost layer of the soil, which plant roots and associated organisms inhabit.

**rhyolite**: A group of extrusive igneous rocks, typically having large grains of quartz and alkali feldspar contained in a fine-grained glassy groundmass; also, any rock in that group; the extrusive equivalent of granite.

**roasting**: Heating an ore to oxidize and drive off sulphur-oxide gases, in preparation for smelting; this may be contrasted with calcination.

**salt marsh:** 1. A coastal march influenced by sea water. 2. An inland marsh in an arid region, with saline water.

**schist:** A strongly foliated crystalline rock, formed by dynamic metamorphism, that has well developed parallelism of more than 50 percent of the minerals present, particularly those of lamellar or elongate prismatic habit, e.g. mica and hornblende.

**scintillation:** A small flash of light produced by an ionization agent, such as radioactive particles, in a phosphor or scintillator.

**sediment diagenesis:** The re-entry of sediment substances into the overlying water through physical, chemical and/or biological action.

**sedimentation:** The settling out of suspended particulate matter in a water body.

**seismic lines:** Lines cleared of major vegetation in order to permit access to the ground surface for seismic surveying purposes.

**semi-autogenous (SAG) milling:** The grinding of ore by means of large pieces of rock and steel balls.

**serpentine:** A group of common rock-forming minerals with the formula $(Mg,Fe)_3Si_2O_5(OH)_4$. Serpentines are always secondary minerals, derived by the alteration of magnesium-rich silicate minerals (esp. olivine), and are found in both igneous and metamorphic rocks; also, a mineral of that group, such as chrysotile or antigorite.

**settling pond:** A natural or artificial water body used to contain effluent for the purpose of removing suspended solids before release to receiving waters.

**shotcrete:** A cement-aggregate mixture pneumatically sprayed on the walls of an underground mine to stabilize the surface.

**siliceous:** Describes a rock or substance containing abundant silica, esp. as free silica rather than silicates.

**silicosis:** A lung disease caused chiefly by prolonged inhalation of high levels of silica dust; it is most common among rock drillers and stone cutters

**siltation:** The deposition of sediments in a water body as fine suspended particulate matter.

**sintering:** The use of heat to fuse some constituents of ores and concentrates, with the purpose of agglomerating them into larger masses for further processing.

**slag:** Material from the smelting process, resulting from the fusion of flux with ash from the coke and impurities from the ore.

**slate:** A compact, fine-grained metamorphic rock that possesses cleavage and hence can be split into slabs and thin plates. Most slate was formed from shale.

**slimes:** Solid hydrospheric residuals from many mineral-processing operations, which are fine enough to remain in suspension or settle very slowly, making removal difficult (see colloids and flocculants).

**sluice:** A long, trough-like box used to separate gold from alluvial material in a hydraulic placer operation.

**smelting:** Chemical reduction of a metal from its ore using pyrometallurgy.

**solvent extraction:** A method of separating substances in a mixture through the use of a solvent to dissolve the required substance, leaving the others.

**source inventory:** The assessment of all of the sources of a given emission (in contrast with emission inventory).

**spectrometer:** An optical device for measuring wavelength of light used, for example, in aerial prospecting for minerals.

**specular hematite:** A variety of hematite $(Fe_2O_3)$ with brilliant black colour and metallic lustre.

**spoil:** Overburden or other waste material removed in mining, quarrying, dredging, or excavating.

**stibine:** Antimony hydride, $SbH_3$, which is a poisonous gas.

**stoma:** A pore in a plant leaf through which transpiration and gas exchange occur. Plural: stomata.

**stratosphere:** The layer of the atmosphere immediately above the troposphere, extending to a height of approximately 50 km; compared with the troposphere, it has very low temperatures and pressure, high levels of ultraviolet radiation, and very little vertical mixing.

**strip mining:** A method of mining shallow, bedded mineral deposits, in which the overlying material is stripped off, the mineral removed, and the overburden replaced.

**stripping ratio**: The ratio of waste rock to ore in a mining operation.

**structural materials**: Those industrial minerals used in constructing buildings, roads, and other structures, e.g. sand and gravel, stone aggregates, cement, and limestone.

**subsidence**: Sinking or downward settling of the earth's surface.

**substrate**: 1. The substance or nutrient on or in which an organism lives and grows. 2. The surface to which a fixed organism is attached. 3. An underlying layer, such as the parent material of a soil.

**succession**: The change in biotic community composition over time.

**swamp**: A wetland containing trees or dominated by trees.

**sylvite**: A white or colourless isometric mineral: KCl. It is the principal ore mineral of potassium compounds. Sylvite occurs in beds as a saline residue with halite and other evaporites.

**synergism**: The combined effect of two or more actions being greater than the sum of the two separate effects.

**synoptic scale**: Spatial scale of the size of a mid-latitude weather system, i. e. about 1000-3000 km.

**taconite**: A local term used in the Lake Superior iron-bearing district of Minnesota for any bedded ferruginous chemical sedimentary silicate rock, especially one that encloses the Mesabi iron ores (granular hematite); an unleached iron formation.

**tailings**: Those portions of washed or milled ore that are regarded as too poor to be treated further, as distinguished from the concentrates, or material of value.

**thickener**: A vessel for reducing the amount of water in pulp (q.v.).

*Thiobacillus ferrooxidans*: Bacteria which catalyze the oxidation of sulphide minerals.

**tidal flat**: A coastal area covered by water at high tides and exposed at low tides.

**till**: Unstratified drift, deposited directly by a glacier without reworking by meltwater, and consisting of a mixture of clay, silt, sand, gravel, and boulders ranging widely in size and shape.

**transmissivity**: The rate at which water is transmitted through a unit width of an aquifer under a unit hydraulic gradient.

**trophic level**: Any of a series of distinct feeding levels in a food chain.

**troposphere**: The layer of the atmosphere closest to the Earth, extending from the surface up to about 10 km, in which there is usually a temperature decrease with height and substantial vertical mixing.

**turbidity**: The extent to which water contains suspended matter. This is often measured as the reduction of light transmission through the water.

**ultramafic**: Said of an igneous rock composed chiefly of mafic minerals, e.g. monomineralic rocks composed of hypersthene, augite, or olivine.

**understorey**: The layer of small trees in a forest below the upper layer, or canopy.

**vat leaching**: Hydrometallurgy carried out in a vat, in contrast to heap and *in-situ* leaching (q.v.).

**ventilation coefficient**: The product of the vertical height of the surface-based layer of the atmosphere in which mixing can take place and the mean horizontal windspeed in that layer.

**venturi scrubbers**: An air-scrubber using venturi tubes to exert suction on the air stream.

**volatilization**: Freeing or giving off a substance as gas or vapour.

**wet deposition**: Fallout of air components scavenged by precipitation.

**wetland**: Land that is saturated with water or submerged, at least during most of the growing season.

**yellowcake**: Ammonium diuranate, the concentrate produced by a uranium mill.

**zeolites**: A class of hydrated aluminosilicate minerals that are used as water softeners, as molecular sieves and in ion-exchange processes.

# BIBLIOGRAPHY

Acott, G.B. 1989. The Creation of a Mountain Lake Sport Fishery at Cardinal River Coals Ltd. *See* Walker et al. 1989.

AECB. Atomic Energy Control Board. 1990. *Canada's radiation scandal?* Ottawa, ON: Atomic Energy Control Board.

Aho, A.E. 1966. Exploration methods in Yukon, with special reference to Anvil District. *Western Miner* 127-148.

Aitken, R. 1991. Sustainable Development— A Canadian Perspective. *See* AMIC 1991, Vol. 2: 28-45.

Al-Hashimi, A. 1989. Raffinate treatment for a pilot uranium mill. *See* Chalkley et al. 1989.

Aldous. P.J., and P.L. Smart. 1988. Tracing ground-water movement in abandoned coal mined aquifers using fluorescent dyes. *Ground Water* 26: 172-178.

Allan, R.J. 1974. Metal contents of lake sediment cores from established mining areas: an interface of exploration and environmental geochemistry. *Geological Survey of Canada Bulletin* 74-1B: 43-49.

Allen, M.F. 1991. *The Ecology of Mycorrhizae.* New York, NY: Cambridge University Press.

Allum, J.A.E., and B.R. Dreisinger. 1987. Remote sensing of vegetation change near Inco's Sudbury mining complexes. *International Journal of Remote Sensing* 8: 399-416.

AMIC. 1991. *Proceedings of the 16th Annual and 2nd International Environmental Workshop.* 7-11 October 1991. 2 vols. Perth, Australia: Australian Mining Industry Council.

Amiro, B.D., and G.M. Courtin. 1981. Patterns of vegetation in the vicinity of an industrially disturbed ecosystem, Sudbury, Ontario. *Canadian Journal of Botany* 59: 1623-1639.

Anderson, D. 1973. *Emission factors for trace substances.* EPA 450 2 73 001. Research Triangle Park, NC: Environmental Protection Agency.

Anderson, D.W., R.E. Redmann, and M.E. Jonescu. 1975. *A soil, vegetation and microclimate inventory of coal strip-mine wastes of the Estevan area, Saskatchewan.* Ottawa, ON: Co-operative Revegetation Project, Energy, Mines and Resources Canada.

Andrews, P. 1993. Clays. 16.1-16.11. *See* Godin 1993.

Aneja, V.P., D.F. Adams, and C.D. Pratt. 1984. Environmental impact of natural emissions—Summary of an APCA International Specialty Conference. *Journal of the Air Pollution Control Association.* 34: 799-803.

Angus Environmental Limited. 1991. *Review and Recommendations for Canadian Interim Environmental Quality Criteria for Contaminated Sites.* (The National Contaminated Sites Remediation Program) (Scientific Series, 197). Ottawa, ON: Inland Waters Directorate, Environment Canada.

Antonovics, J., A.D. Bradshaw, and R.G. Turner. 1971. Heavy metal tolerance in plants. *Advances in Ecological Research.* 7: 1-85.

Archer, A. 1970. Sub-sea minerals and the environment. *New Scientist* 3: 372-373.

Archibold, O.W. 1978. Vegetation recovery

following pollution control at Trail, B.C. *Canadian Journal of Botany* 56: 1625-1637.

——. 1985. The metal content of windblown dust from uranium tailings in northern Saskatchewan. *Water, Air, and Soil Pollution* 24: 63-76.

Arkay, K.E. 1975. Exploration and abandonment—environmental aspects. 36. *See* Canada. Environment Canada 1975.

Asklof, S. 1974. New system of quarry plant dust control. *Quarry Managers Journal* 58: 24-27.

Aston, T. 1989. That sinking feeling: Scientists monitor seafloor subsidence in Nova Scotia's Sydney Coalfield. *Geos* 18: 18-21.

Atkins, A.S., et al. 1986. An alternative method of surface disposal and stabilization of coal mine tailings. In *Proceedings of the 8th annual symposium on Geotechnical and Geohydrological Aspects of Waste Management, 5-7 February*. Rotterdam, Netherlands: A.A. Balkema. 277-286.

Axtmann, E.V., and S.N. Luoma. 1987. Trace metal distributions in floodplain and fine bed sediments of the Clark Fork River, Montana. *See* Lindberg and Hutchinson 1987.

Aylsworth, J., and L. Shapiro. 1993. Coal. 17.1-17.9. *See* Godin 1993.

Aylsworth, J. L. Shapiro, and B. Lomas. 1994. Coal. 17.1-17.9. *See* Godin 1994.

Bagatto, G. 1992. Concentrations of minerals (Cu, Ni, Fe, Mn, Mg, and Ca) in tissues of lowbush blueberry (*Vaccinium angustifolium*) and an insect gall induced by *Hemadas nubilipennis* (*Hymenoptera: Pteromalidae*) and its inhabitants in metal-contaminated sites. Ph.D. Thesis. Queen's University, Kingston, Ontario.

Bagatto, G., and M.A. Alikhan. 1987a. Copper, cadmium and nickel accumulation in crayfish populations near copper-nickel smelters at Sudbury, Ontario, Canada. *Bulletin of Environmental Contamination and Toxicology* 38: 540-545.

——. 1987b. Zinc, iron, manganese, and magnesium accumulation in crayfish populations near copper-nickel smelters at Sudbury, Ontario, Canada. *Bulletin of Environmental Contamination and Toxicology* 38: 1076-1081.

Bagatto, G., A.A. Crowder, and J.D. Shorthouse. 1993. Concentrations of metals in tissues of lowbush cranberry (*Vaccinium augustifolium*) near a copper-nickel smelter at Sudbury, Ontario, Canada: A factor analytic approach. *Bulletin of Environmental Contamination and Toxicology* 57: 600-604.

Baker, J.P., and S.W. Christensen. 1991. Effects of Acidification on Biological Communities in Aquatic Ecosystems. *See* Charles 1991.

Baker, L.A., et al. 1991. Interregional comparisons of surface water chemistry and biogeochemical processes. *See* Charles 1991.

Ballantyne, A.R. 1975. Water Pollution Problems in the Canadian Mining and Milling Industry. *See* Environment Canada 1975.

Balsberg-Pahlsson, A.-M. 1989. Toxicity of Heavy Metals (Zn, Cu, Cd, Pb) to Vascular Plants. *Water, Air, and Soil Pollution* 47(3-4): 287-319.

Barcelo, J., and C. Poshenreider. 1990. Plant-water relations as affected by heavy-metal stress—A review. *Journal of Plant Nutrition* 13: 1-37.

Barker, L., and L. Curtis. 1991. *The realities of environmental compliance—how to stay green and clean in mineral exploration.* Paper presented at the 59th Annual Meeting and Convention, Prospectors and Developers Association, Toronto, 26 March. Toronto, ON: Platinova Resources Ltd.

Barrie, L.A., and R.S. Schemenauer. 1989. Wet deposition of heavy metals. In *Control and Fate of Atmospheric Trace Metals*, eds. Pacyna, J.M. and B. Ottar. (NATO Advanced Science Institutes Series, Series C: Mathematical and Physical Sciences—268). Dordrecht: Kluwer Academic Publishers.

Barrie, W.B., and C.H. Carr. 1988. *Water use in the coal mining industry and its influence on ground water in western Canada.*

IWD Technical Bulletin No. 154, NHRI Paper, No. 38. Saskatoon, SK: Inland Waters Directorate, National Hydrology Research Institute.

Barth, H. (ed.). 1987. *Reversibility of acidification*. Environmental Research Programme of the Commission of the European Communities, Air Pollution Report Series, 8. London, UK: Elsevier.

Barton, B.J. 1993. *Canadian Law of Mining*. Calgary, AB: The Canadian Institute of Resources Law.

Barton, B., B. Roulston, and N. Strantz. 1988. *A Reference Guide to Mining Legislation in Canada*. Calgary, AB: The Canadian Institute of Resources Law.

Beach, R.A., and G. Cloutier. 1987. *Summary of spill events in Canada, 1974-1983*. Environmental Protection Series, EPS 5/SP/1. Ottawa, ON: Environment Canada.

Beak Associates Consulting Ltd. 1988. *An evaluation of persistent toxic contaminants in the Bay of Quinte*. Technical Report No. 1, Bay of Quinte Remedial Action Plan. Toronto, ON: Environment Ontario.

———. 1990. *Claude Resources Inc. Seabee project environmental impact statement*. Regina, SK: Environmental Assessment Branch, Saskatchewan Environment.

Beamish, R.J. 1974. Loss of fish populations from unexploited remote lakes in Ontario, Canada, as a consequence of atmospheric fallout of acid. *Water Research* 8: 85-95.

Beanlands, G.E., and P.N. Duinker. 1983. *An Ecological Framework for Environment Impact Assessment in Canada*. Hull, Québec: Federal Environmental Assessment Review Office.

Beaton, J.D. 1974. Soil-plant relationships. *Land Reclamation Short Course Notes*. Vancouver, BC: Centre for Continuing Education, University of British Columbia.

Beckett, P.J., et al. 1982. Distance dependent uranium and lead accumulation patterns. In *Lichens and mosses as monitors of industrial activity associated with uranium mining in northern Ontario, Canada*. Environmental Pollution Series B 4: 91-107.

Beggs, G.L., and J.M. Gunn. 1986. Response of lake trout (*Salvelinus namaycush*) and brook trout (*S. fontinalis*) to surface water acidification in Ontario. *Water, Air, and Soil Pollution* 30: 711-717.

Begon, M., J.L. Harper, and C.R. Townsend. 1986. *Ecology: Individuals, populations and communities*. Sunderland, MA: Sinauer Associates.

Beilke, S., and H.W. Georgii. 1968. Investigation on the Incorporation of Sulphur Dioxide into Fog and Rain Droplets. *Tellus (Sweden)* 20: 435-445.

Bell, J.C., W.L. Daniels, and C.E. Zipper. 1989. The practice of "approximate original contour" in the central Appalachians. 1. Slope stability and erosion potential. *Landscape and Urban Planning* 18: 127-138.

Berberi, G. 1977. Un autre mineral canadien sur le marché mondial. *Geos* 6: 2-4.

Bergeron, M. 1992. Peat. 36.1-36.11. *See* Godin 1992.

———. 1993. Peat. 36.1-36.9. *See* Godin 1993.

Bergeron, M., and P. Andrews. 1993. Talc, steatite and pyrophyllite. 48.1-48.10. *See* Godin 1993.

Bernstein, J.W., and S.M. Swanson. 1989. Hematological parameters and parasite load in wild fish with elevated radionuclide levels. In *Aquatic Toxicology and Water Quality Management*, ed. J.A. Nriagu. New York, NY: John Wiley & Sons.

Bhowmik, N.G. 1985. Erosion. Chapter 19 in *Handbook of Applied Meteorology*, ed. D.D. Houghton. John Wiley and Sons, New York.

Bietz, B.F. 1989. *Sediment impact on fisheries*. Technical and Research Committee on Reclamation. Vancouver, BC: Mining Association of British Columbia.

Blakeman, B., et al. (eds.). 1990. *Proceedings of Acid Mine Drainage Seminar/Workshop*. Halifax, Nova Scotia. 23-26 March 1987. Ottawa, ON: Environment Canada.

Bloom, P.A., J.H. Maysilles and H. Dolezal. 1982. *Hydrometallurgical treatment of arsenic-containing lead-smelter flue dust*. Report No. 8679. Washington, DC: Bureau of Mines, U.S. Department of the Interior.

Blowes, D.W., C.J. Ptacek, and J.L. Jambor. 1994. Remediation and Prevention of Low-Quality Drainage from Tailings Impoundments. *See* Jambor and Blowes 1994.

Boffey, P.M. 1975. Energy: Plan to use peat as fuel stirs concern in Minnesota. *Science* 190: 1066-1070.

Boivin, D.J. 1981. Les excavations abandonnées: Un sérieux problème, une tâche urgente. *Geos* 10: 12-14.

———. 1989. Geological and environmental considerations in abandoned underground mine reuse and rehabilitation. [French, with English abstract.] 459-470. *See* Walker et al. 1989.

Boldt, J.R. 1967. *The Winning of Nickel.* Toronto, ON: Longmans.

Bond, W. 1989. Heat recovery from the manufacture of sulphuric acid. *See* Centre in Mining and Mineral Exploration Research 1989.

Bosserman, R.W., and P.L. Hill. 1986. Community ecology of three wetland ecosystems impacted by acid mine drainage. *See* Brooks et al. 1986.

Bouchard, G., and D. Cranstone. 1994. Canadian mineral exploration. 6.1-6.24. *See* Godin 1994.

Boucher, M.A. 1992. Graphite. 22.1-22.12. *See* Godin 1992.

———. 1993a. Graphite. 24.1-24.10. *See* Godin 1993.

———. 1993b. Silica. 43.1-43.9. *See* Godin 1993.

———. 1994. Graphite. 24.1-24.9. *See* Godin 1994.

Boyd, B. 1976. *Potash.* Mineral Bulletin MR 156. Ottawa, ON: Energy, Mines and Resources Canada.

———. 1992. Iron ore. 24.1-24.12. *See* Godin 1992.

———. 1993. Iron ore. 26.1-26.11. *See* Godin 1993.

———. 1994. Iron ore. 26.1-26.9. *See* Godin 1994.

Boyd, B.W., and M.-C. Campeau. 1988. *Iron ore in Canada 1886-1986.* Mineral Policy Sector Internal Report MRI8812. Ottawa, ON: Energy, Mines and Resources Canada.

Boyd, B.W., and F. Johnson. 1991. *Canadian iron ore industry statistics—1991 and 1990.* Ottawa, ON: Mineral Policy Sector, Energy, Mines and Resources Canada.

Boyd, B.W., and L. Perron. *Canadian iron ore industry statistics 1992.* Mining Sector, Energy, Mines and Resources Canada. Ottawa, ON: Minister of Supply and Services.

Boyle, R., B. Coker, and G. Hall. 1992. Geochemical exploration research at the GSC: 40 years of progress. *Geos* 21: 14-15.

Boyle, R.W. 1987. *Gold: History and genesis of deposits.* New York, NY: Van Nostrand Reinhold.

Bradshaw, A.D. 1990. The reclamation of derelict land and the ecology of ecosystems. *Restoration ecology,* eds. W.R. Jordan, M.E. Gilpin and J.D. Aber. Cambridge, UK: Cambridge University Press. 53-74.

Bradshaw, A.D., and M.J. Chadwick. 1980. *The Restoration of Land. The ecology of reclamation of derelict and degraded land.* Oxford: Blackwell.

Brady, N.C. 1974. *The Nature and Properties of Soils.* 8th edition. New York, NY: Macmillan.

Bragg, K. 1975. Tailings pond and lagoon treatment systems. *See* Canada. Environment Canada 1975.

Brandt, C.J., and R.W. Rhoades. 1972. Effects of limestone dust accumulation on composition of a forest community. *Environmental Pollution* 3: 217-225.

Bratton, D.L. 1987. Regulatory response to changing reclamation demands. 5-12. *See* Powter 1987.

Brawner, C.O. 1986. Groundwater and coal mining. *Mining Science and Technology* 3: 187-198.

Brehaut, H. 1991. It Isn't Easy Being—or Becoming—Green. *CRS Perspectives,* No. 38, August 1991. Kingston, ON: Centre for Resource Studies, Queen's University.

Briggs, G.A. 1975. Plume Rise Predictions.

Chapter 3 in *Lectures on Air Pollution and Environmental Impact Analyses,* ed. D.A. Haugen. Boston, MA: American Meteorological Society.

Briggs, R.A. 1973. Heaps of Durham disappear. *Surveyor* 142: 34-35.

Brimblecombe, P., and A.Y. Lein (eds.). 1989. *Evolution of the global biogeochemical sulphur cycle.* International Council of Scientific Unions. SCOPE 39. Chichester, UK: John Wiley & Sons.

Brink, S., et al. 1991. *Application of petrographic techniques to assess in situ leach mining potential.* Circular IC 9295. Washington, DC: Bureau of Mines, U.S. Department of the Interior.

Brooks, R.P., D.E. Samuel, and J.B. Hill (eds.). 1986. *Wetlands and Water Management on Mined Lands.* Proceedings of a conference, 23-26 October 1985, Pennsylvania State University, J.O. Keller Conference Center, University Park, PA.

Brooks, R.R.. 1983. *Biological Methods of Prospecting for Minerals.* New York, NY: John Wiley & Sons.

———. 1987. *Serpentine and its vegetation.* Portland, OR: Dioscorides Press.

Brown, A., and D.J. Kushner. 1987. Feasibility of assessing metal pollution in Canadian Shield mining areas through analyses of peat soils. *Symposium 87: Wetlands/ Peatlands,* eds. C. Rubec and R.P. Overend. Edmonton, AB.

Brown, M.R. 1990. High cost means deep mines' days are numbered. *Northern Miner* 76(41): 4.

Browning, G. 1987. Successful landscape manipulation in the urban fringe—from a gravel pit to a lake. 24-29. *See* Powter 1987.

Brownridge, J.D. 1985. Use of tree bark to monitor radionuclide pollution. *Bulletin of Environmental Contamination and Toxicology* 35: 193-201.

Bruno, G.A. et al. 1984. *U.S. uranium mining industry: Background information on economics and emissions.* Contract DE-AC06-76RLO 1830. Richland, WA: Batelle Pacific Northwest Laboratory.

Bukovac, M.J., and S.H. Witener. 1957. Absorption and Mobility of Foliar-Applied Nutrients. *Plant Physiology* 8: 428-435.

Burckle, J.O., G.H. Marchant, and R.L. Meek. 1981. Arsenic emissions and control: Gold roasting operations. *Environment International* 6: 443-451.

Burnett, J.A. 1991. Where We Live: The Ecosphere. Chapter 1 in *The State of Canada's Environment,* ed. S. Burns. Ottawa, ON: Minister of Supply and Services.

Burton, J. 1991. Mining Systems and the Environment. In *Proceedings of the Curtin University of Technology Conference on the Management and Rehabilitation of Mined Lands.* Perth, Australia: Brodie-Hall Research and Consultancy Centre.

Cairns, J. (ed.). 1988. *Rehabilitating Damaged Ecosystems.* Vol. I and II. Boca Raton, FL: CRC Press.

Cameco. 1990. *Environmental impact statement: Modification to tailings placement procedure: Rabbit Lake in-pit tailings disposal system.* Regina, SK: Environment and Public Safety Saskatch-ewan.

Campbell, D., and I.B. Marshall. 1991. Mining: Breaking New Ground. Chapter 11 in *The State of Canada's Environment.* ed. S. Burns. Ottawa, ON: Minister of Supply and Services.

Campbell, P.G.C., and P.M. Stokes. 1985. Acidification and toxicity of metals to aquatic biota. *Canadian Journal of Fisheries and Aquatic Sciences.* 42:2034-2049.

Campbell, P.G.C., et al. 1988. *Biologically available metals in sediments.* No. 27694. Ottawa, ON: National Research Council of Canada.

Canada. Department of Fisheries and Oceans. 1991. Fisheries: Taking Stock. Chapter 8 in *The State of Canada's Environment,* ed. S. Burns. Ottawa, ON: Minister of Supply and Services.

———. Department of Indian and Northern Affairs. 1982. *Environmental guidelines pits and quarries,* prepared by MacLaren Plansearch. Cat. No. R72-180/1982E. Ottawa, ON: Minister of Supply and Services.

———. 1983. *Land use guidelines mineral exploration: Yukon and Northwest Terri-*

*tories*. QS-8321-000-EE-A1. Ottawa, ON: Minister of Supply and Services.

———. 1986. *Investigation of the magnitude and extent of sedimentation from Yukon placer mining operations*, prepared by Envirocon Limited. Environmental Studies No. 45. Ottawa, ON: Minister of Supply and Services.

———. Department of Natural Resources. 1994. Mineral production of Canada in 1992 with estimates for 1993. *Canadian Minerals* [a supplement to *Canadian Mining Journal*] April: 10-11.

———. Department of Regional Industrial Expansion. 1985. *Canada-Québec subsidiary agreement on mineral development 1985-1990*. Québec, PQ: Ministère des communications du Québec, direction générale des publications gouvernementales.

———. Energy, Mines and Resources Canada. 1977. *A summary view of Canadian reserves and additional resources of iron ore*. Mineral Bulletin MR 170. Ottawa, ON: Minister of Supply and Services.

———. 1982. *Potash—a proposed strategy*. Mineral Bulletin MR 194. Ottawa, ON: Minister of Supply and Services.

———. 1984a. *Canada's nonferrous metals industry: Nickel and copper. A special report*. Ottawa, ON: Minister of Supply and Services.

———. 1984b. *The Canadian iron ore industry: Current and emerging problems*. Mineral Policy Sector Internal Report MRI 8412. Ottawa, ON: Energy, Mines and Resources Canada.

———. 1989a. *Gold deposits and occurrences in Canada*. Ottawa, ON: Minister of Supply and Services.

———. 1989b. *Uranium in Canada—1989 assessment of supply and requirements*. Report EP 89-3. Ottawa, ON: Uranium Resource Appraisal Group.

———. 1992. *Mining and mineral processing operations in Canada 1991*. Mineral Bulletin MR 229. Ottawa, ON: Minister of Supply and Services.

———. 1993a. *Mining and mineral processing operations in Canada 1992*. Mineral Bulletin MR 232. Ottawa, ON: Minister of Supply and Services.

———. 1993b. *The production of primary iron in Canada: Managing the air quality challenges*. Coal and Ferrous Division. Ottawa, ON: Minister of Supply and Services.

———. Energy, Mines and Resources Canada, State of the Environment Reporting, and Environment Canada. 1993. *Canada: Terrestrial Ecoregions*. Ottawa: Energy, Mines and Resources Canada.

———. Environment Canada. 1973. *A nationwide inventory of air pollutant emissions 1970*, prepared by the Pollution Data Analysis Division. Technical Appraisal Report EPS 3-AP-73-2. Ottawa, ON: Minister of Supply and Services.

———. 1974. *National inventory of sources and emissions of asbestos, beryllium, lead, and mercury. Summary of emissions for 1970*. Data Analysis Division. Economic and Technical Review Report EPS 3-AP-74-1. Ottawa, ON: Minister of Supply and Services.

———. 1975. *Technology transfer seminar on mining effluent regulations/guidelines and effluent treatment technology as applied to the base metal, iron ore, and uranium mining and milling industry*. Ottawa, ON: Minister of Supply and Services.

———. 1976. *A nationwide inventory of air pollutant emissions. Summary of emissions for 1972*. Pollution Data Analysis Division. Economic and Technical Review Report EPS 3-AP-75-5. Ottawa, ON: Minister of Supply and Services.

———. 1981. *Copper-nickel smelter complexes in Canada: $SO_2$ emissions, historical and projected (1950-2000)*. EPS 3-AP-80-5. Data Analysis Division, Programs Branch, Air Pollution Control Directorate. Ottawa, ON: Environment Canada.

———. 1982a. *Air pollution emissions and control technology: Primary copper industry*. Economic and Technical Review Report. EPS 3-AP-82-4. Ottawa, ON: Air Pollution Control Directorate.

———. 1982b. *Environmental aspects of the extraction and production of nickel. Economic and Technical Review Report.* EPS 3-AP-82-5. Ottawa, ON: Air Pollution Control Directorate.

———. 1986. *Western Canadian low-sulphur coal: Its expanded use in Ontario.* Summary Report of the Federal/Provincial Task Force. Ottawa, ON: Minister of Supply and Services.

———. 1987. *Environmental aspects of nickel production: Sulphide pyrometallurgy and nickel refining.* Environmental Protection Service. Report EPS 2/MM/2. Ottawa, ON: Environment Canada.

———. 1988. *Status Report on Water Pollution Control in the Canadian Metal Mining Industry (1986)* Environmental Protection Series. EPS Report 1/MM/3, May 1988.

———. 1990a. *Natural urban air quality trends 1978-1987.* Report EPS 7/UP/3. Ottawa, ON: Environment Canada.

———. 1990b. *A framework for discussion on the environment: The Green Plan.* Ottawa, ON: Minister of Supply and Services.

———. 1992. *Status Report on Water Pollution Control in the Canadian Metal Mining Industry (1990 and 1991)* Environmental Protection Series. EPS Report 1/MM/4, December 1992.

———. Environment Canada and Department of Health and Welfare. 1989. *Canadian Environmental Protection Act.* Proceedings of the priority substances science forum. Burlington, February 22-23, 1989, ed. J. Waterston. Ottawa, ON: Rawson Academy of Aquatic Science.

Canadian Environmental Assessment Research Council. 1987. *The Place of Negotiation in Environmental Assessment.* Ottawa, ON: Minister of Supply and Services.

———. 1988. *Evaluating environmental impact assessment: an action prospectus.* Ottawa, ON: Minister of Supply and Services.

Cannon, H.L. 1960. Botanical Prospecting for Ore Deposits. *Science* 132: 591-598.

Capes, C.E., A.E. McIchinney, and R.D. Coleman. 1970. Beneficiation and balling of coal. *Transactions of the Society of Mining Engineers.* American Institute of Mining, Metallurgical and Petroleum Engineers. 247: 233-237.

Carter, T.R. 1984. *Marmoraton iron mine.* Report 5515. Metallogeny of the Grenville Province, Ontario. Toronto, ON: Ontario Geological Survey.

Casagrande, D.J., R.B. Finkelman, and F.T. Caruccio. 1989. The non-participation of organic sulphur in acid-mine drainage generation. *Environmental Geochemistry and Health* 11: 187-192.

Cassidy, N.G. 1968. The effect of cyclic salt in a maritime environment, II. The absorption by plants of colloidal atmospheric salt. *Plant and Soil* 28: 1397-1402.

Cathcart, G. (ed.). 1988. *1987 Canadian Minerals Yearbook—Review and Outlook.* Ottawa, ON: Energy, Mines and Resources Canada.

CEARC. *See* Canadian Environmental Assessment Research Council.

Centre in Mining and Mineral Exploration Research (ed.). 1989. *Energy Efficient Technologies in the Mining and Metals Industry.* Proceedings of a conference held at Laurentian University, Sudbury, 29 March 1989. Toronto, ON: Ontario Ministry of Energy.

Chadwick, J. 1994. Mining zinc and lead at the top of the world. *Mining Magazine* 171 (4): 205-212.

Chalkley, M.E., et al. (eds.). 1989. *Tailings and effluent management.* Proceedings of the International Symposium on Tailings and Effluent Management, 20-24 August, Halifax. New York, NY: Pergamon Press.

Chamberlin, P.D. 1986. Agglomeration: Cheap insurance for good recovery when heap leaching gold and silver ores. *Mining Engineering* 38: 1105.

Chan, W.H., and D.H.S. Chung. 1986. Regional-scale precipitation scavenging of $SO_2$, $SO_4$, $NO_3$, and $HNO_3$. *Atmospheric Environment.* 20: 1397-1402.

Chan, W.H., and M.A. Lusis. 1986. Smelting operations and trace metals in air and precipitation in the Sudbury Basin. Report A.R.B.-TDA 40-86. Toronto, ON: Ontario Ministry of the Environment.

Chan, W.H., et al. 1980. An analysis of the impact of Inco emissions on precipitation quality in the Sudbury area. Report A.R.B.-TDA 35/80. Toronto, ON: Ontario Ministry of the Environment, .

Chan, W.H., et al. 1984. Long-term precipitation quality and wet deposition fields in the Sudbury Basin. *Atmospheric Environment.* 18:1175-1188.

Charles, D. (ed). 1991. *Acidic Deposition and Aquatic Ecosystems. Regional Case Studies.* New York, NY: Springer-Verlag.

Chepil, W.S., and N.P. Woodruff. 1963. The Physics of Wind Erosion and Its Control. In *Advances in Agronomy.* Vol. 15, ed. A.G. Norman. New York, NY: Academic Press.

Chepil, W.S., F.H. Siddoway, and D.V. Armbruster. 1962. Climatic factors for estimating wind erodibility of farm fields. *Journal of Soil and Water Conservation* 17: 162-165.

Churcher, D. 1989. Design, implementation and cost effectiveness of backfill plants. *Canadian Mining Journal* 110: 13-14.

Cienski, T., and D. Doyle. 1992. Energy conservation in the comminution of industrial minerals. *CIM Bulletin* 85: 101-109.

Clair, T.A., J.P. Witteman, and S.H. Whitlow. 1982. *Acid Precipitation Sensitivity of Canada's Atlantic Provinces.* Technical Bulletin No. 124. Moncton, NB: Inland Waters Directorate, Environment Canada.

Clulow, F.V., et al. 1986. Radium-226 concentrations in faeces of snowshoe hares, *Lepus americanus,* established near uranium mine tailings. *Journal of Environmental Radioactivity* 3: 305-314.

Coates, D.F. 1972. Tentative Design Guide for Mine Waste Embankments in Canada. Technical Bulletin 145. Ottawa, ON: Energy, Mines and Resources Canada.

Coates, D.F., and M. Gyenge. 1972. Pit slopes angled to the future. *Geos* 1: 10-11

Coates, D.F., and Y.S. Yu. 1977. Waste embankments. *Pit Slope Manual.* CANMET Report 77-1. Ottawa, ON: Energy, Mines and Resources Canada.

Coates, W.E. 1975. Can surface mining be compatible with urbanization? *CIM Bulletin* 68: 41-47.

Colborn, T.E., et al 1990. *Great Lakes: Great Legacy?* The Institute for Research on Public Policy, Ottawa, and the Conservation Foundation, Washington. Baltimore, MD.

Collerson, K.D., et al. 1991. *Effect of coal dewatering and coal use on the water quality of the East Poplar River, Saskatchewan: A literature review.* Scientific Series, 177. Regina, SK: Environment Canada.

Collings, R.K. 1977. *Mineral waste resources of Canada. Report no. 2—mining wastes in Quebec.* CANMET Report 77-55. Ottawa, ON: Mineral Sciences Laboratories, Energy, Mines and Resources Canada.

———. 1978. *Mineral waste resources of Canada. Report no. 3—mining wastes in British Columbia.* CANMET Report 79-22. Ottawa, ON: Mineral Sciences Laboratories, Energy, Mines and Resources Canada.

———. 1980a. *Mineral waste resources of Canada. Report no. 4—mining wastes in the Atlantic provinces.* CANMET Report 80-12E. Ottawa, ON: Mineral Sciences Laboratories, Energy, Mines and Resources Canada.

———. 1980b. *Mineral waste resources of Canada. Report no. 6—mineral wastes as potential fillers.* CANMET Report 80-13E. Ottawa, ON: Mineral Sciences Laboratories, Energy, Mines and Resources Canada.

Committee on Mineral Resources and the Environment. 1975. *Mineral resources and the environment.* Washington, DC: National Academy of Sciences.

Constable, T.W., and W.J. Snodgrass. 1987. *Leachability of radioactive constituents from uranium mine tailings—final report.* Report EPS 3/MM/2. Ottawa, ON: Environmental Protection Service, Environment Canada.

Cook, B., and A.K. Biswas. 1974. *Beneficial*

*Uses for Thermal Discharges.* Report No. 2. Ottawa, ON: Environment Canada.

COSEWIC. Committee of the Status of Endangered Wildlife in Canada. 1994. Summary sheets. Environment Canada, Canadian Wildlife Service.

Costello, J.M., et al. 1982. A review of the environmental impact of mining and milling of radioactive ores, upgrading processes, and fabrication of nuclear fuels. In *Nuclear energy and the environment,* ed. E.E. El-Hinnawi. Environmental Sciences and Applications, Volume 11. Oxford, UK: Pergamon Press.

Couch, W.J., (ed.). 1988. *Environmental assessment in Canada.* 1988 summary of current practice. Canadian Council of Resource and Environment Ministries. Ottawa, ON: Minister of Supply and Services.

Coupal, B. 1985. La tourbe: Un agent dépolluant. *Geos* 14: 10-13.

Couper, A.S. 1990. Dust monitoring overcomes a burning Bing thought too hot to handle, Ramsay Bing, Loanhead. 209-221. *See* Rainbow 1990.

Coutant, C.C., and S.S. Talmage. 1976. Thermal Effects. *Journal of the Water Pollution Control Federation* 48: 1486-1544.

Couturier, G. 1994. Gold. 23.1-23.18. *See* Godin 1994.

Cox, L., et al. (eds.). 1989. *Proceedings: Gold mining effluent treatment seminars.* Burlington, ON: Environment Canada.

Cox, R.A. 1973. Predictions of fog formation due to a warm water lagoon proposed for power station cooling. *Atmospheric Environment* 7: 363-368.

Craigen, W.J.S. 1991. CANMET ferric chloride process. *See* Lakshmanan. 1991.

Cranstone, D., and G. Bouchard. 1994. Diamond discoveries promising. *Canadian Minerals* [a supplement to *Canadian Mining Journal*] April: 5-6.

Cristovici, M.A., and G.W. Leigh. 1986. Source of gold: Abandoned tailing dumps. *Geos* 15: 19-20.

Crowder, A.A. 1991. Acidification, metals, and macrophytes. *Environmental Pollution* 71: 171-204.

Crowder, A.A., et al. 1982. Site factors affecting semi-natural herbaceous vegetation on tailings at Copper Cliff, Ontario. *Reclamation and Revegetation Research* 1: 177-193.

Crowder. A.A., et al. 1989. Metal contamination in sediments and biota of the Bay of Quinte, Lake Ontario, Canada. *Hydrobiologia* 188/189: 337-343.

Curry, R.R. 1975. Biogeochemical limitations on Western Reclamation. *See* Wali 1975.

Darley, E.F. 1966. Studies on the effects of cement kiln dust on vegetation. *Journal of the Air Pollution Control Association* 16: 145-150.

Das, B.M., F. Claridge, and V.K. Garga. 1990. The use of rock drains in surface mine waste dumps. *CIM Bulletin* 83(942):78-83.

Davé, N.K., and A.J. Vivurka. 1994. Water Cover on Acid Generating Uranium Tailings—Laboratory and Field Studies. *See* USBM 1994. Vol. 1: 297-306.

Davé, N.K., T.P. Lim, and N.R. Cloutier. 1985. *Ra-226 concentrations in blueberries* (Vaccinium angustifolium) Ait. near *an inactive uranium tailings site in Elliot Lake, Ontario, Canada.* Environmental Pollution Series B 10. 301-314.

David, Y., et al. 1991. Coal 1990. *Mining Engineering* 43: 519-528.

Davidson, C.I., and Y.L. Wu. 1988. Dry Deposition of Trace Elements. In *Control and Fate of Atmospheric Trace Metals,* eds. J.M. Pacyna and B. Ottar. (NATO Advanced Science Institutes Series, Series C: Mathematical and Physical Science—268: 147-202). Dordrecht: Kluwer Academic Publishers.

Dawson, K.R. 1985. *Geology of barium, strontium and fluorine deposits in Canada.* Economic Geology Report #34. Geological Survey of Canada. Ottawa, ON: Minister of Supply and Services.

De Nie, H. 1987. The decrease in aquatic vegetation in Europe and its consequences for fish populations. European Inland Fisheries Advisory Commission, Food and Agricultural Organization of the United Nations. Rome. Italy. Occasional Paper No. 19.

De Ruiter, H., L. Desjardin, and G. Champion. 1989. Heat recovery from mine waste water. *See* Centre in Mining and Mineral Exploration Research 1989.

Deniseger, J., A. Austin, and W.P. Lucey. 1986. Periphyton communities in a pristine mountain stream above and below heavy metal mining operations. *Freshwater Biology* 16: 209-128.

Dennis, W.H. 1965. *Extractive metallurgy: Principles and application.* New York, NY: Philosophical Library Inc.

Devuyst, E.A., et al. 1989. The Inco SO$_2$/Air process. 103-116. *See* Cox et al. 1989.

Diamond, M.L. 1990. Modelling the fate and transport of arsenic and other inorganic chemicals in lakes. Ph.D. Thesis. Toronto, ON: Institute for Environmental Studies, University of Toronto.

Diamond, M.L., and D. Mackay. 1991. *A mass balance model of the fate of toxic substances in the Bay of Quinte.* Remedial Action Plan Report. Toronto, ON: Environment Ontario.

Diamond, M.L., and J.A. Meech. 1984. An environmental investigation of the Kognak River at Cullaton Lake Gold Mines. Kingston, ON: Department of Mining Engineering, Queen's University.

DIAND. *See* Canada. Department of Indian and Northern Affairs.

Dibbs, H.P., and P. Marier. 1973. *Methods for the removal of sulphur from coal.* Report EPS 3-AP-73-3. Ottawa, ON: Air Pollution Control Directorate.

DiLabio, R.N.W. 1990. Drift prospecting. Geologists use glacial sediments to find ore deposits. *Geos* 19: 7-15.

Dillon, P.J., and P.J. Smith. 1984. Trace metals and nutrient accumulation in the sediments of lakes near Sudbury, Ontario. *See* Nriagu 1984.

Dillon, P.J., R.A. Reid, and R. Girard. 1986. Changes in the chemistry of lakes near

Sudbury, Ontario following reductions of SO$_2$ emissions. *Water, Air, and Soil Pollution* 31: 59-66.

Dirschl, H.J., N.S. Novakowski, and L.C.N. Burgess. 1992. *An overview of the biophysical environmental impact of existing uranium mining operations in northern Saskatchewan.* Prepared for the Saskatchewan Uranium Mine Development Review Panel. Ottawa, ON: Environmental-Social Advisory Services (ESAS) Inc.

Dixit, S.S., A.S. Dixit, and J.P. Smol. 1991. Multivariable environmental inferences based on diatom assemblages from Sudbury (Canada) lakes. *Freshwater Biology* 26: 251-266.

Dixit, S.S. 1986. *Algal Microfossils and Geochemical Reconstructions of Sudbury Lakes: A Test of the Paleoindicator Potential of Diatoms and Chrysophytes.* Ph.D. Thesis. Kingston, ON: Queen's University.

Doggett, M.D., and B.W. Mackenzie. 1995. *The Changing Economic Climate for Minerals Supply in Canada.* Kingston, ON: Centre for Resource Studies, Queen's University.

Down, C.G., and J. Stocks. 1977. *Environmental impact of mining.* New York, NY: Halsted Press.

Dreesen, D.R., et al. 1984. Thermal stabilization of uranium mill tailings. *Environmental Science and Technology* 18: 658-667.

DRIE. *See* Canada. Department of Regional Industrial Expansion.

Dubnie, A. 1972. *Surface mining practice in Canada.* Mines Branch Information Circular 292. Ottawa, ON: Energy, Mines and Resources Canada.

Dufour, M.F. 1982. Le recyclage à grande échelle. *Geos* 11: 16-18.

Dunbavan, M. 1990. Utilizing coarse refuse for tailings dam construction. 267-274. *See* Rainbow 1990.

Duncan, W.F.A., M.J.S. Tevesz, and R.L.R. Towns. 1987. Use of fingernail clams (*Pisidiidae*) and x-ray fluorescence spectrometry for monitoring metal pollution in

Contwoyto Lake, Northwest Territories. *Water Pollution Journal of Canada* 22: 270-279.

Dunn, R.W., L.A. Melis, and A.J. Vivurka. 1972. The pollution control program of Rio Algom Mines Ltd. in the Elliot Lake region. Part I: Tailing revegetation. In *Proceedings Ontario Industrial Waste Conference*. No. 19. 85-119.

Dupuis, J. 1993. Cobalt dogged by supply problems. *Northern Miner Magazine* 79(1): 6.

Durham, R.W., and T. Pang. 1976. *Asbestiform fibre levels in Lakes Superior and Huron.* Scientific Series No. 67. Burlington, ON: Inland Waters Directorate, Canada Centre for Inland Waters.

Dushenko, W.T. 1990. Physical and chemical factors affecting nearshore aquatic vegetation in the Bay of Quinte, Lake Ontario. Ph.D. Thesis. Kingston, ON: Queen's University.

Earl, T.A. 1983. Mine-site monitoring: A systems approach. *Ground Water Monitoring Review* 3: 52-56.

Ecoregions Working Group. 1989. *Ecoclimatic Regions of Canada, First Approximation.* Ecological Land Classification Series, No. 23. Ottawa, ON: Environment Canada.

Edmonds, P. 1965. Solution mining and refining of potash. In *Proceedings of Saskatchewan Potash Show.* Saskatoon, SK.

Edwards, C.R. 1992. Uranium extraction process alternatives. *CIM Bulletin* 85(958): 112-136.

Edwards, R.P., and A.L. Mular. 1992. An expert system supervisor of a flotation circuit. *CIM Bulletin* 85(959): 69-76.

Eedy, W.K., et al. 1979. Environmental feasibility study of an Arctic Mine Site. *See* Rubec 1979.

Ellis, D. 1989. *Environments at Risk.* Berlin: Springer-Verlag.

Ellis, D.V. (ed.). 1982. Marine Tailings Disposal. *Ann Arbor Science.* Ann Arbor, MI.

Ellis, D.V. 1986. Problems of metal bioaccumulation arising from mining industry disposal of tailings to the sea. In *Proceedings of the 11th Annual Aquatic Toxicity Workshop, Vancouver, British Columbia, Canada, November 13-15, 1984,* eds. G.H. Green and K.L. Woodward. Canadian Technical Report on Fisheries and Aquatic Science. 1-17.

———. 1987. A decade of environmental impact assessment at marine and coastal mines. *Marine Mining* 6: 385-417.

Ellis, D.V., and P.M. Hoover. 1990. Benthos on tailings beds from an abandoned coastal mine. *Marine Pollution Bulletin* 21: 477-480.

Emery, J.J. 1975. New uses of metallurgical slags. *CIM Bulletin* 68: 60-68.

Emery, J.J., and C.S. Kim. 1974. Trends in the utilization of wastes for highway construction. In *Proceedings of the Fourth Mineral Waste Utilization Symposium, Chicago, Illinois.* 22-32.

EMR. *See* Canada. Energy, Mines and Resources Canada.

Environment Ontario. *See* Ontario. Ministry of the Environment.

Ericksson, M.O.G. 1984. Acidification of lakes: effects on waterbirds in Sweden. *Ambio* 13:260-262.

Ernst, W. 1985. Accumulation in aquatic organisms. In *Appraisal of tests to predict the environmental behaviour of chemicals,* eds. P. Sheehan, et al. Somerset, NJ: John Wiley & Sons. 243-255.

Ernst, W.H.O. 1975. Physiology of Heavy Metal Resistance in Plants. *See* Hutchinson 1975.

———. 1989. Mine vegetation in Europe. *See* Shaw 1989.

Etter, H.M. 1971. *Preliminary report of water quality measurements and revegetation trials on mined land at Luscar, Alberta.* Rep. NOR-3. Edmonton, AB: Northern Forest Research Centre, Environment Canada.

———. 1973. Mined land reclamation studies on bighorn sheep range in Alberta, Canada. *Biological Conservation* 5: 191-195.

Faulkner, B.B., and J.G. Skousen. 1994. Treatment of Acid Mine Drainage by Passive Treatment Systems. *See* USBM 1994. Vol. 1: 250-259.

F/P/TDBWG. Federal/Provincial/Territorial Biodiversity Working Group. 1994. Draft Canadian biodiversity strategy discussion paper. June 1994. Biodiversity Convention Office, Hull, PQ.

Feenstra, S., et al. 1985. *Uranium tailings sampling manual.* NUTP-1E. Ottawa, ON: CANMET.

Ford, D.C. 1993. Karst in Cold Environments. *See* French and Slaymaker 1993.

FPACAQ. Federal-Provincial Advisory Committee on Air Quality. 1987. *Review of the ambient air quality objectives for sulphur dioxide: Desirable and acceptable levels.* Ottawa, ON: Federal-Provincial Advisory Committee on Air Quality.

Frankling, F.T. 1984. Ecology of plants growing on abandoned uranium mill tailings in Saskatchewan. M.Sc. Thesis. Saskatoon, SK: University of Saskatchewan.

Franz, J., B. Stenberg, and J. Strongman. 1986. *Iron ore—global prospects for the industry, 1985-95.* Industry Department. Industry and Finance Series, Volume 12. Washington, DC: World Bank.

Franzin, W.G., G.A. Mcfarlane, and A. Lutz. 1979. Atmospheric fallout in the vicinity of a base metal smelter at Flin Flon, Manitoba, Canada. *Environmental Science and Technology* 13(12): 1513-1522.

Franzin, W.G. 1984. Aquatic Contamination in the Vicinity of the Base Metal Smelter at Flin Flon, Manitoba, Canada: A Case Study. *See* Nriagu 1984.

Fraser, W.W., and J.D. Robertson. 1994. Subaqueous Disposal of Reactive Mine Waste: An Overview and Update of Case Studies—MEND Canada. *See* USBM 1994. Vol. 1: 250-259.

Freedman, B., and T.C. Hutchinson. 1980a. Effects of smelter pollutants on forest leaf litter decomposition near a nickel-copper smelter at Sudbury, Ontario. *Canadian Journal of Botany* 58: 1722-1736.

———. 1980b. Long-term effects of smelter pollution at Sudbury, Ontario, on forest community composition. *Canadian Journal of Botany* 58: 2123-2140.

———. 1980c. Pollutant inputs from the atmosphere and accumulations in soils and vegetation near a nickel-copper smelter at Sudbury, Ontario, Canada. *Canadian Journal of Botany* 58: 108-132.

Freeze, R.A., and J.A. Cherry. 1979. *Groundwater.* Englewood Cliffs, NJ: Prentice Hall.

French, H.M. 1976. *The Periglacial Environment.* London and New York: Longmans.

———. 1993. Cold-Climate Processes and Landforms. *See* French and Slaymaker 1993.

French, H.M., and O. Slaymaker (eds.). 1993. *Canada's Cold Environmets.* Montreal and Kingston: McGill-Queen's Press.

Friedland, A.J. 1989. The movement of metals through soils and ecosystems. *See* Shaw 1989.

Frohliger, J.O., and R. Kane. 1975. Precipitation: Its acidic nature. *Science* 189: 455-457.

Fulkerson, W., et al. (eds. ). 1973. *Cadmium the dissipated element.* U.S. Department of Commerce ORNL-NSF-EP-21. Oak Ridge, TN: Oak Ridge National Laboratory.

Gagan, E.W. 1977. *Air pollution emissions and control technology: Asbestos mining and milling industry.* EPS 3-AP-76-6. Ottawa, ON: Environmental Protection Service, Environment Canada.

Gagnon, J. 1987. Natural revegetation of the Beattie Mine tailings, Duparquet, Quebec. M.Sc. Thesis. Kingston, ON: Queen's University.

Gardner, J.S. 1986. Snow as a Resource and Hazard in Early 20th Century Mining, Selkirk Mountains, B.C. *Canadian Geographer* 30: 217-28.

———. Mountain hazards. See French and Slaymaker 1993.

Gerard, R. 1976. Environmental effects of deep-sea mining. *MTS Journal* 10(7 September): 7-16.

Gersberg, R.M., et al. 1984. The removal of heavy metals by artificial wetlands. *Proceedings of the Water Reuse Symposium.* San Diego, CA: American Water Resources Association Research Foundation.

Giblin, A.E. 1985. Comparison of the processing of elements by ecosystems. II metals. In *Wetland Treatment of Municipal Waste Waters, eds.* P.J. Godfrey, E.R. Kaynor, and S. Pelczarshi. New York, NY: Van Nostrad Reinhold.

Gignac, L.D,. and P.J. Beckett. 1986. The effect of smelting operations on peatlands near Sudbury, Ontario, Canada. *Canadian Journal of Botany* 64: 1138-1147.

Gilbert, G.W. 1989. *A brief history of placer mining in the Yukon.* QS-Y002-000-EE-A1. Ottawa, ON: Indian Affairs and Northern Development Canada.

Gilbert, G., et al. 1985. *Ecosystem Sensitivity to Acid Precipitation for Québec.* Ecological Land Classification Series, No. 20. Ottawa, ON: Environment Canada.

Gilbert R. 1994. Canada looks at Elliot Lake's uranium waste. *Great Lakes United* 9:5 (Spring 1994).

Gilchrist, J.D. 1989. *Extraction Metallurgy.* Third ed. Oxford, UK: Pergamon Press.

Gillson, J.L. (ed.). 1960. *Industrial Minerals and Rocks.* Third ed. New York, NY: American Institute of Mineral and Petroleum Engineers.

Glooschenko, W., and W. Stevens. 1985. Sources of acidity in wetlands at Sudbury, Ontario. *Science of the Total Environment.* 54: 53-59.

Goblot, R. 1981. Les nodules: la dernière richesse minière vierge. *Geos* 10: 11-13.

Godin, E. (ed. ). 1990. *1989 Canadian minerals yearbook—review and outlook.* Mineral Report No. 38. Ottawa, ON: Energy, Mines and Resources Canada.

———. 1991. *1990 Canadian minerals yearbook—review and outlook.* Ottawa, ON: Energy, Mines and Resources Canada.

———. 1992. *1991 Canadian minerals yearbook—review and outlook.* Ottawa, ON: Energy, Mines and Resources Canada.

———. 1993. *1992 Canadian minerals yearbook—review and outlook.* Ottawa, ON: Minister of Supply and Services.

———. 1994. *1993 Canadian minerals yearbook—review and outlook.* Ottawa, ON: Natural Resources Canada, Minister of Supply and Services.

Godwin, R.C., and Z.M. Abouguendia. 1986. *Revegetation of coal spoils, evaluation of field trials: 1985 season.* Saskatoon, SK: Saskatchewan Research Council.

Goodier, J.L., and S. Soehle. 1971. Protecting the environment during marine mining operations. *Oceanology International* 6: 25-27.

Gordon, A.G., and E. Gorham. 1963. Ecological aspects of air pollution from an iron-sintering plant at Wawa, Ontario. *Canadian Journal of Botany* 41: 1063-1078.

Gorham, E., and A.G. Gordon. 1960. Some effects of smelter pollution northeast of Falconbridge, Ontario. *Canadian Journal of Botany* 383:307-312.

———. 1963. Some effects of smelter pollution upon aquatic vegetation near Sudbury, Ontario. *Canadian Journal of Botany* 41: 371-378.

Gorham, E., et al. 1987. The natural and anthropogenic acidification of peatlands. In *Effects of atmospheric pollutants on forests, wetlands, and agricultural ecosystems*, eds. T.C. Hutchinson and K.M. Meema. Berlin: Springer-Verlag.

Goyette, D.E. 1975. Marine Tailings Disposal Case Study. *Technology Transfer Seminar on Mining Effluent Regulations/Guidelines and Effluent Treatment Technology as Applied to the Base Metal, Iron, Ore, and Uranium Mining and Milling Industry.* Ottawa, ON: Environment Canada.

Green, W.R. 1991. *Exploration with a computer—geoscience data analysis applications.* Toronto, ON: Pergamon Press.

Gregory, C.E. 1983. *Rudiments of Mining Practice.* Houston, TX: Gulf Publishing Company.

Greig, J. 1989. Distribution of selected elements in substrates, plants and snails from marshes in the Bay of Quinte, Ontario. M.Sc. Thesis. Kingston, ON: Queen's University.

Grime, J.P. 1977. Evidence for the existence of three primary strategies in plants and its relevance to ecological and evolutionary theory. *American Naturalist* 3: 1169-1174.

Grimshaw, P.N. 1986. Environmental benefits of surface mining. *Mining Magazine* 155: 581-585.

Guild, P.W. 1976. Discovery of natural resources. *Science* 191: 709-713.

Guillet, G.R. 1985. *Industrial mineral resources of the north Clay Belt.* Kapuskasing, ON: North Clay Belt Development Association.

Habashi, F., and M.I. Ismail. 1975. Health hazards and pollution in the metallurgical industry due to phosphine and arsine. *CIM Bulletin* 68: 99-103.

Hackbarth, D. 1981. *The effects of surface mining of coal on water chemistry near Grande Cache, Alberta.* Bulletin 40. Edmonton, AB: Alberta Research Council.

Hall, D.L. 1987. Reclamation planning for coal strip-mined lands in Montana. *Landscape and Urban Planning* 14: 45-55.

Hamilton, R., and W.W. Fraser. 1978. A case history of natural underwater revegetation: Mandy Mine high sulfide tailings. *Reclamation Review* 1: 61-65.

Hammer, U.T. 1980. *Acid Rain.* Regina, SK: Environmental Advisory Council.

Hardcastle, S., and A. Sheikh. 1988. Applying tracer gas techniques to evaluate the air distribution in flood leaching stopes. *CIM Bulletin* 81.913: 53-58.

Hare, F.K., and M.K. Thomas. 1979. *Climate Canada.* 2nd edition. Toronto, ON: John Wiley & Sons.

Hart, P.A. 1990. An assessment of the marine application of minestone. *See* Rainbow 1990.

Hart, R.T. 1983. The effect of potash dust emissions on vegetation. *See* McKercher 1983.

———. 1985. *A review of waste management in the potash industry, and options for decommissioning and abandonment of potash tailings piles.* CANMET Project No. 350204. Saskatoon, SK: Potash Corporation of Saskatchewan.

———. 1986. *Investigation into weathering of the surface of potash tailings in Saskatchewan.* Contract report no. OSQ85-00126. Ottawa, ON: CANMET.

———. 1989. Salt tailings—the ultimate saline reclamation challenge. *See* Walker et al. 1989.

Hart, M. 1992. Aggregate mining and wetlands: An overview. *Great Lakes Wetlands* 3(3): 1-5.

Hartley, D.M., and G.E. Schuman. 1984. Soil Erosion Potential of Reclaimed Mined Lands. *Transactions of the American Society of Agricultural Engineers* 27: 1067-1073.

Haryett, C.R. 1983. Potash—our world class reserve. *Geos* 12: 19-21.

Haw, V.A. 1982. The Canadian research programme into the long-term management of uranium mine tailings. In *Management of Wastes from Uranium Mining and Milling.* Vienna, Austria: International Atomic Energy Agency. 663-677.

Hawley, J.R. 1972a. *The Problem of Acid Mine Drainage in the Province of Ontario.* Toronto, ON: Industrial Wastes Branch, Ontario Ministry of the Environment.

———. 1972b. *The Use, Characteristics and Toxicity of Mine-Mill Reagents in the Province of Ontario.* Toronto, ON: Industrial Wastes Branch, Ontario Ministry of the Environment.

———. 1974. *Mine waste control in Ontario.* Toronto, ON: Environment Ontario.

———. 1979. *The Chemical Characteristics of Mineral Tailings in the Province of Ontario.* Toronto, ON: Ontario Ministry of the Environment.

Hay, A.E., and D.V. Ellis. 1982. The Effects of Submarine Channels on Mine Tailings Disposal in Rupert Inlet, British Columbia. *See* Ellis 1982.

Heffernan, V. 1992. Matachewan Awaits Verdict in Tailings Trial. *Northern Miner Magazine* 78(16).

Heginbottom, J.A. 1973. *Some effects of surface disturbance on the permafrost active layer at Inuvik, N.W.T.* Info. Can. Cat. no. R72-9573. Ottawa, ON: Information Canada.

Helming, E.M. 1976. *Symposium on fugitive emissions measurement and control.* EPA-600/2-76-246. Washington, DC: Office of Research and Development, Environmental Protection Agency.

Hills, L.V., and W. Jones. 1981. *Trace elements in coal.* Ed. Science Advisory Committee. Edmonton, AB: Environment Council of Alberta.

Hiskey, J.B. 1986. Technical innovations spur resurgence of copper solution mining. *Mining Engineering* 38: 1036-1039.

Hocking, D. 1974. *The Forest Impact of Sulphur Dioxide Fumes from Underground Combustion of Sulphide Ores near Kimberley, B.C.* Edmonton, AB: Canadian Forestry Service.

Hocking, D., and W.R. MacDonald (eds.). 1974a. *Proceedings of a workshop on reclamation of disturbed lands in Alberta.* Information Report NOR-X-116. Edmonton, AB: Northern Forest Research Centre, Environment Canada. 184-188.

Hodder, R.L. 1973. Surface-Mined Land Reclamation Research in Eastern Montana. Chapter 6 in *Research and Applied Technology Symposium on Mined-Land Reclamation,* ed. J.R. Garvey. Monroeville, PA: Bituminous Coal Research Inc.

Hoddinott, P.J. 1993. Energy minerals: Coal. In *Metals and minerals annual review: 1993.* London, UK: Mining Journal. 121-135.

Hogg, J.L.E. 1971. Mined land reclamation in British Columbia: The importance of institutional and environmental factors in the formulation of policy. *The Forestry Chronicle* 47: 1-4.

Hoos, R.A.W. 1975. Pacific region mine effluent chemistry and acute toxicity survey, 1973. 15. *See* Canada. Environment Canada 1975.

Hopkins, J.L. 1987. Ready for Eagle Point. *Canadian Mining Journal* 108(5): 17-23.

Horn, W.R. 1966. *Problems and progress in abatement and control of industrial wastes in the mining and non-ferrous smelting industries.* Toronto, ON: The Mining Association of Canada.

Hoskin, W.M.A. 1991. Cadmium. 15.1-15.7. *See* Godin 1991.

———. 1993. *Controlled Use: A Case Study of Asbestos and Possible Future Application to Potentially Dangerous Industrially Important Materials.* Ottawa, ON: Minister of Supply and Services Canada.

Hossner, L.R. (ed.). 1988. *Reclamation of Surface-Mined Lands.* Vol. I and II. Boca Raton, FL: CRC Press.

Howard, P.H., and R.S. Datta. 1977. Desulfurization of coal by use of chemical comminution. *Science* 197: 668-669.

Howard, R., D. Wilcox, and M. Cahill. 1991. Environmental Management in the WA Goldfields: Examples from the Kalgoorlie Area. *See* AMIC 1991, Vol. 1: 188-202.

Howells, G. 1990. *Acid Rain and Acid Waters.* Ellis Horwood Series in Environmental Management, Science and Technology. New York, NY: Ellis Horwood.

Humphrey, C.L., K.A. Bishop, and V.M. Brown. 1990. Use of biological monitoring in the assessment of effects of mining wastes on aquatic ecosystems of the Alligator Rivers region, tropical northern Australia. *Environmental Monitoring and Assessment* 14: 139-181.

Hunter, R.A., et al. 1982. A biophysical estuarine habitat mapping and classification system for British Columbia. *See* Stelfox and Ironside 1982.

Hutcheson, M.S. 1983. Toxicological effects of potash brine on Bay of Fundy marine organisms. *Marine Environmental Research* 9: 237-255.

Hutcheson, J.R.M. 1971. Environmental control in the asbestos industry of Quebec. *CIM Bulletin* 64: 83-89.

Hutchinson, G.E. 1993. *A Treatise on Limnology.* 4 volumes. (Vol. 4 edited by Y. Edmonson). New York, NY: John Wiley & Sons.

Hutchinson, T.C. (ed.). 1975. *International Conference on Heavy Metals in the Environment* (Abstracts). Toronto, ON: National Research Council of Canada.

Hutchinson, T.C., and R.M. Cox. 1984. Tolerances of the native grass *Deschampsia caespitosa* for possible mine tailings revegetation. *Environmental Studies No. 21.* Ottawa, ON: Indian Affairs and Northern Development Canada.

Hutchinson, T.C., and M. Havas. 1986. Recovery of previously acidified lakes near Coniston, Canada, following reductions in atmospheric sulphur and metal emissions. *Water, Air, and Soil Pollution* 28: 319-333.

Hutchinson, T.C., and L.M. Whitby. 1974. Heavy metal pollution in the Sudbury mining and smelting region of Canada. I. Soil and vegetation contamination by nickel, copper and other metals. *Environmental Conservation* 1: 123-132.

———. 1977. The effects of acid rainfall and heavy metal particulates on a boreal forest ecosystem near the Sudbury smelting region of Canada. *Water, Air, and Soil Pollution* 7: 421-438.

Hutton, M. 1984. Impact of Airborne Metal Contamination on a Deciduous Woodland System. In *Effects of Pollutants at the Ecosystem Level*, eds. P.S. Sheehan et al. Chichester, UK: John Wiley & Sons.

Hynes, T.P. 1990. *The Impacts of the Cluff Lake Uranium Mine and Mill Effluents on the Aquatic Environment of Northern Saskatchewan.* M.Sc. thesis. Saskatoon, SK: University of Saskatchewan.

Hyslop, C., and D. Brunton. 1991. Wildlife: Maintaining Biological Diversity. Chapter 6 in *The State of Canada's Environment*, ed. S. Burns. Ottawa, ON: Minister of Supply and Services.

Ibrahim, S.A., and F.W. Whicker. 1988. Plant/soil concentration ratios of $^{226}$Ra for contrasting sites around an active U mine-mill. *Health Physics* 55: 903-910.

IEC Beak Consultants Ltd. 1986. *An assessment of the radiological impact of uranium mining in northern Saskatchewan.* Report EPS 2/MM/1. Ottawa, ON: Environmental Protection Programs Directorate.

Ignatow, A., et al. 1991. Specialty metals. 58.1-58.38. *See* Godin 1991.

IJC. 1981. *Water quality in the Poplar River Basin.* Ottawa, ON: International Joint Commission.

———. 1994. *Source Investigation for Lake Superior.* Windsor, ON: International Joint Commission.

Indicators Task Force. 1991. *A Report on Canada's Progress Towards a National Set of Environmental Indicators.* State of the Environment Report, 91-1. Ottawa, ON: Environment Canada.

Ingles, J., and J.S. Scott. 1987. *State-of-the-art processes for the treatment of gold mill effluents.* Ottawa, ON: Industrial Programs Branch, Environment Canada.

Jackson, L.M. 1995. *Environmental Management by Canadian Mining Companies: Two Case Studies.* Kingston, ON: Centre for Resource Studies, Queen's University.

Jambor, J.L. 1994. Mineralogy of Sulfide-Rich Tailings and Their Oxidation Products. *See* Jambor and Blowes 1994.

Jambor, J.L., and D.W. Blowes (eds.). 1994. *Short Course Handbook on Environmental Geochemistry of Sulfide Mine-Wastes.* Waterloo, ON: Mineralogical Association of Canada.

Jaques, A.P. 1987. *Summary of emissions of antimony, arsenic, cadmium, chromium, copper, lead, manganese, mercury, and nickel in Canada.* Ottawa, ON: Environment Canada.

Jasper, D., et al. 1991. Ensuring Diversity and Growth in Revegetation—A Program to Develop Techniques for Inoculation with VA Mycorrhizal Fungi. *See* AMIC 1991, Vol. 1: 172-174.

Jaworski, J.F. (ed.). 1980. *Executive reports: Effects of chromium, alkali halides, arsenic, asbestos, mercury, and cadmium in the Canadian environment.* NRCC No. 17585. Ottawa, ON: National Research Council of Canada.

Jaworski, J.F. 1982. Data sheets on selected toxic elements. Publication No. 19252 Ottawa, ON: National Research Council of Canada.

Jeffries, D.S. 1990. Snowpack storage of pollutants, release during snowmelt and impact on receiving surface waters. In *Acid Precipitation*. Vol. 4., eds. S.A. Norton, S.E. Lindber and A.L. Page. New York, NY: Springer-Verlag.

———. 1991. Southeastern Canada: an overview of the effect of acidic deposition on aquatic resources. *See* Charles 1991.

Jen, Lo-Sun, and A. Cadieux. 1992. Principal Canadian nonferrous and precious metal mine production in 1990, with highlights for 1991. 53.1-53.7. *See* Godin 1992.

Jen, Lo-Sun, and B. McCutcheon. 1993. Principal Canadian nonferrous precious metal mine production in 1991. 57.1-57.9. *See* Godin 1993.

———. 1994. Principal Canadian nonferrous precious metal mine production in 1992. 57.1-57.7. *See* Godin 1994.

John, R.D. 1987. Review of the National Uranium Tailings Program research into acid generation. *See* Blakeman et al. 1990.

Johnson, M.G. 1987. Trace element loadings to sediments of fourteen Ontario lakes and correlations with concentrations in fish. *Canadian Journal of Fisheries and Aquatic Sciences* 44: 3-13.

Johnson, M.G., M.F.P. Michalski, and A.E. Christie. 1970. Effects of acid mine wastes on phytoplankton communities in two northern Ontario lakes. *Journal of the Fisheries Research Board of Canada* 27: 425-444.

Johnson, R.A., J.H. Chapman, and R.M. Lipchak. 1987. Recycling to reduce sediment discharge in placer mining operations. *Journal of the Water Pollution Control Federation* 59: 274-283.

Johnston, W.B., and R.E. Ruff. 1975. Observational Systems and Techniques in Air Pollution Meteorology. Chapter 8 in *Lectures on Air Pollution and Environmental Impact Analyses*, ed. C.A. Haugen. Boston, MA: American Meteorological Society.

Jones, H.G. 1983. *Plants and Microclimate*. Cambridge, MA: Cambridge University Press.

Jones, C.E., and J.Y. Wong. 1994. Shotcrete as a Cementitious Cover for Acid Generating Waste Rock Piles. *See* USBM 1994. Vol. 2: 104-112.

Jonescu, M.E. 1974. *Natural Vegetation and Environmental Aspects of Strip-Mined Lands in the Lignite Coal Fields of Southeastern Saskatchewan*. M.Sc. Thesis. Saskatoon, SK: University of Saskatchewan.

Jordan, W.R., M.E. Gilpin, and J.D. Aber (eds.). 1990. *Restoration Ecology: A Synthetic Approach to Ecological Research*. Cambridge, MA: Cambridge University Press.

Joyner, W.M. (ed.). 1985. *Compilation of air pollutant emission factors. Fourth ed. Vol. I: Stationary point and area sources*. AP-42. Research Triangle Park, NC: U.S. Environmental Protection Agency.

Kalin, M. 1982. *Environmental conditions of two abandoned uranium mill tailings sites in northern Saskatchewan*. Regina, SK: Saskatchewan Environment.

———. 1983. 1. Synoptic survey and identification of invading biota. In *Long-term ecological behaviour of abandoned uranium mill tailings*. Technological Development Report, EPS 4-ES-83-1. Ottawa, ON: Environmental Strategies Directorate.

———. 1987. *Ecological engineering for gold and base metal mining operations in the Northwest Territories: Final report*. Northern Affairs Program Environmental Studies No. 59. Ottawa, ON: Indian Affairs and Northern Development Canada.

———. 1988. 3. Radionuclide concentrations and other characteristics of tailings, surface waters, and vegetation. In *Long-term ecological behaviour of abandoned uranium mill tailings*. EPS 3/HA/4. Ottawa, ON: Environmental Strategies Directorate.

———. 1990. Ecological engineering applied to base metal mining wastes. In *Proceedings of the Canadian Land Reclamation Association, 15th Annual Meeting: Acidity and alkalinity in terrestrial and aquatic reclamation*, eds. D.J. Boivin and N. Trepanier. Ottawa, ON: Canadian Land Reclamation Association.

———. 1991. Biological alkalinity generation in acid mine drainage. Chapter 9 in *Proceedings of the 23rd Annual Meeting of the Canadian Mineral Processors.* 22-24 January. Ottawa, ON: Canadian Mineral Processors.

Kalin, M., and P.M. Stokes. 1981. Macrofungi on uranium mill tailings—associations and metal content. *The Science of the Total Environment* 19: 83-94.

Kalin, M., and R.O. van Everdingen. 1987. Ecological engineering: biological and geochemical aspects, Phase I experiment. *See* Blakeman et al. 1990.

Kalin, M., R.O. van Everdingen, and R.G.L. McCready. 1990. Ecological engineering: Interpretation of hydrogeochemical observations in a sulphide tailings deposit. In *Proceedings of the 92nd Annual General Meeting of the CIM.* Toronto, ON: Boojum Research Limited.

Karvinen, W.O., and M.L. McAllister. 1994. *Rising to the Surface: Emerging Trends in Groundwater Policy in Canada.* Kingston, ON: Centre for Resource Studies, Queen's University.

Kay, B.H. 1989. *Pollutants in British Columbia's marine environment.* State of the Environment Report, 89-1. Ottawa, ON: Environment Canada.

Keating, J. 1992. Silver. 40.1-40.14. *See* Godin 1992.

———. 1993. Silver. 44.1-44.14. *See* Godin 1993.

Keating, J., and P. Wright. 1994. Lead. 27.1-27.19. *See* Godin 1994.

Keen, A.J. 1992. Polaris Update. *CIM Bulletin* 85 (861): 51-57.

Keith, R.F. 1994. The ecosystem approach: Implications for the North. *Northern Perspectives* 22: 3-6.

Keller, W. ,and J. Pitblado. 1986. Water quality changes in Sudbury area lakes: A comparison of synoptic surveys in 1974-1976 and 1981-1983. *Water, Air, and Soil Pollution* 29: 285-296.

Keller, W., J.R. Pitblado, and N.I. Conroy. 1986. Water quality improvements in the Sudbury, Ontario, Canada area related to reduced smelter emissions. *Water, Air, and Soil Pollution* 31: 765-776.

Kellogg, H.H. 1981. Trends in non-ferrous pyrometallurgy. In *Metallurgical Treatise,* eds. J.K. Tien and J.F. Elliott. Warrendale, PA: The Metallurgical Society of AIME.

Kellogg, W.W., et al. 1972. The sulfur cycle. *Science* 175: 597-596.

Kelly, G.J., and R.J. Slater. 1994. The Mining Environment Database on Abandoned Mines, Acid Mine Drainage and Land Reclamation. *See* USBM 1994 Vol. 4: 352-356.

Kelly, E.G., and D.J. Spottiswood. 1982. *Introduction to Mineral Processing.* New York, NY: John Wiley & Sons.

Kelly, M., et al. 1988. *Mining and the freshwater environment.* Barking, Essex, UK: Elsevier.

Kelso, J.R.M., and D.S. Jeffries. 1988. Response of headwater lakes to varying atmospheric deposition in north-central Ontario, 1979-85. *Canadian Journal of Fisheries and Aquatic Sciences* 45: 1905-1911.

Kemmer, F.N., and J.A. Beardsley. 1971. Environmental challenge: Wastewater. *Engineering and Mining Journal* 172: 92-99.

Kennedy, A.J., and B.M. Kovach. 1987. The use of unconsolidated runoff material in coal waste dump reclamation. In *Acid mine drainage: Proceedings of the Eleventh Annual British Columbia Mine Reclamation Symposium,* ed. Technical and Research Committee on Reclamation. Vancouver, BC: Mining Association of British Columbia.

Kennedy, P. 1992. Newmont getting the bugs out of bioleaching. *The Northern Miner Magazine* 77(47): 1.

Kennett, S.A. (ed.). 1993. *Law and Process in Environmental Management.* Calgary, AB: Canadian Institute of Resources Law.

Kent, D.D., and A.W. Clifton. 1986. Potash industry: Containing the wastes. *Canadian Mining Journal* 107(2): 88-89.

Ketel, D.H., W.G. Dirske, and A. Ringoet. 1972. Water uptake from foliar-applied drops and its further distribution in the oat-leaf. *Acta Botanica Neerlandia* 21: 155-166.

Report Series No. 39. Burlington, ON: Inland Waters Directorate, Canada Centre for Inland Waters.

Lyons, T., and B. Scott. 1990. *Principles of Air Pollution Meteorology*. Boca Raton, FL: CRC Press.

Maathuis, H., and G. van der Kamp. 1986. *Theory of groundwater flow in the vicinity of brine ponds and salt tailings piles*. SRC Technical Report No.152. Saskatoon, SK: Saskatchewan Research Council.

———. 1994. *Subsurface brine migration at potash waste disposal sites in Saskatchewan: Final report*. Saskatchewan Research Council Report No. R-1220-10-E-94.

MacCallum, B. 1989. Seasonal and spatial distribution of bighorn sheep at an open pit coal mine in the Alberta foothills. *See* Walker et al. 1989.

MacDonald, D.D., and L.E. McDonald. 1987. The influence of surface coal mining on potential salmonid spawning habitat in the Fording River, British Columbia. *Water Pollution Research Journal of Canada* 22: 584-595.

MacIsaac, H.J., et al. 1986. Natural changes in the planktonic rotifera of a small acid lake near Sudbury, Ontario, following water quality improvements. *Water, Air, and Soil Pollution* 31: 791-797.

MacLaren Plansearch. 1982. *Environmental Guidelines Pits and Quarries*. No. R72-180/1982E. Ottawa, ON: Indian and Northern Affairs Canada.

MacLatchy, J. 1994. *SO$_2$ Emission and Acid Production Data for Base Metal Smelters from 1970 to 1992, Expressed in Terms of Metal Production*. Ottawa, ON: Environment Canada.

MacWilliams, W.A., and R.G. Reynolds. 1973. Solution mining of sodium sulphate. *CIM Bulletin* 66: 115-119.

Macyk, T.M. 1974. Revegetation of strip mined land at No. 8 Mine Grande Cache, Alberta. 184-188. *See* Hocking and MacDonald 1974.

Magrab, E.B. 1972. Noise control. *Chemical Rubber Company Critical Reviews in Environmental Control* 3: 69-83.

Malhotra, S.S., and R.A. Blauel. 1980. *Diag-nosis of air pollutant and natural stress symptoms on forest vegetation in western Canada*. Report NOR-X-228. Edmonton, AB: Environment Canada.

Maltby, E. 1992. Peatlands—dilemmas of use and conservation. *See* Maltby et al. 1992.

Maltby, E., P.J. Dugan, and J.C. Lefevre (eds.). 1992. *Conservation and development: the sustainable use of wetland resources*. Gland, Switzerland: International Union for Conservation of Nature.

Mance, G. 1987. *Pollution threat of heavy metals in aquatic environments*. Pollution Monitoring Series, ed. K. Mellanby. London, UK: Elsevier.

Maneval, D.R. 1975. Reclaiming land for recreational development. *Coal Mining and Processing* 12: 84-86.

Marmorek, D.R., et al. 1989. A protocol for determining lake acidification pathways. *Water, Air and Soil Pollution* 44: 235-257.

Marshall, I.B. 1982. *Mining, land use, and the environment: Part I, a Canadian overview*. Land Use in Canada Series, No. 22. Ottawa, ON: Lands Directorate, Environment Canada.

———. 1983. *Mining, land use, and the environment: Part II, a review of mine reclamation activities in Canada*. Land Use in Canada Series, No. 23. Ottawa, ON: Lands Directorate, Environment Canada.

Martinez, J.D. 1971. Environmental significance of salt. *American Association of Petroleum Geologists (Bulletin)* 55: 810-825.

Martinez, E., and D.E. Spiller. 1991. Gravity-magnetic separation—the recovery of magnetic and weakly magnetic minerals by combined gravity-magnetic separation. *Engineering and Mining Journal* 192: 16EE, 16GG.

Mathews, W.H., F.E. Smith, and E.D. Goldberg (eds.). 1971. *Man's Impact on Terrestrial and Oceanic Ecosystems*. Cambridge, MA: The MIT Press.

Mathieu, J. 1981. Nouvelle méthode de nettoyage de charbon. *Geos* 10: 17-19.

Maxwell, C.D. 1991. Floristic changes in soil algae and cyanobacteria in reclaimed metal-contaminated land at Sudbury, Canada. *Water, Air, and Soil Pollution* 60: 381-393.

McCarthy, R.E. 1971. Surface mine siltation control. *Mining Congress Journal* 59: 30-35.

McCleary, E.C., and D.A. Kepler. 1994. Ecological Benefits of Passive Wetland Treatment Systems Designed for Acid Mine Drainage: With Emphasis on Watershed Restoration. *See* USBM 1994. Vol. 3: 111-119.

McClelland, G.E. 1986. Agglomerated and unagglomerated heap leaching behavior is compared in production heaps. *Mining Engineering* 38: 500-503.

McCorkell, R.H., and M. Silver. 1987. Effect of compost and soil covers on radon emanation from uranium tailings. *CIM Bulletin* 80(908): 43-45.

McCready, R.G. 1976. *Microbial assessment of the vegetated test plots on the pyrite uranium tailings in the Elliot Lake area.* CANMET Report MRP/MRL 76-161 TR. Elliot Lake, ON: CANMET.

————. 1986a. *Current biotechnology research in mining.* CANMET Report MRP/MSL 86-69 OP. Ottawa, ON: CANMET.

————. 1986b. Underground bioleaching: Extracting from low-grade ore. *Geos* 15: 9-10.

————. 1987. A Review of the Physical, Chemical and Biological Measures to Prevent Acid Mine Drainage: An Application to the Pyritic Halifax Shale. *See* Blakeman et al. 1990.

McDonald, J.D., and J. Dick. 1973. *Reclamation guidelines for exploration.* Victoria, BC: Reclamation Branch, Mines and Mineral Resources British Columbia.

McKee, P.M., et al. 1987. Sedimentation rates and sediment core profiles of $^{238}$U and $^{232}$Th decay chain radionuclides in a lake affected by uranium mining and milling. *Canadian Journal of Fisheries and Aquatic Sciences* 44: 390-398.

McLaughlin, B.E., and A.A. Crowder. 1988. The distribution of *Agrostis gigantea* and *Poa pratensis* in relation to some environmental factors on a mine-tailings area at Copper Cliff, Ontario. *Canadian Journal of Botany* 66: 2317-2322.

McLeay, D.J., et al. 1987. Responses of Arctic grayling (*Thymallus arcticus*) to acute and prolonged exposure to Yukon placer mining sediment. *Canadian Journal of Fisheries and Aquatic Sciences* 44: 658-673.

McLellan, A.G. 1975. The aggregate dilemma. *Bulletin of the Conservation Council of Ontario* 22(4): 12-20.

McNamara, V.N. 1989. The AVR process for cyanide recovery, and cyanogen control for barren recycle and barren bleed. *See* Cox et al. 1989.

McNicol, D.K., B.E. Bendell, and R.K. Ross. 1987. *Studies of the effects of acidification on aquatic wildlife in Canada: Waterfowl and trophic relationships in small lakes in northern Ontario.* Occasional Paper 62. Ottawa, ON: Canadian Wildlife Service.

McTavish, S.F. 1990. History and evolution of rubber liner systems for grinding mills. *Mining Engineering* 42: 1249-1251.

Mehling, P., and L. Broughton. 1989. Fate of cyanide in abandoned tailings ponds. 269-279. *See* Chalkley et al. 1989.

Melis, L.A., P.I. Tones, and S.M. Swanson. 1987. *Identification and evaluation of impacts resulting from the discharge of gold mill effluents.* Publication E-901-18-E-87. Saskatchewan: Saskatchewan Research Council. .

Merritt, G.J. 1994. The Reserve Mining Case. *Focus* 19(3): 1-4.

Merritt, R.C. 1971. *The extractive metallurgy of uranium.* Boulder, CO: Johnson Publishing Co.

Meyers, R.A., J.W. Hamersma, and M.L. Kraft. 1975. Sulphur dioxide pressure leaching—a new pollution-free method to process copper ore. *Environmental Science and Technology* 9: 70-71.

Mikhail, M.W., L.C. Bird, and N.T.L. Landgren. 1981. *Feasibility study on recovery of thermal coal from waste dumps in Nova Scotia.* CANMET Report 81-3E. Ottawa, ON: Energy Research Laboratories, Energy, Mines and Resources Canada.

Mikhail, M.W., A.I.A. Salama, and O.E.

Humeniuk. 1992. The role of a pilot scale faility in performing R & D and helping Canadian coal and mineral industries. *CIM Bulletin* 85(961): 58-64.

Millar, J.B. 1976. *Wetland Classification in Western Canada*. Report No. 37. Saskatoon, SK: Canadian Wildlife Service, Environment Canada.

Miller, C.G. 1990. *Guide for environmental practice*. Ottawa, ON: The Mining Association of Canada.

———. 1991. *The four Rs of mining in Canada*. Ottawa, ON: The Mining Association of Canada.

Miller, G.T. 1979. *Living in the environment*. Belmont, CA: Wadsworth Publishing Co.

Miller, J.D. 1981. Solution concentration and purification. 95-117. *See* Tien and Elliott 1981.

Miller, J.R. 1976. The direct reduction of iron ore. *Scientific American* 235(1): 68-80.

Milliken, J.G., and H.E. Mew. 1970. *Recreational impact of reclamation reservoirs*. Washington, DC: Bureau of Reclamation, U.S. Department of the Interior.

Mining Association of British Columbia. 1972. *Mining and Milling*. Brief submitted to public inquiry pursuant to section 14 of the *Pollution Control Act*, 1967. Vancouver, BC: Mining Association of British Columbia.

*Mining Engineering*. 1990. Pressure oxidation will cut emissions at Campbell Red Lake. 1990. *Mining Engineering* 42: 1234.

*Mining Review*. 1993. Canada's north: A diamond in the rough. *Mining Review* 13(3): 10-18.

Minish, B., and K. Hurley. 1991. Fresh Water: Liquid Assets. Chapter 3 in *The State of Canada's Environment*, ed. S. Burns. Ottawa, ON: Minister of Supply and Services.

Minns, C.K. ,and J.R.M. Kelso. 1986. Estimates of existing and potential impact of acidification on the freshwater fishery resources and their use in eastern Canada. *Water, Air, and Soil Pollution* 31: 1079-1090.

Mitchell, E.R. 1973. Environmental restraints on energy conversion. *CIM Bulletin* 66: 65-73.

MOE. *See* Ontario. Ministry of the Environment.

Moffett, D. 1977. *The disposal of solid wastes and liquid effluents from the milling of uranium ores*. CANMET Report 76-19. Ottawa, ON: CANMET.

Moffett, D., and M. Tellier. 1977. Uptake of radioisotopes by vegetation growing on uranium tailings. *Canadian Journal of Soil Science* 57: 417-424.

Montreal Engineering Co. Ltd. 1973. *Base metal mine waste management in northeastern New Brunswick*. Report No. EPS 8-WP-73-1. Ottawa, ON: Water Pollution Control Directorate, Environment Canada.

Moore, J.N. 1987. Processes of metal deposition and remobilization in a contaminated, high-gradient river system: the Clark Fork River, Montana. *See* Lindberg and Hutchinson 1987.

Moore, P.D., and D.J. Bellamy. 1974. *Peatlands*. New York, NY: Springer-Verlag.

Moore, T.R., and R.C. Zimmermann. 1972. Establishment of vegetation on serpentine asbestos mine wastes, southeastern Quebec, Canada. *Journal of Applied Ecology* 14: 589-599.

———. 1979. Follow-up studies of vegetation establishment on asbestos tailings, southeastern Quebec, Canada. *Reclamation Review* 2: 143-146.

Morel-à-l'Huissier, P. 1992. Sodium sulphate. 41.1-41.7. *See* Godin 1992.

———. 1993. Sodium sulphate. 45.1-45.6. *See* Godin 1993.

———. 1994a. Salt. 41.1-41.13. *See* Godin 1994.

———. 1994b. Asbestos. 10.1-10.14. *See* Godin 1994.

Morel-à-l'Huissier, P., and W.M.A. Hoskin. 1993. Asbestos. 10.1-10.15. *See* Godin 1993.

Morin, K.A., and J.A. Cherry. 1988. Migration of acidic groundwater seepages from uranium-tailings impoundments. 3. Simulation of the conceptual model with application to seepage area A. *Journal of Contaminant Hydrology* 2: 323-342.

Morin, K.A., et al. 1988a. Migration of acidic groundwater seepages from uranium-tailings impoundments. 1. Field study and conceptual hydrogeochemical model. *Journal of Contaminant Hydrology* 2: 271-303.

———. 1988b. Migration of acidic groundwater seepages from uranium-tailings impoundments. 2. Geochemical behavior of radionuclides in groundwater. *Journal of Contaminant Hydrology* 2: 305-322.

MRAZ Project Consultants Ltd. 1989. *Use of backfill in New Brunswick potash mines phase II.* Saskatoon, SK: MRAZ Project Consultants Ltd.

Mrowaka, J.P. 1974. Man's Impact on Stream Regimen and Quality. Chapter 4 in *Perspectives on Environment*, eds. I.R. Manners and M.W. Mikesell (Commission of College Geography, Publication No. 13). Washington, DC: Association of American Geographers.

Mudroch, A. 1991. *Geochemical composition of the nepheloid layer in Lake Ontario.* Burlington,ON: National Water Research Institution.

Mudroch, A., and J.A. Capobianco. 1980. Impact of past mining activities on aquatic sediments in Moira River Basin, Ontario. *Journal of Great Lakes Research* 6: 121-128.

Mudroch, A., and T.A. Clair. 1986. Transport of arsenic and mercury from gold mining activities through an aquatic system. *The Science of the Total Environment* 57: 205-216.

Mudroch, A., L. Sarazin, and T. Lomas. 1988. Summary of surface and background concentrations of selected elements in the Great Lakes sediments. *Journal of Great Lakes Research.* 14: 241-251.

Munn, R.E., and B. Bolin. 1971. Global Air Pollution—Meteorological Aspects, A Survey. *Atmospheric Environment* 5: 363-402.

Murphy, J.E., and M.F. Chambers. 1991. *Production of lead metal by molten-salt electrolysis with energy-efficient electrodes.*

Report of Investigations 9335. Washington, DC: Bureau of Mines, U.S. Department of the Interior.

Murray, D. 1972. Part II. Chemical conditions. In *Factors affecting revegetation of uranium tailings in the Elliot Lake area.* CANMET Report 72/55. Elliot Lake, ON: CANMET.

———. 1977. *Pit slope manual. Vol. 2—Mine waste inventory by satellite imagery.* Supplement 10-1: Reclamation by Vegetation. Ottawa, ON: CANMET.

———. 1978. The influence of uranium mine tailings on tree growth at Elliot Lake, Ontario. *CIM Bulletin* December, 1978: 79-81.

Murray, D., and D. Moffett. 1977. Vegetating the uranium mine tailings at Elliot Lake, Ontario. *Journal of Soil and Water Conservation* 32: 171-174.

Murray, D.R. 1973. *Vegetation of Mine Waste Embankments in Canada.* Information Circular IC301. Energy, Mines and Resources Canada.

Murray, M.L. et al. 1987. Estimation of long-term risk from Canadian uranium mill tailings. *Risk Analysis* 7: 287-298.

Murray, F., and S. Wilson. 1989. The relationship between sulfur dioxide concentration and yield of five crops in Australia. *Clean Air* 23: 51-55.

Murtha, P.A. 1974. Detection of $SO_2$ fume damage to forests on ERTS-1 imagery. *Canadian Surveyor* 28: 167-170.

Muthuswami, S.V., I. Nirdosh, and M. Selamat. 1989. Simultaneous leaching of uranium, $^{230}$Th, and $^{226}$Ra from Saskatchewan ores by nitric and hydrochloric acids. *Minerals & Metallurgical Processing* 6: 186-189.

Nakatsu, C. et al. 1987. Metal Tolerances of Bacteria from Sudbury Soils and from an Unpolluted Site. *See* Lindberg and Hutchinson 1987.

Nathwani, J.S., and C.R. Phillips. 1979a. Absorption of $^{226}$Ra by soils (I). *Chemosphere* 5: 285-291.

———. 1979b. Absorption of $^{226}$Ra by soils in the presence of $Ca^{2+}$ ions. Specific absorption (II). *Chemosphere* 5: 293-299.

National Academy of Sciences. 1975. *Underground disposal of coal mine wastes.* Washington, DC: National Academy of Sciences.

N.R.C.C. 1981a. *Acidification in the Canadian Aquatic Environment.* NRCC 18475. National Research Council of Canada: Ottawa, ON.

———. 1981b. *Effects of Nickel in the Canadian Environment.* No. 18568. National Research Council of Canada: Ottawa, ON.

———. 1986. *Aluminum in the Canadian Environment.* No. 24759. National Research Council of Canada: Ottawa, ON.

National Wetlands Working Group. 1988. *Wetlands of Canada.* Ecological Land Classification Series No. 24. Sustainable Development Branch, Environment Canada, and Montreal, Polyscience Publications Inc.

Nawrot, J.R. 1986. Wetland development on coal mine slurry impoundments. *See* Brooks et al. 1986.

Nebel, B.N. 1987. *Environmental Science: The Way the World Works.* Englewood Cliffs, NJ: Prentice-Hall.

Nicholas, A.K., and R.J. Hutnik. 1971. *Ectomycorrhizal Establishment and Seedlings Response on Variously Treated Deep-Mine Coal Refuse.* Special Research Report CR-80. University Park, PA: Pennsylvania State University.

Nicholson, K.W. 1988. The dry deposition of small particles: a review of experimental measurements. *Atmospheric Environment* 22: 2653-2666.

Nieman, T.J., and D. Meshako. 1990. The Star Fire Mine reclamation experience. *Journal of Soil and Water Conservation* 45: 529-532.

*The Northern Miner.* 1991a. In situ leaching key to Arimetco's aggressive plans as copper miner. *The Northern Miner* 77(17): 17.

———. 1991b. Newmont tests bioleaching for Carlin refractory ores. *The Northern Miner* 77(33): 19.

———. 1991c. United Reef asked to clean up waste rock dumped 80 years ago. *The Northern Miner* 76(49): 1.

Nriagu, J.O. (ed.). 1984. *Environmental Impacts of Smelters.* New York, NY: John Wiley & Sons.

Nriagu, J.O. 1988. Natural Versus Anthropogenic Emissions of Trace Metals to the Atmosphere. In *Control and Fate of Atmospheric Trace Metals*, eds. J.M. Pacyna and B. Ottar. NATO Advanced Science Institute Series, Series C, Mathematical and Physical Sciences 268. Dordrecht: Kluwer Academic Publishers.

Nriagu, J.O., and J.F. Gaillard. 1984. The speciation of pollutant metals in lakes near the smelters at Sudbury, Ontario. *See* Nriagu 1984.

Nriagu, J.O., and H.K. Wong. 1983. Selenium pollution of lakes near the smelters at Sudbury, Ontario. *Nature* 301: 55-57.

Odum, E.P. 1983. *Basic Ecology.* Philadelphia, PA: Saunders College Publications.

———. 1985. Trends expected in stressed ecosystems. *Bioscience* 35: 419-422.

Ontario. Ministry of Energy and the Environment and Ministry of Natural Resources. 1994. *Fish habitat protection guidelines for developing areas.* Toronto, ON: Environment Ontario.

———. Ministry of the Environment. 1990. *Bay of Quinte Remedial Action Plan. Stage 1—environmental setting + problem definition*, prepared by the Remedial Action Plan Coordinating Committee. Toronto, ON: Environment Ontario.

———. 1991. *Site rehabilitation reduces arsenic loading to Moira River.* Environment Information. Toronto, ON: Environment Ontario.

———. 1992a. *Air Quality in Ontario.* Toronto, ON: Environment Ontario, Air Resources Branch.

———. 1992b. *Bay of Quinte Remedial Action Plan. Stage 2. Time to Act*, prepared by the Remedial Action Plan Coordinating Committee. Toronto, ON: Environment Ontario.

———. 1992c. *Status report on the metal mining sector effluent monitoring data.* Toronto, ON: Environment Ontario.

———. Ministry of Northern Development

and Mines. 1990. *Ontario gold.* Toronto, ON: Mines and Minerals.

———. Water Resources Commission. 1971. *Water Pollution from the Mining Industry in the Elliott Lake and Bancroft Areas.* Vol. 1, Summary. Toronto, ON: Ontario Water Resources Institute.

Osiensky, J.L., and R.E. Williams. 1990. Factors affecting efficient aquifer restoration at in situ uranium mine sites. *Ground Water Monitoring Review* 10: 107-112.

Owen, R.M. 1977. An assessment of the environmental impact of mining on the continental shelf. *Marine Mining* 1: 85-103.

Ozvacic, V. 1982. *Emissions of Sulphur Oxides and Trace Elements in the Sudbury Basin.* Report #ARB-ETRD-09-82. Toronto, ON: Ontario Ministry of the Environment.

Paine, P.J., and W.B. Blakeman. 1987. *The national coal wastewater survey: 1982—1984.* Report IP-94, December. Ottawa, ON: Environment Canada.

Palmer, S.D., and M. Murphy. 1984. *Cumulative hydrologic impact assessment: The effects of coal mining on the hydrologic systems of the Raton Coal Field in Central New Mexico.* Santa Fe, NM: New Mexico Energy and Minerals Department.

Panu, U.S. 1989. Hydrological assessment of peat mining operations in domed bogs: A case study. *Canadian Water Resources Journal* 14: 54-65.

Parekh, B.K., et al. 1991. Mineral processing 1990. *Mining Engineering* 43: 529-541.

Pearse, P.H., F. Bertrand, and J.W. MacLaren. 1985. *Currents of Change.* Final report of the inquiry on federal water policy. En 37/71/19855555-I E. Ottawa, ON: Environment Canada.

Pearson, J.S. 1975. *Ocean floor mining.* Park Ridge, NJ: Noyes Data Corporation.

Pedersen, T.F., and A.J. Losher. 1988. Diagenetic processes in aquatic mine tailings deposits in British Columbia, Canada. In *Chemistry and Biology of Solid Waste: Dredged Material and Mine Tailing,* eds. W. Salomons and V. Forstner. Berlin: Springer-Verlag.

Pedersen, T.F., et al. 1994. Geochemistry of Submerged Tailings in Anderson Lake, Manitoba: Recent Results. *See* USBM 1994. Vol. 1: 288-296.

Penrose, W.R. 1974. Arsenic in the marine and aquatic environments: Analysis, occurrence and significance. *Chemical Rubber Company Critical Reviews in Environmental Control* 4: 465-482.

Peters, T.H. 1984. Rehabilitation of mine tailings: A case of complete ecosystem reconstruction and revegetation of industrially stressed lands in the Sudbury area, Ontario, Canada. In *Effects of pollutants at the ecosystem level.* SCOPE 22, eds. P.J. Sheehan et al. New York, NY: John Wiley & Sons. 403-421.

———. 1989. Overview of tailings reclamation in eastern Canada. *See* Walker et al. 1989.

Peterson, E.G., and H.M. Etter. 1970. *A background for disturbed land reclamation and research in the Rocky Mountain region of Alberta.* Info. Rep. A-X-34. Edmonton, AB: Forest Research Laboratory, Canadian Forestry Service.

Peterson, J.B., and R.F. Nielson. 1973. Toxicities and Deficiencies in Mine Tailings. In *Ecology and Reclamation of Devastated Land.* Vol. 2, eds. R. Hutnik and G. Davis. New York, NY: Gorden and Breach.

Pierce, W.G., et al. 1994. Composted Organic Wastes as Anaerobic Reducing Covers for Long Term Abandonment of Acid-Generating Tailings. *See* USBM 1994. Vol. 2: 148-157.

Platzoder, R. 1979. Deep seabed mining and the law of the sea. In *The Mineral Resources Potential of the Earth.* Proceedings of the 2nd International Symposium, Hannover, April 18-20, ed. F. Bender. Stuttgart: E. Schweizerbart'sche Verlagsbuchhandlung. 118-131.

Poling, G.W. 1973. Sedimentation of mill tailings in fresh water and in sea water. *CIM Bulletin* 66: 97-102.

Powell, R.J., and L.M. Wharton. 1982. Development of the Canadian Clean Air Act.

*Journal of the Air Pollution Control Association* 32: 62-65.

Pretty, K.M., R.R. Sentis, and R.L. Thomas. 1977. The movement of alkali halides into the environment. In *The Effects of Alkali Halides in the Canadian Environment.* NRCC No. 15019, ed. D.A. Rennie. Ottawa, ON: National Research Council. 37-61.

Prud'homme, M. 1986. Graphite. 29.1-29.13. *1985 Canadian Minerals Yearbook—Review and Outlook*, ed. G. Cathcart. Ottawa, ON: Energy, Mines and Resources Canada.

———. 1990. Fluorspar. 27.1-27.11. *See* Godin 1990.

———. 1991. Sulphur. 61.1-61.20. *See* Godin 1991.

———. 1992. Potash. 35.1-35.26. *See* Godin 1992.

———. 1993. Potash. 38.1-38.19. *See* Godin 1993.

———. 1994. Graphite. 24.1-24.9. *See* Godin 1994.

Prud'homme, M., and D. Francis. 1986. *The clay products industry.* Mineral Policy Sector Internal Report MRI 8613. Ottawa, ON: Energy, Mines and Resources Canada.

Purves, D. 1977. *Trace-Element Contamination of the Environment.* Amsterdam: Elsevier.

Rabbitts, F.T., et al. 1971. *Environmental control in the mining and metallurgical industries in Canada.* Ottawa, ON: National Advisory Committee on Mining and Metallurgical Research.

Radford, D.S., and D.N. Graveland. 1978. *The water quality of some coal mine effluents and their effect on stream benthos and fish.* Fisheries Pollution Report No. 4. Lethbridge, AB: Ministry of Recreation, Parks and Wildlife.

Rainbow, A.K.M. (ed.). 1990. Reclamation, treatment and utilization of coal mine wastes. *Proceedings of the Third International Symposium on the Reclamation, Treatment and Utilization of Coal Mining Wastes*, 3-7 September, in Glasgow, UK. Rotterdam, Netherlands: A.A. Balkema.

Ranta, W., F.D. Tomassini, and E. Nieboer. 1978. Elevation of copper and nickel levels in primaries from black and mallard ducks collected in the Sudbury district, Ontario. *Canadian Journal of Zoology* 56: 581-586.

Rauser, W.E., and E.K. Winterhalder. 1985. Evaluation of copper, nickel and zinc tolerances in four grass species. *Canadian Journal of Botany* 63: 58-63.

Raymont, M.E.D. 1972. Sulphur sources and uses—past, present and future. *CIM Bulletin* 65: 49-54.

Redmann, R.E. 1983a. *Potash dust and vegetation—pathways and impacts.* Saskatoon, SK: Department of Crop Science and Plant Ecology, University of Saskatchewan.

———. 1983b. Resistance of plant species to airborne residual salts from potash refineries. 823-828. *See* McKercher 1983.

Redmann, R.E., and F.T. Frankling. 1982. *Revegetation of abandoned uranium mill tailings near Uranium City, Saskatchewan—plant species selection.* Report to the Saskatchewan Department of the Environment. Saskatoon, SK: University of Saskatchewan.

Redmann, R.E., and J. Haraldson. 1986. *Annual report 1985—vegetation research.* Saskatoon, SK: Department of Crop Science and Plant Ecology, University of Saskatchewan.

Redmann, R.E., and M.K.A. Ryan. 1988. *Annual report 1987 — vegetation research.* Saskatoon, SK: Department of Crop Science and Plant Ecology, University of Saskatchewan.

Redmann, R.E., J. Haraldson, and L.V. Gusta. 1986. Leakage of UV-absorbing substances as a measure of salt injury in leaf tissue of woody species. *Physiologia Plantarum* 67: 87-91.

Reeder, S.W. 1971. *Cross-sectional study of the effects of smelter waste-water disposal on water quality of the Columbia River downstream from Trail, British Columbia.* Technical Bulletin No. 39. Ottawa, ON: Inland Waters Branch, Energy, Mines and Resources Canada.

Reichenbach, I. 1994. Review of Canadian Legislation Relevant to Decommissioning

Acid Mine Drainage Sites. *See* USBM 1994. Vol. 4: 275-284.

Reimers, J.H. and Associates. 1980. *A study of sulphur containment technology in the non-ferrous metallurgical industry.* EPS 3-AP-79-8. Ottawa, ON: Environment Canada.

Reiter, E.R . 1974. Dispersion of Radioactive Material on Small, Meso- and Global Scales. *Physical Behavior of Radioactive Contaminants in the Atmosphere.* Proceedings of a Symposium, Vienna, 12-16 November 1973, Jointly Organized by the IAEA and WMO. Vienna, Austria: International Atomic Agency.

Rennie, D.A. 1983. Fate and impact of potash dust on the soil environment in Saskatchewan. 841-846. *See* McKercher 1983.

Reynoldson, T.B. 1987. Interactions between sediment contaminants and benthic organisms. *Hydrobiologia* 149: 53-66.

Ripley, E.A. 1987. Climatic change and the hydrological regime. Chapter 6 in *Canadian Aquatic Resources. Canadian Bulletin of Fisheries and Aquatic Sciences,* eds. M.C. Healey and R.R. Wallace. 215: 137-178. Department of Fisheries and Oceans, Ottawa, ON.

Ripley, E.A., R.E. Redmann, and J. Maxwell. 1978. *Environmental Impact of Mining in Canada.* First printing. Kingston, ON: Centre for Resource Studies, Queen's University.

———. *Environmental Impact of Mining in Canada.* Second printing. Kingston, ON: Centre for Resource Studies, Queen's University.

Ritchie, J. 1987. *Postglacial vegetation of Canada.* London, UK: Cambridge University Press.

Ritter, J. 1993. *Deloro mine site rehabilitation.* Toronto, ON: Ontario Ministry of the Environment.

Roberts, J.M., and A.L. Masullo. 1986. Pneumatic stowage becomes affordable. *Coal Age* 21: 52-56.

Roberts, T.M. 1984. Effects of air pollutants in agriculture and forestry. *Atmospheric Environment* 18: 629-652.

Roberts-Pichette, P. 1994. Canada's ecosystem monitoring and assessment initiative: Building a framework of ecological science centres. *Northern Perspectives* 22: 6-9.

Robertson, A.M. 1986. *The technology of uranium tailings covers.* Contract report OSQ85-00226. Ottawa, ON: National Uranium Tailings Program, CANMET.

Robertson, J.L. 1986. Owl Rock builds land while mining. *Rock Products* 89(9): 66-69.

Robinsky, E.I. 1975. Thickened discharge—a new approach to tailings disposal. *CIM Bulletin* 68: 47-53.

Roe, N.A., and A.J. Kennedy. 1989. Moose and deer habitat use and diet on a reclaimed mine in west-central Alberta. 127-135. *See* Walker et al. 1989.

Rolia, E., and K.G. Tan. 1985. The generation of thiosalts in mills processing complex sulphide ores. *Canadian Metallurgical Quarterly* 24: 293-302.

Romaine, M.J., and G.R. Ironside. 1979. *Canada's Northlands.* Ecological Land Classification Series. Ottawa, ON: Environment Canada.

Romaniuk, A.S., and H.G. Naidu. 1987. *Coal mining in Canada: 1986.* CANMET Report 87-3E. Ottawa, ON: CANMET.

Rose, D., and J.R. Marier. 1977. *Environmental fluoride 1977.* NRCC No. 16081. Ottawa, ON: National Research Council of Canada.

Rose, G.A., and G.H. Parker. 1982. Effects of smelter emissions on metal levels in the plumage of ruffed grouse near Sudbury, Ontario, Canada. *Canadian Journal of Zoology* 60: 2659-2667.

Roy, M., P. LaRochelle, and C. Anctil. 1973. *Stability of Dykes Embankments at Mining Sites in the Yellowknife Area.* Arctic Land Use Research 1972-73-31. Ottawa, ON.: Indian and Northern Affairs Canada.

Royex Gold Mining Corporation. 1988. Executive Summary. Vol. 1. In *Royex Gold Mining Corporation Jolu Project environmental impact statement.* Regina, SK: Royex Gold Mining Corporation.

Rubec, C.D.A. (ed.). 1979. *Application of Eco-*

logical (Biophysical) Land Classification in Canada. Ecological Land Classification Series, No. 7. Ottawa, ON: Environment Canada.

Rumble, M.A., and A.J. Bjugstad. 1986. Uranium and radium concentrations in plants growing on uranium mill tailings in South Dakota. Reclamation and Revegetation Research 4: 271-277.

Ryerson, R., and J. Cihlar. 1990. Remote Sensing Techniques Monitor Global Change. Geos 19: 1-6.

Sage, R. 1976. Summary. See Murray 1977.

Salomons, W., et al. 1987. Sediments as a source for contaminants? Hydrobiologia 149: 13-30.

Saskatchewan. Ministry of Energy and Mines. Energy Division. 1984. Update of energy supply and demand. Misc. Report 84-5. Regina, SK: Minister of Supply and Services.

———. Ministry of Environment and Public Safety. 1984. Reclamation guidelines for the Estevan mining area. Regina, SK: Saskatchewan Environment.

———. 1991. Saskatchewan state of the environment report. Regina, SK: Saskatchewan Environment.

Scales, M. 1986. Base metal mastery: Brunswick Mining gets the Pb out (also Cu, Zn and Ag). Canadian Mining Journal 107(12): 33-5, 38.

———. 1992. Floating new ideas. Canadian Mining Journal 113 (April): 10-15.

Scheider, W.A., D.S. Jeffries, and P.J. Dillon. 1980. Bulk deposition in the Sudbury and Muskoka-Haliburton areas of Ontario during the shutdown of Inco Ltd. in Sudbury. Rexdale, ON: Water Resources Branch, Environment Ontario.

Schelske, C.L. 1991. Historic nutrient enrichment of Lake Ontario: palaeolimnological evidence. Canadian Journal of Fisheries and Aquatic Science 48: 1529-1538.

Scheuhammer, A.M. (ed). 1991. Environmental Pollution: pH-related changes in metal biochemistry and effects on aquatic systems. Environmental Pollution 71: 1-380.

Schindler, D.W., et al. 1985. Long-term ecosystem stress: the effects of years of experimental acidification on a small lake. Science 228 (4706): 1395-1401.

Schmidt, F.H. 1963. Atmospheric Pollution. Chapter 10 in Physics of Plant Environment. W.R. Van Wijk (ed). North-Holland Publishing Company, Amsterdam.

Schmidt, J.W. 1988. Evaluation of the full scale Inco SO₂/Air system treating gold mill effluent at Carolin Mines. WTC-PC-2-1988. Burlington, ON: Wastewater Technology Centre, Environment Canada.

Schramm, J.R. 1966. Plant Colonization Studies on Black Wastes from Anthracite Mining in Pennsylvania. Transactions of the American Philosophical Society 56 (Part 1): 1-194.

Schreier, H., and J. Taylor. 1980. Asbestos fibres in receiving waters. Technical Bulletin No. 117. Inland Waters Directorate, Pacific and Yukon Region, Water Quality Branch. Vancouver, BC: Environment Canada.

———. 1981. Variations and mechanisms of asbestos fibre distribution in stream water. Technical Bulletin No. 118. Inland Waters Directorate, Pacific and Yukon Region, Water Quality Branch. Vancouver, BC: Environment Canada.

Schultz, V., and F.W. Whicker. 1974. Radiation ecology. Chemical Rubber Company Critical Reviews in Environmental Control 4: 423-464.

Schutzman, R.J. 1983. Size distributions of particulates in potash dryer exhausts. See McKercher 1983.

Scollan, M. 1994. Mining and mineral processing operations in Canada 1993. Mineral Bulletin MR 235. Ottawa, ON: Energy, Mines and Resources Canada.

Scott, D. 1991. Potash mined by different methods. Northern Miner Magazine 77(23): 2.

———. 1992a. Dickenson close to decision on using bioleaching. Northern Miner Magazine 77(51): 1-2.

———. 1992b. Autoclaving, bioleaching replacements for roasting. Northern Miner Magazine 78(2): 16.

Scott, J.S. 1987. *Mine and mill wastewater treatment*. EPS 2/MM/3. Industrial Programs Branch. Mining, Mineral and Metallurgical Processes Division. Ottawa, ON: Environment Canada.

———. 1989a. An overview of gold mill effluent treatment. *See* Cox et al. 1989.

———. 1989b. The treatment of gold mill effluents—an update. *See* Chalkley et al. 1989.

Scott, J.S., and K. Bragg (eds.). 1975. *Mine and mill wastewater treatment*. Economic and Technical Review Report EPS 3-WP-75-5. Ottawa, ON: Water Pollution Control Directorate.

Seip, H.M. 1986. Surface water acidification. *Nature* 322: 118.

Semkin, R.G., and J.R. Kramer. 1976. Sediment geochemistry of the Sudbury-area lakes. *Canadian Mineralogist* 14: 79-90.

Sengupta, M. 1994. *Environmental Impacts of Mining*. Boca Raton, FL: Lewis Publishers.

Shaw, A.J. (ed.). 1989. *Heavy Metal Tolerance in Plants: Evolutionary Aspects*. Boca Raton, FL: C.R.C. Press.

Shaw, D. 1988. Mica. 43.1-43.9. *See* Cathcart 1988.

Shaw, D.J., and M.A. Boucher. 1988. Barite and celestite. 12.1-12.8. *See* Cathcart 1988.

Shaw, G. 1982. Concentrations of 28 elements in fruiting shrubs downwind of the smelter at Flin Flon, Manitoba. *Environmental Pollution* 25: 197-210.

Sheard, J.W. 1986. Distribution of uranium series radionuclides in upland vegetation of northern Saskatchewan. I: Plant and soil concentrations. *Canadian Journal of Botany* 64: 2446-2452.

Shearer, S.D. 1964. Water characteristics at uranium mills and associated environmental aspects. In *11th Ontario Industrial Wastes Conference*. 23-41.

Sheffield, A. 1983. *National inventory of sources and emissions of mercury (1978)*. Economic and Technical Review Report, EPS 3-AP-81-1. Ottawa, ON: Air Pollution Control Directorate.

Sherwood, D.R., B.E. Opitz, and R.J. Serne.

1985. Evaluation of neutralization treatment processes and their use for uranium tailings solutions. *Nuclear Safety* 26: 334-345.

Shreffler, J.H. 1976. A model for the transfer of gaseous pollutants to vegetational surface. *Journal of Applied Meteorology* 15: 744-746.

Shubert, L.E. 1976. Soil Algae from Surface-Mined Areas of Western North Dakota. In *Proceedings of North Dakota Academy of Science*. Vol. 30. (Abstract). Valley City, ND: North Dakota Academy of Science.

Shugar, S. 1979. *Effects of asbestos in the Canadian environment*. NRCC No.16452. Ottawa, ON: National Research Council of Canada.

Shugart, H.H., R. Leemans, and G.B. Bonan. 1992. *A Systems Analysis of the Global Boreal Forest*. Cambridge, UK: Cambridge University Press.

Sileo, L., and W.N. Beyer. 1985. Heavy metals in white-tailed deer (*Odocoileus virginianus*) living near a zinc smelter in Pennsylvania. *Journal of Wildlife Diseases* 21(3): 289-296.

Simon, S.L., and E.J. Deming. 1986. Time-dependent leaching of radium from leaves and soil. *Journal of Environmental Quality* 15: 305-308.

Simovic, L., and W.J. Snodgrass. 1989. Tailings pond design for cyanide control at gold mills using natural degradation. *See* Cox et al. 1989.

Sindelar, B.W. 1979. Successful development of vegetation on surface mined land in Montana. In *Ecology and Coal Resource Developments*, ed. M.K. Wali. New York, NY: Pergamon Press.

Singhal, R.K., and H. Sahay. 1986. Canadian coal industry: Developments in resource conservation and productivity improvement. *Mining Engineering* 38: 879-884.

Sirois, L.L., and R.J.C. MacDonald (eds.). 1983. *Minerals, metals, and mining technologies*. Minerals Processing—Research and Development. Ottawa, ON: CANMET.

Sjörs, H. 1980. Peat on Earth: Multiple use or conservation? *Ambio* 9: 303-308.

Sladen, J.A., and R.C. Joshi. 1988. Mining subsidence in Lethbridge, Alberta. *Canadian Geotechnical Journal* 25: 768-777.

Slagle, S.E., B.D. Lewis, and R.W. Lee. 1985. *Ground-water resources and potential hydrologic effects of surface coal mining in the Northern Powder River Basin, Southeastern Montana*. Water-Supply Paper 2239. Washington, DC: U.S. Geological Survey.

Slancy, F.F. 1971. Environmental impact of surface coal mining operations in Alberta. In *Proceedings of the public hearings on the impact on the environment of surface mining in Alberta*. Edmonton, AB: Environmental Conservation Authority. 1-59.

Slivitsky, M. (ed.). 1978. *Hydrological Atlas of Canada*. Ottawa, ON: Environment Canada and Fisheries and Oceans Canada.

Sly, P.G. 1991. The effects of land use and cultural development on the Lake Ontario system since 1750. *Hydrobiologia* 213: 1-75.

Smecht, L.M., D. Laguitton, and Y. Berube. 1975. *Control of arsenic level in gold mine waste waters*. ALUR 74-74-351, INA Pub. No. QS-8035-000-EE-A1. Ottawa, ON: Indian and Northern Affairs Canada.

Smith, P.R., D.W. Bailey, and R.E. Doane. 1972. Minerals processing. *Engineering and Mining Journal* 173: 175-176.

Smith, R.L. 1986. *Elements of Ecology*. New York, NY: Harper and Row.

————. 1990. *Elements of Ecology*. 2nd edition. New York, NY: Harper-Collins College.

Smol, J.P. 1992. Paleolimnology, an important tool for effective ecosystem management. *Journal of Aquatic Ecosystem Health* 1:49-58.

Smol, J.P., and J.R. Glew. 1992. Paleolimnology. In *Encyclopedia of Earth System Science* Vol. 3, ed. W.A. Nierenberg. New York, NY: Academic Press. 551-564.

Snelgrove, W.R.N. and J.C. Taylor. 1989. Environmental controls and energy consumption in non-ferrous metallurgy. *See* Centre in Mining and Mineral Exploration Research 1989.

Somcynsky, T. 1986. Un avenir prometteur pour l'amiante. *Geos* 15: 5-8.

Sonntag, N.C., et al. 1987. *Cumulative effects assessment: a context for further research and development*. En 106-7/1987E. Ottawa, ON: Canadian Environmental Assessment Research Council.

Sopko. M. 1994. Environmental Programs at Sudbury. In *Proceedings Volume for the 96th Annual General Meeting of the Canadian Institute of Mining, Metallurgy and Petroleum and the 1994 Mineral Outlook Conference*. 1-4 May 1994. Toronto, ON.

Sopper, W.E. 1992. Rapid ecological restoration of mined land using municipal sewage sludge. *Land Reclamation: Advances in Research and Technology*, eds. T. Younos, P. Diplas and S. Mostaghimi. St. Joseph, MI: American Society of Agricultural Engineers.

SPAPOP. Study Panel on Assessing Potential Ocean Pollutants. 1975. *Assessing potential ocean pollutants*. Washington, DC: National Academy of Sciences.

Sparrow, H. 1984. *Soil at risk: Canada's eroding future*. Standing Senate Committee on Agriculture, Fisheries and Forestry. Ottawa, ON: Senate of Canada.

Spear, P.A., and R.C. Pierce. 1979. *Copper in the aquatic environment: chemistry, distribution and toxicology*. No. 16454. Ottawa, ON: National Research Council of Canada.

St-Arnaud, L. 1994. Water Covers for the Decommissioning of Sulfidic Mine Tailings Impoundments. *See* USBM 1994. Vol. 1: 279-287.

St-Cyr, L. 1989. *Iron Plaque of Phragmites australis (Cav.) Trin. ex Steudel and Bioavailability of Iron, Manganese, Copper, Zinc, and Nickel*. Ph.D. Thesis. Kingston, ON: Queen's University.

St-Cyr, L., and A.A. Crowder. 1990. Manganese in the root plaque of *Phragmites australis* (Cav.) Trin. ex Steudel. *Soil Science* 149: 191-198.

Stelfox, H.A., and G.R. Ironside (eds.). 1982. *Land/Wildlife Integration No. 2*. Ecological Land Classification Series, No. 17. Ottawa, ON: Environment Canada, .

————. 1988. *Land/Wildlife Integration No. 3.* Ecological Land Classification Series, No. 22. Ottawa, ON: Environment Canada.

Stern, A.C. 1968. *Air Pollution.* London, UK: Academic Press.

Stern, A.C., et al. 1984. *Fundamentals of Air Pollution.* Second ed. San Diego, CA: Academic Press.

Stewart, B.M., R.R. Backer, and R.A. Busch. 1990. Flocculation of fine coal waste: At discharge. *See* Rainbow 1990.

Stiller, D., and R.G. Reider. 1979. Impacts of cement manufacturing on soils in a semiarid climate. *Journal of Soil and Water Conservation* 34: 279-281.

Stilley, B., et al. 1971. The environmental challenge. *Engineering and Mining Journal* 172: 71-128.

Stoermer, E.F., et al. 1985. An assessment of ecological changes during the recent history of Lake Ontario based on siliceous algal microfossils preserved in the sediments. *Journal of Phytology* 21: 257-276.

Stonehouse, D.H. 1973. *Cement in Canada.* Mineral Bulletin MR 133. Mineral Resources Branch, Energy, Mines and Resources Canada. Ottawa, ON: Information Canada.

Stoner, J.D. 1983. Probable hydrologic effects of subsurface mining. *Ground Water Monitoring Review* 3: 128-137.

Strathdee, G. 1994. Potash production in Saskatchewan. *Phosphorus & Potassium* 191 (May-June 1994): 19-25.

Struthers, G. 1991. Brine solution from mine destroys 40 acres of farmland. *Saskatoon Star-Phoenix,* September 27: 1.

Sullivan, T.J., et al. 1988. Atmosphere wet sulphate deposition and lakewater chemistry. *Nature* 331: 607-609.

Swanson, S.M. 1983. Levels of $^{226}$Ra, $^{210}$Pb and total U in fish near a Saskatchewan uranium mine and mill. *Health Physics* 45: 67-80.

————. 1985. Food-chain transfer of U-series radionuclides in a northern Saskatchewan aquatic system. *Health Physics* 49: 747-770.

————. 1990. *Cluff Lake Status of the Environment Report.* A review of predicted and observed environmental impacts at the AMOK Ltd. Cluff Lake mine and mill site. Saskatoon, SK: Beak Associates.

Swanson, S.M., and Z. Abouguendia. 1981. *The problem of abandoned uranium tailings in northern Saskatchewan: An overview.* No. C-805-48-81. Saskatoon, SK: Saskatchewan Research Council.

Swanson, S.M., and R.J.W. Bernstein. 1984. *Physiological parameters in three fish species from the Beaverlodge area, Saskatchewan.* No. E-90-40-E-84. Saskatoon, SK: Saskatchewan Research Council.

Sweigard, R.J., and R.V. Ramani. 1986. Site planning process: Application to land use potential evaluation for mined land. *Mining Engineering* 38: 427-432.

Sweigard, R.J., and L.W. Saperstein. 1991. Innovative Techniques to Alleviate Soil Compaction on Reclaimed Land in the United States. *See* AMIC 1991. Vol. 1: 1-12.

Tang, A.J.S., et al. 1987. An analysis of the impact of the Sudbury smelters on wet and dry deposition in Ontario. *Atmospheric Environment* 21: 813-824.

Tarjan, G. 1981. *Mineral processing—fundamentals, comminution, sizing and classification.* Vol. 1. Translated by B. Balkay. Budapest: Akademiai Kiado.

Tarnocai, C. 1980. Canadian Wetland Registry. In *Proceedings of Workshop on Canadian Wetlands,* eds. C.D.A. Rubec and F.C. Pollett. Ecological Land Classification Series, No. 12. Ottawa, ON: Environment Canada.

Task Force on Water Quality Guidelines. 1987. *Canadian water quality guidelines.* Ottawa, ON: Canadian Council of Resource and Environment Ministers.

Tate, D.M., and D.N. Scharf. 1989a. *1986 Industrial water use survey tables. Vol. I: Water use in manufacturing.* Ottawa, ON: Water Planning and Management Branch, Environment Canada.

————. 1989b. *1986 Industrial water use survey tables. Vol. II: Water use in mineral extraction, thermal power and hydro power.* Ottawa, ON: Water Planning and Management Branch, Environment Canada.

Taylor, G.J., and A.A. Crowder. 1983a. Accumulation of atmospherically-deposited metals in wetland soils of the Sudbury, Ontario, region. *Water, Air, and Soil Pollution* 19: 29-42.

————. 1983b. Uptake and accumulation of heavy metals by *Typha latifolia* in wetlands of the Sudbury, Ontario, region. *Canadian Journal of Botany* 61: 63-73.

Thirgood, J.V. 1969. Land disturbance and revegetation in Canada. *Canadian Mining Journal* 90: 33-37.

Thomas, D.J., and S. Metikosh. 1984. The effect of a lead-zinc mining operation on the marine environment of Strathcona Sound, NWT, Canada. *International Conference on Environmental Contamination*, London, July 1984. Edinburgh: CEP Consultants.

Thomas, L.J. 1973. *An introduction to mining: Exploration, feasibility, extraction, rock mechanics.* New York, NY: John Wiley & Sons.

Thomas, P.A., J.W. Sheard, and S. Swanson. 1994. Transfer of $^{210}$Po and $^{210}$Pb through the lichen-caribou-wolf food chain of northern Canada. *Health Physics* 66(6): 666-677.

Thompson, D.N., and R.J. Hutnik. 1971. *Environmental Characteristics Affecting Plant Growth on Deep-Mine Banks.* Special Report SR-88. University Park, PA: Pennsylvania State University.

Thompson, J.A.J., and F.T. McComas. 1974. *Copper and zinc levels in submerged tailings at Brittania Beach.* Technical Report 437. B.C. Fish Research Board, Canada.

Thompson, J.V. 1991. Silver recovery by older methods. *Engineering and Mining Journal* 192: 39-41.

Thorpe, M.B. 1989. *Rehabilitation studies on saline land caused by potash mining activity.* Ph.D. Thesis. Saskatoon, SK: University of Saskatchewan.

Thrush, P.W. (ed.). 1968. *A dictionary of mining, mineral, and related terms.* Washington, DC: Bureau of Mines, Department of the Interior.

Tien, J.K., and J.F. Elliott (eds.). 1971. *Metallurgical Treatises.* Warrendale, PA: The Metallurgical Society of AIME.

Tipping, E., et al. 1986. Effects of pH on the release of metals from naturally-occurring oxides of Mn and Fe. *Environmental Technology Letters* 7: 109-114.

Tomkins, R.J. 1954. *Natural sodium sulphate in Saskatchewan.* Regina, SK: Saskatchewan Department of Mineral Resources.

Townley-Smith, L.J., and R.E. Redmann. 1980. Injury to aspen (*Populus tremuloides*) exposed to airborne salt from potash mines in Saskatchewan. *Canadian Journal of Botany* 58: 2616-2623.

Tremblay, R.L. 1994. Controlling Acid Mine Drainage Using an Organic Cover: The Case of the East Sullivan Mine, Abitibi, Québec. *See* USBM 1994. Vol. 2: 122-127.

Treshow, M. (ed.). 1984. *Air Pollution and Plant Life.* New York, NY: John Wiley & Sons.

Trinh, D.T. 1981. *Exploration camp wastewater characterization and treatment plant assessment.* Report No. EPS 4-WP-81-1. Ottawa, ON: Water Pollution Directorate.

Trost, W.R. (ed.). 1972. *The impact on the environment of surface mining in Alberta.* Report and recommendations. Edmonton, AB: Environment Conservation Authority.

Trudell, M.R. 1986. *Post-mining groundwater supply potential at the Highvale site: Plains hydrology & reclamation project.* Report #RRTAC 86-7. Edmonton, AB: Alberta Land Conservation and Reclamation Council.

Tucker, C.J., J.R.G. Townshend and T.E. Goff. 1985. African land-cover classification using satellite data. *Science* 227: 367-375.

Tyler, G., et al. 1989. Heavy-Metal Ecology of Terrestrial Plants, Microorganisms, and Invertebrates. *Water, Air and Soil Pollution* 47 (3-4): 189-215.

Udd, J.E. 1989. Mining in Canada's north—a technical overview. In *Mining in the Arctic.* Proceedings of the 1st International Symposium on Mining in the Arctic, Fairbanks, 17-19 July, eds. S. Bandopadhyay and F.J. Skudrzyk. Rotterdam, Netherlands: A.A. Balkema.

Ulman, N. 1994. Arctic hunters on a collision

course. Toronto, ON: *Globe & Mail, Metro Edition.* May 31: B13.

United Nations Commission on Environment and Development. 1987. *Our Common Future* (The Brundtland Report). New York, NY: Oxford University Press.

United Nations Department of International Economic and Social Affairs. 1986. *Methodologies for assessing the impact of deep sea-bed minerals on the world economy.* ST/ESA/168. New York, NY: United Nations.

UNSCEAR. United Nations Special Committee on Atomic Radiation. 1988. *Sources, effects and risks of ionizing radiation.* New York, NY: United Nations.

USBM. 1994. *International Land reclamation and Mine Drainage Conference and Third International Conference on the Abatement of Acidic Drainage.* 4 volumes. Proceedings of a conference, Pittsburgh, PA, 24-29 April 1994. United States Department of the Interior, Bureau of Mines Special Publications SP 06A-94, SP 06A-94, SP 06B-94, SP 06C-94, SP 06D-94.

Vagt, O. 1976. *Asbestos.* Mineral Policy Series. Mineral Bulletin MR 155. Ottawa, ON: Energy, Mines and Resources Canada.

———. 1994a. Cement. 14.1-14.11. *See* Godin 1994.

———. 1994b. Stone. 46.1-46.27. *See* Godin 1994.

———. 1994c. Mineral aggregates. 32.1-30.14. *See* Godin 1994.

———. 1994d. Gypsum and anhydrite. 25.1-25.9. *See* Godin 1994.

Van Arsdale, G.D. 1953. *Hydrometallurgy of base metals.* New York, NY: McGraw-Hill.

Van der Hoven, I. 1985. Environmental Impacts. Chapter 40 in *Handbook of Applied Meteorology,* ed. D.D. Houghton. New York, NY: John Wiley & Sons.

Van Kalsbeek, L.P., A.R. Waroway, and D.A. Latoski, (eds.). 1991. *Yukon Placer Mining Industry 1989-1990.* Whitehorse, YT: Northern Affairs Program, Placer Mining Section.

Van Nieuwenhuyse, E.E., and J.D. LaPerriere. 1986. Effects of placer gold mining on primary production in subarctic streams of Alaska. *Water Resources Bulletin* 22: 91-99.

Vandergaast, G. et al. 1988. The application of ecological engineering at a uranium mining facility in northern Saskatchewan. In *Proceedings of International Symposium on Uranium and Electricity: The complete nuclear fuel cycle.* Saskatoon, SK: Canadian Nuclear Society.

Varney, R., and B.M. McCormac. 1971. Atmospheric Pollutants. *Introduction to the Scientific Study of Atmospheric Pollution,* ed. B.M. McCormack. Dordrecht: Reidel.

Veldhuizen, H., D.W. Blowes, and R.S. Siwik. 1987. The effectiveness of vegetation in controlling acid generation in base metal tailings. *See* Blakeman et al. 1987.

Venkatram. A., P.K. Karamchandani, and P.K. Misra. 1988. Testing a comprehensive acid deposition model. *Atmospheric Environment* 22: 737-747.

Vickell, G.A., D.T. Davies, and R. Gec. 1989. Hydrogen peroxide treatment of gold mill wastes. 145-154. *See* Cox et al. 1989.

Vonhof, J.A. 1983. *Hydrogeological and hydrochemical investigation of the waste disposal basin as I.M.C.C. K2 potash plant, Esterhazy, Saskatchewan.* NHRI Paper No.13. IWD Scientific Series, 116. Ottawa, ON: Inland Waters Directorate.

Vroom, A.H. 1971. *Sulphur utilization, a challenge and an opportunity.* Report of a Study for the National Research Council of Canada. Ottawa, ON: National Research Council.

W & W Radiological and Environmental Consultant Services Inc. 1978. *Monitoring program design recommendations for uranium mining localities.* EPS 3-EC-78-10. Ottawa, ON: Environment Canada.

Waite, D.T., S.R. Joshi, and H. Sommerstad. 1988. The effect of uranium mine tailings on radionuclide concentrations in Langley Bay, Saskatchewan, Canada. *Archives of Environmental Contamination and Toxicology* 17: 373-380.

Waite, D.T., et al. 1990. A toxicological examination of whitefish (*Coregonus clupeaformis*) and Northern Pike (*Esox lucius*) exposed to uranium mine tailings. *Archives of Environmental Contamination and Toxicology* 19: 578-582.

Wali, M.K. (ed.). 1975. *Practices and Problems of Land Reclamation in Western North America*. Grand Forks, ND: University of North Dakota Press.

Walker, D.G., C.B. Powter, and M.W. Pole (eds.). 1989. *Proceedings of the conference: Reclamation, A Global Perspective. Vol. 1*. Report # RRTAC 89-2. Edmonton, AB: Alberta Conservation and Reclamation Council.

Ward, J.R. 1983. Well design and construction for in situ leach uranium extraction. *Ground Water Monitoring Review* 3: 79-85.

Watson, G.P., A.N. Rencz, and G.F. Bonham-Carter. 1989. Computers assist prospecting. *Geos* 18: 8-15.

Watson, L.E., R.W. Parker, and D.F. Polster. 1980. *Manual of Species Suitability for Reclamation in Alberta*. 2 vols. Report #RRTAC 80-5. Edmonton, AB: Alberta Land Conservation and Reclamation Council.

Watt, W.D., C.D. Scott, and W.J. White. 1983. Evidence of acidification of some Nova Scotian rivers and its impact on Atlantic salmon, *Salmo salar. Canadian Journal of Fisheries and Aquatic Sciences* 40: 462-473.

Webster, H.J., M.J. Lacki, and J.W. Hummer. 1994. Biotic Development Comparisons of a Wetland Constructed to Treat Mine Water Drainage with a Natural Wetland System. *See* USBM 1994. Vol. 3: 102-110.

Wedel, J.H., and R.L. Wedel. 1991. *Hydrometric Network Evaluation: An Assessment of User Needs*. Ottawa, ON: Northern Waters Resources Studies. Indian and Northern Affairs Canada.

Welch, D.E., and F.W. Firlotte. 1989. Tailings management in the gold mining industry. *See* Chalkley et al. 1989.

Welch, D.W. 1978. Land-Water Classification. *Proceedings of Land/Water Integra-tion Working Groups*. Ottawa, ON: Environment Canada.

Werniuk, J. 1987. Taking another look: Dredging up profits from tailings ponds. *Canadian Mining Journal* 108.3: 31-33.

Westlake, G.F., et al. 1992. *Acute lethality data for Ontario's industrial minerals sector effluents covering the period from May 1990 to May 1991*. Municipal/Industrial Strategy for Abatement, Ontario Ministry of the Environment. Toronto, ON: Queen's Printer for Ontario.

Wetzel, R. 1983. *Limnology*. New York, NY: Saunders College Publishers.

Whelpdale, D.M. 1992. An overview of the atmospheric sulphur cycle. In *Sulphur cycling on the continents: wetlands, terrestrial ecosystems and associated water bodies* (SCOPE 48), eds. R.W. Howarths, J.W.B. Stewart and N.V. Ivanov. Chichester, UK: John Wiley & Sons.

Whillans, R.T. 1992. Uranium. 50.1-50.21. *See* Godin 1992.

———. 1994. Uranium. 53.1-53.19. *See* Godin 1994.

Whitby, L.M., and T.C. Hutchinson. 1974. Heavy metal pollution in the Sudbury mining and smelting region of Canada, II: Soil toxicity tests. *Environmental Conservation* 1: 191-200.

Whitby, L.M., et al. 1976. Ecological consequences of acidic and heavy metal discharges from the Sudbury smelters. *Canadian Mineralogist* 14: 47-57.

Whitby-Costescu, L., J. Shillabeer, and D.F. Coates. 1977. Environmental planning. In *Pit Slope Manual*. CANMET Report 77-2. Ottawa, ON: Energy, Mines and Resources Canada.

White, L. 1985. Boliden improves processing economics. *Engineering and Mining Journal* 186: 32-36.

Whitehead, A.J., B.W. Kelso, and J.G. Malick. 1989. Nitrogen removal from coal mine wastewater using a pilot scale wetland: Year 1 results. 37-49. In *Water Management at Minesites: Proceedings of the Thirteenth Annual British Columbia Mine Reclamation Symposium,* ed. Technical and

Research Committee on Reclamation. Vancouver, B.C.: Mining Association of British Columbia.

Whiteway, P.( ed.). 1990. *Mining Explained: A Guide to Prospecting and Mining.* Toronto, ON: The Northern Miner Press, Inc.

Whiting, J., et al. 1982. *Summary report on the surface water and groundwater investigations of the Lorado Mill area and meteorology data for the Gunnar uranium tailings. Vol. 1.* No. E-820-4-E-82. Saskatoon, SK: Saskatchewan Research Council.

Whittaker, R.H. 1975.*Communities and Ecosystems.* 2nd edition. New York, NY: Macmillan.

Whittaker, B.N., D.J. Reddish, and X.L. Yao. 1992. Environmental issues arising from subsidence occurring at underground operational and abandoned mine sites. *Environmental Issues in Energy and Minerals Production*, eds. Singhal et al. Rotterdam, Netherlands: Balkema.

Whittle, L. 1992. The piloting of Vitrokele for cyanide recovery and waste management at two Canadian gold mines. In *Randol gold forum '92,* ed. H. von Michaelis. Vancouver, BC: Randol International Ltd.

Wieder, R.K., G.E. Land, and A.E. Whitehouse. 1986. Metal removal in sphagnum-dominated wetlands: experience with a man-made wetland system. *See* Brooks et al. 1986.

Wiken, E. 1986. *Terrestrial Ecozones of Canada.* Ecological Land Classification Series No. 19. Ottawa, ON: Lands Directorate, Environment Canada.

Wilder, A.L., and S.N. Dixon. 1989. Heap leach solution application at Coeur-Rochester. *Minerals & Metallurgical Processing* 6: 57-59.

Williams, G.P. 1973. Changed Spoil Dump Shape Increases Stability on Contour Strip Mines. Chapter 21 in *Research and Applied Technology Symposium on Mined Land Reclamation*, ed. J.R. Garvey. Monroeville, PA: Bituminous Coal Research Inc.

Williams, R.E. 1975. *Waste Production and Disposal in Mining, Milling, and Metallurgical Industries.* San Francisco, CA: Miller Freeman Publications, Inc.

Williamson, N.A., M.S. Johnson, and A.D. Bradshaw. 1982. *Mine waste reclamation.* London, UK: Mining Journal Books Ltd.

Wilson, H.R. 1989. Tailings management program: An operating success, Echo Bay's Lupin mine. 83-102. *See* Cox et al. 1989.

———. 1991. Tailings management in the Canadian Arctic: Echo Bay's Lupin mine. *Mining Engineering* 43: 213-214.

Winkler, M.G., and C.B. DeWitt. 1985. Environmental impacts of peat mining in the United States: Documentation for wetland conservation. *Environmental Conservation* 12: 317-330.

Winterhalder, E.K. 1984. Environmental degradation and rehabilitation in the Sudbury area. *Laurentian University Review* 16: 15-47.

Wischmeier, W.H., and L.D. Meyer. 1973. Soil Erodibility on Construction Areas. In *Soil Erosion: Causes and Mechanisms, Prevention and Control*, ed. Highway Research Board. Washington, DC: National Academy of Sciences.

Wolfe, J. 1994. International developments in environmental management systems. In *CRS Perspectives* 49: 22-26. Kingston, ON: Centre for Resource Studies, Queen's University.

Wood, P.A. 1983. *Underground stowing of mine waste.* Report Number ICTIS/TR23. London, UK: IEA Coal Research.

Wren, C.D., and G.L. Stephenson. 1991. Effect of acidification on metal accumulation and toxicity of metals to freshwater invertebrates. *Environmental Pollution* 71: 205-242.

Yamartino, R.J. 1985. Atmospheric pollutant deposition modeling. Chapter 27 in *Handbook of Applied Meteorology*, ed. D.D. Houghton. New York, NY: John Wiley & Sons.

Yan, N.D., and G.E. Miller. 1982. *Characterization of lakes near Sudbury, Ontario.* Sudbury Environmental Study, Report SES

009/82. Toronto, ON: Ontario Ministry of the Environment.

———. 1984. Effects of deposition of acids and metals in chemistry and biology of lakes near Sudbury, Ontario. *See* Nriagu 1984.

Yan, N.D., and P.J. Dillon. 1984. Experimental neutralization of lakes near Sudbury, Ontario. *See* Nriagu 1984.

Yanful, E.K., et al. 1991. Overview of Noranda Research on Prevention and Control of Acid Mine Drainage. *See* AMIC 1991, Vol. 1: 62-75.

Yundt, S.E. 1975. The problems of urban mining—social pressures and the aggregate industry. *CIM Bulletin* 68(11): 1-4.

Ziemkiewicz, P.F., and R. Gallinger. 1989. Overview of tailings reclamation in western Canada. *See* Walker et al. 1989.

Zoltai, S.C. 1988. Wetland Environments and Classification. In *Wetlands of Canada*. National Wetlands Working Group, Ottawa, ON: Environment Canada.

Zonneveld, I.S. 1983. Principles of Bio-Indication. *Environmental Monitoring and Assessment* 3: 207-217.

# INDEX